T0251111

PLANNING AND THE INTELLIGENCE OF INSTITUTIONS

Planning and the Intelligence of Institutions

Interactive approaches to territorial policy-making between institutional design and institution-building

ENRICO GUALINI
AME - Amsterdam Study Centre for the Metropolitan Environment
University of Amsterdam

Routledge
Taylor & Francis Group

LONDON AND NEW YORK

Contents

PART II AN EXPERIENCE IN INSTITUTIONAL INNOVATION: THE CROSS-ACCEPTANCE PROCESS IN NEW JERSEY

List of Figures

Preface

The more humans act according to plans, the more effectively they are struck by chance.

Friedrich Dürrenmatt

And yet, how else is it possible to break the hold of the vicious circle through which a high affectivity of ideas and a low ability to control threats to humans by humans reinforce one another? [...] There might be after all the possibility for humans to hit upon a way of getting out of a critical situation more by accident than by design. But even this may only occur if the structure of the process offers the humans involved loopholes for escape.

Norbert Elias

The role of social interaction and of its institutional patterns in realizing an effective linkage between *knowledge* and *action* is a classic issue in planning theory. As such, it is linked to shifting conceptions of the 'effectiveness' of planning and of its normative aims.

For a long time, in the social-welfarist tradition, the question of the effectiveness of planning has been identified with the prospects of institutionalization of its practices, and its 'ineffectiveness' (i.e. the evidence of a mismatch between the aims pursued by its form of knowledge and expertise and their actual outcomes) has been accordingly interpreted as a result of a lack of institutionalization, as an unfulfilled prospect of social reform.

Since however the awareness has been reached of the need – in order for knowledge to be 'usable', and for the link between knowledge and action to be effective – to combine the multiplicity of knowledge-bases which affect policy-making processes – 'scientific', 'expert', and 'ordinary' forms of knowledge – through their constructive interplay (Lindbolm and Cohen 1979), planning theory has increasingly recognized that 'how (relevant) knowledge is *used*' and 'how (relevant) knowledge is *created*' are hermeneutically linked issues, tied by their mutual belonging to a dimension of *enactment* and by the interactive, situational and relational foundations of symbolic-cognitive processes. Knowledge, as a key for effective social action, has been hence itself acknowledged as a social construct, and its construction conceived as a result of processes of *interpretive mediation*, contributing to the constitution and to the ongoing renewal of 'common knowledge-bases' as contingent frames of reference for action. Planning practices have been thus reconceptualized according to their possible institutional role in realizing this kind of mediation, introducing to new perspectives of normative and ethical commitment.

Precisely this kind of reconceptualization and this redefinition of aims, however, have also represented a source of major dilemmas for planning theory. In its understanding, the institutional embeddedness and the degree of institutionalization of planning as a specific set of social practices – once held as an indicator of social recognition for the 'progressive' character of the discipline – have progressively turned into challenges to its 'effectiveness' and to its very definition. The mismatch between the normative aims and the actual outcomes of planning has been increasingly perceived as emerging from an actual background of conflict, incommensurability, and mutual exclusion among forms of social knowledge set at different levels of institutionalization. The challenge to the effectiveness of planning has been thus no more identified with the alleged constraints to the enforcement of its expert form of knowledge due to a troubled path towards institutionalization, but rather with the constraints in realizing the mediation between expert and common forms of knowledge due to its very patterns of institutionalization. 'Ineffectiveness', in this sense, has been more and more seen as a distinctive mark of the institutionalization of planning, as a reflex of the very contradictions inherent in its institutionalized dimension. In a critical-pragmatic tradition of planning theory – particularly in approaches focussing on the social position as well as on the subjective condition of the planner as a mediator and as an activist – the interactive and communicative mediation between different forms of knowledge has been accordingly interpreted – in more or less 'radical' ways – as the means for turning the social-institutional role of the planner into a potential for working across and against social patterns of institutionalization, 'in the face' of their connotations of power.

It is as a result of a rather recent theoretical turn – favored by a growing awareness of issues of culture and identity as well as of the symbolic-cognitive dimension of policy-making – that attention has been shifting, from a rather subjectivist focus on the planner's role and conduct in mediating social practices in an institutionalized environment, towards a more pluralist and desubjectivized understanding of the dynamics of institutionalization that affect the field of social practices to which planning belongs. Planning (and policy-making) are increasingly interpreted not only as practices set in a framework of institutional constraints – a perspective that would stress the inherent contradictions of any aim of planning, as an 'institutionalized' practice, to pursue an idealtype of effectiveness intended as the fair, integrative and inclusionary mediation of plural forms of social knowledge. Planning (and policy-making) are no more only seen as 'institutionalized' practices, amenable to be scrutinized as such. They are seen as practices embedded in an *institutional field*, defined by the interplay of a plurality of social practices set at different, varying, and coevolutive degrees of institutionalization.

As part of an institutional field, planning and policy-making participate in processes of institutionalization. This means – in other words – that they bear themselves the features of institutionalization processes. The forms taken by their practices are entangled in the duality of processes of structuration which are ongoingly reproduced through everyday interactions. Thus, both their patterned, 'institutionalized', and their contingent, rather 'weakly institutionalized' forms of enactment, are constantly set in a tension between their constraining and enabling potential relative to other forms and to other degrees of institutionalization of social practice involved in their field.

Planning, as a knowledge-based practice, may be interpreted as a form of institutionalization inasmuch as the heuristic focus is placed on its embeddedness within a pluralist field of social practices, and on its ability to contribute in generating and consolidating within this field new situated forms of collective action, through the interpretive, symbolic-cognitive and strategic-communicative mediation realized in framing processes of social interaction.

The relative degree of generative potential with which planning practice is endowed, in this view, resides in the ability of achieving an effective and legitimate balance between the enabling and the constraining dimensions of institutionalization: in the ability of defining patterns of collective action open to the emergence of social innovation, to the democratic renewal of combinations between social claims, normative values, and public aims. In institutionalist terms, the experimental challenge of planning may be thus defined as the achievement of an effective and legitimate balance between purposive inputs of *institutional design* and coevolutive, path-dependent processes of *institution-building*. In this endeavor, yet, the pursuit of *effectiveness* becomes itself part of a social-constructive and coevolutive experience: its definition shares in a dimension of *innovation* and *discovery*.

The assumption on which this book is grounded is that such questions are gaining a particular relevance today, as research on new forms of territorial governance hints at processes that entail a radical redefinition of conditions for the effectiveness, representativeness, and legitimation of planning, and points to the need for experimental attitudes towards the innovation of its institutional settings.

This book is an attempt in addressing such questions from an interactionist, actor- and action-oriented perspective: it is an invitation to think of planning – in an attitude of critical pragmatism – not merely, or primarily, as an 'institutionalized' practice, as a 'function' of institutions, but as a factor of institutionalization: as a form of *generative action,* potentially constitutive of new forms of relationship between social practices and, as such, of resources for their experimentation.

In such a perspective, focussing on actual institutionalization processes – rather than on formal institutions which, all to often, do not account for the reality of policy processes nor for the normative values which stood behind their foundation –, a new pragmatic direction may be envisioned for combining the normative aim of planning – as an expression of democratic life – with an understanding of how its practices affect the 'social construction of reality'. And it is possibly in such perspective that a better understanding may be achieved of the actual social mechanisms at play in experiments in social innovation, and of their possible contribution in redefining politics, in reconstructing the public sphere.

The book is subdivided into three parts.

The *first part* is devoted to tracing the reasons for a renewed focus on the institutional dimension of planning, with reference to the imperatives as well as to the challenges for institutional innovation inherent in today's tasks of territorial governance.

The reasons for such a focus – and the reasons behind its assumption – are further contrasted with the recent and increasingly explicit attention for institutional issues in planning theory. In doing so, however, an attempt is made in establishing a critical distance between our line of argumentation and the ever-present temptation of acknowledging – or even of 'establishing' – a new planning-theoretical 'paradigm'. The way in which the dimension of institutional embeddedness of planning is discussed here points rather to the *longue durée* of planning-theoretical debate in coping with institutional issues – through its 'shifting involvement' with issues like the planner's role and expertise, social interaction, argumentation, communicative action, and the like – stressing to the need to overarch the critical lessons coming from these important contributions with a new *experimental* commitment towards issues of institutionalization and institutional change.

Among the positions which are discussed in the course of our arguments, approaches based on consensus-building are identified as the possibly most promising and fruitful attempts in pursing this experimental direction. The conceptual reference to interactive and consensual forms of policy-making which supports their practices is therefore further scrutinized – from a theoretical, policy-analytical point of view – in order to identify both their limits and opportunities in addressing the challenges of effectiveness, representativeness and innovativeness which planning practices are increasingly called to face.

The *second part* of the book is devoted to an extensive case-study of a planning experience based on practices of consensus-building, the *cross-acceptance* approach to state planning adopted in New Jersey in the late 1980s. It could be easily agreed upon – albeit from a rather different perspective – that the case chosen might be seen, to a certain extent, "as only modestly 'out of the ordinary'" (Mandelbaum 2000, p. xi): the more so, as experimental planning initiatives inspired by principles of consensus-building are rapidly spreading, in different versions and across different contexts, calling for extensive comparisons and fine-tuned categorizations in order to allow for 'generalizable' interpretive results. The New Jersey experience, however, remains in our view quite 'out of the ordinary' for a very specific reason, which underlines its fairly exceptional position among US state-sponsored growth management programs: for the consistency by which it combines the response to a need for institutional change with the adoption of an experimental commitment to a new style of policy-making, based on patterned forms of interaction and on reflexive feedbacks on their very design and conduct. For this reason, the subject and the methodological format of the case-study have been assumed as an occasion for an interpretive evaluation of processes of institutional change specifically focussed on their *experiential* dimension, on their internal coevolutive dynamics, and on the social mechanisms involved in their interactive course of development. Inquiring into the New Jersey *cross-acceptance* process has represented as such for the author an enlightening hermeneutic experience in understanding the dynamics of institutionalization phenomena as well as a fascinating opportunity for experiencing reflexive loops between theoretical questioning and the empirical conduct of research. To a certain extent, the exceptionality of the case may perhaps even lessen the main shortcoming of the analysis presented, i.e. its temporal restriction to the very first phase of policy development (1988-1994) and, hence, its necessarily limited extension towards the evolutive dimension of the institutional experience.

The *third part* of the book, finally, develops and generalizes the theoretical assumptions which have guided the research. Its core aim is an attempt in addressing issues concerning the role of social interaction and of its institutional patterns of conduct in an action-oriented perspective. The background is given by a discussion of new-institutionalist contributions to a critical understanding of the institutional embeddedness of planning and of the constraining and enabling duality of its processes of institutionalization. A crucial key for addressing the challenge these dimensions represent in a normative perspective towards interactive planning approaches is identified in a revision of theoretical assumptions on collective action. The possibility of effective institutional patterns of action for identifying, framing, and handling collective issues, and the nature of institutionalization processes involved in the emergence of such patterns of action, are reinterpreted in a perspective of experimental and constructive interplay between inputs of *institutional design* and coevolutive processes of *institution-building*.

Needless to say, most of the arguments presented in this book are very much indebted to the research and to the theoretical efforts of many acknowledged scholars. The importance of their influence – although not always made explicit – should emerge clearly out of the discussion of the ideas to which they have contributed. This personal interpretation of some of these ideas is intended as a tribute to their path-breaking work.

Acknowledgements

The research presented in this book is the result of work done mainly at Dipartimento di Scienze del Territorio, Politecnico di Milano, between 1994 and 1997. I am grateful to Pier Carlo Palermo, Luigi Mazza, Giorgio Ferraresi and Alessandro Balducci for their support during its conduct.

The case study has been made possible by the hospitality and support of the New Jersey Office of State Planning and of its staff, which I was able to benefit from during my stay in Trenton between June and September 1995; thanks to the director, Herbert Simmens, to the vice-directors, Robert A. Kull and Charles P. Newcomb, and to their assistants, Mary Housel, Sheila Bogda and Wendy McVicker, for their help and kindness. My thanks extend to all the interviewees, whose names can be found in the appendix. For their contribution in making my stay in New Jersey and my first visit to the USA enjoyable, I would like to thank Foster and all members of the family of Pat and Ed Krupa, Lawrenceville NJ.

My research has advanced further during a visiting scholarship at the Institute of Urban and Regional Development, University of California at Berkeley, between June and September 1996; thanks to its director, Judith E. Innes, to her assistants, Barbara Hadenfeldt, Kate Blood and Christine Amado, and to the scholars who devoted time to discussing my research project: Eugene Bardach, Frederick Collignon, Bernard Frieden, Judith Gruber, Judith E. Innes, Michael Neuman, Michael Teitz, Melvin M. Webber and, moreover, Seymour J. Mandelbaum.

Parts of this work have been previously presented and discussed at various international conferences, and in particular at the Oxford Planning Theory Conference (Oxford, April 2-5, 1998), at the XII AESOP Congress (Aveiro, July 22-25, 1998), at the seminar 'Institutional Capacity, Social Capital and Urban Governance' (Newcastle, April 15-16, 1999), and at the 41st ACSP Conference (Chicago, October 21-24, 1999).

Graham Cass has greatly helped by revising most of the English text.

Finally, I am grateful to the institutions that have contributed more or less directly to making my research and this book possible: the Italian Ministry of University and Scientific Research; the John F. Kennedy Institute for North-American Studies at the Free University in Berlin; the Italian Regional Science Association; and the Department of Spatial Planning in Europe of the University of Dortmund.

Figure 2.1 is adapted from: Nielsen, R.P. (1993), 'Woolman's "I Am We" Triple-Loop Action Learning: Origin and Application in Organizational Ethics', *Journal of Applied Behavioral Science*, Vol. 29, No. 1, pp. 117-38, ©1993 by NTL Institute; reprinted in adapted form by permission of the NTL Institute for Applied Behavioral Science, Alexandria VA. The source is an adaptation of figures taken from: Argyris, C. (1990), *Overcoming Organizational Defense: Facilitating Organizational Learning*, Prentice Hall, Englewood Cliffs NJ, ©1990 by Prentice Hall, a Pearson Education company; reprinted in adapted form by permission of Pearson Education, Inc., Upper Saddle River NJ; and from: Argyris, C. (1993), *Knowledge for Action: A Guide to Overcoming Barriers to Organizational Change*, Jossey-Bass, San Francisco CA, ©1993 by Jossey-Bass, Inc. Publishers; reprinted in adapted form by permission of Jossey-Bass, Inc. Publishers, San Francisco CA, a subsidiary of John Wiley & Sons, Inc., New York NY.

Figures 8.1, 12.1, and 12.2 are taken from: Ostrom, E. (1990), *Governing the Commons: The Evolution of Institutions for Collective Action*, Cambridge University Press, Cambridge, ©1990 by Cambridge University Press; reprinted by permission of Cambridge University Press, Cambridge.

Figure 11.1 is taken from: Lanzalaco, L. (1995), *Istituzioni organizzazioni potere. Introduzione all'analisi istituzionale della politica*, La Nuova Italia Scientifica, Roma (new edition: Carocci Editore, Roma 1995) ©1995 by Carocci Editore; reprinted by permission of Carocci Editore, Roma.

Figure 11.2 is taken from: Powell, W. W. and DiMaggio, P.J. (eds.) (1991), *The New Institutionalism in Organizational Analysis*, University of Chicago Press, Chicago IL, ©1991 by The University of Chicago; reprinted by permission of The University of Chicago Press, Chicago IL.

Figures 13.1 and 13.2 are taken from: Keohane, R.O. and Ostrom, E. (1994), 'Introduction', in: 'Local Commons and Global Interdependence: Heterogeneity and Cooperation in Two Domains', *Journal of Theoretical Politics*, Vol. 6, No. 4, pp. 403-28, ©1994 by Sage Publications Ltd.; reprinted by permission of Sage Publications Ltd., London.

PART I
AN 'INSTITUTIONAL TURN' IN PLANNING THEORY?

1 Governance and the Challenges to Institutional Innovation

Introduction

Changing principles of territorial sovereignty and emerging rationales of territorial governance represent a challenge for both research and action. A crucial aspect in this challenge is the increasingly problematic connection between the aims of *effectiveness*, *representativeness*, and *innovativeness* which new forms of territorial governance are called to face.

Two interpretive keys frame our approach to a discussion of this challenge and of its implications on rethinking the institutional dimension of territorial policy-making and planning. The first points to the conceptual assumptions entailed in reframing discourse on territorial policy as an issue of *governance*; the second points to the changing meaning of issues and efforts of *institutional innovation*. The reason for the paradigmatic significance attributed to these notions resides primarily in the identification of a mutual problematic relationship, which renders them jointly a privileged way of interpreting empirical evidence on the renewal of forms of territorial governance.

The introductory sections of this chapter focus on a discussion of the notion of governance, aiming not so much at isolating it in a dimension of theoretical purity, but rather at identifying its heuristic fertility in the interaction with complementary notions as well as with contributions from different disciplinary traditions which are undergoing – and contributing to – a revision in interpreting the territorial dimension of social, economic and policy processes. A discussion is further introduced of the implications of such perspectives on territorial processes in terms of conceiving 'paths towards innovation' and of the possible institutional responses to their imperative. What is stressed by this is the importance of an experimental, interactive and social-constructivist attitudes in addressing the challenge of governance and institutional innovation, as may be critically recognized in the diffusion of concerted and consensus-building policy-making approaches, in the framework of which institutional rationales of territorial policy-making are being reframed, and new modes of institutionalization of territory-bound practices are emerging.

3

Interpretive Perspectives of the Notion of Governance

Governance is a term which seems recently to have acquired the role of a summary concept for the evolution of forms of public action in mature liberal-democratic systems. Its power of generalization implies some risks, of course. In fact, it is not difficult to agree on the popularity and, at the same time, on the imprecision and polysemy of the word: 'governance' has entered common use from time to time signifying "a change in the meaning of government, referring to a *new* process of governing; or a *changed* condition or ordered rule; or the *new* method by which society is governed" (Rhodes 1996, pp. 652-3). As has been nonetheless noted, "contrary to the suggestion of some authors, the attempt (or attempts) to conceptualize in terms of governance is not always due to the effects of methods but is rather the result of comparative research which aims to highlight similarities and differences and which makes new conceptualizations necessary" (Le Galès 1998, p. 495). It cannot be furthermore ignored that the fortune of the notion of governance is an expression of the constitution of a new discourse, which hermeneutically involves analysts and producers of public policies, redefining modes of attention and relevance criteria for action. Hence, beyond ambitions of rigor, our interest here is mainly in identifying elements common to a redefinition of the problems of territorial governing and development which is exerting a *constitutive* influence on policies.

Reframing Governability between State, Markets, and Civil Society

The transition from a traditional approach to questions of territorial *government* towards the notion of *governance*, which is gaining momentum as an interpretative paradigm in the field of urban and regional studies, refers to an evolution in practices pertaining to effective forms in the management of territorial resources at the regional and local level, progressively involving translevel and interjurisdictional dimensions, which tend to include forms of action, of knowledge and of organization non-reducible to the statutes of political-administrative institutions and into formal definitions of their prerogatives.[1]

Policy orientations are thus developing complex governance settings throughout various territorial and socioeconomic contexts which may be traced back to three main dimensions:
- the broadening of the field of actors and organizational forms involved in the development and implementation of policies, in the framework of scattered, dispersed, but tendentially interrelated decision-making arenas;

- the broadening of the field of relations between policy areas pertaining to territorial management (structural political-economic strategies, active labor and human capital policies, research and technologic development policies, locational and environmental policies, territorial marketing);
- the narrowing role of governmental actors at the formal level of institutional competencies and the broadening of opportunities and arenas for informal involvement. Representatives of organizations and stakeholders, collective actors and associations crowd the intermediate and more and more continuous space between the market and the state, often intervening in filling effectiveness and legitimation gaps in public action, and contributing to the definition of a relatively fragmented and volatile arena of problem- and solution-setting.

In this analytical perspective, as has been noted, "the various approaches to governance share a rejection of the conceptual trinity of market-state-civil society which has tended to dominate mainstream analyses of modern societies" (Jessop 1995a, p. 310). At their basis stands a dissatisfaction with top-down explanations of the exercise of power, as well as an interest in inquiry into forms of sociopolitical coordination which broaden the field of traditional relations between the public and the private, constituting new forms of interdependencies in the public sphere: hence an emphasis on the shift from a restricted conception of *government* to an enlarged conception of *governance* as *governing activity*, characterized by the rejection of the presumption that connotations of public action be grounded on decision-making mechanisms anchored in the sovereignty of the state.

Starting from an interest in alternative coordination approaches in business relations, like "relational contracting, organized markets in group enterprises, clans, networks, trade associations and strategic alliances" (Jessop 1995a, p. 310), which has developed in the field of new institutional economics, around the notion of governance contributions from social and political sciences have progressively clustered which move from an interest in the constitution and the functioning of forms of 'regimes', and sometimes of 'governance without government', extending to the application to the logic of organizational behavior, to a revision of normative models of public action, and to the definition of new strategies.[2]

The notion of governance which emerges from the rather diversified profile of disciplinary contributions to its definition, both as a theoretical concept and as a concept-in-use, is therefore also nurtured by an ambiguous tension between analytical-interpretive intentions and normative ambitions, as far as to render impossible – perhaps useless – a clear distinction between these two spheres: *governance*, as previously noted, constitutes a discursive construct which redefines parameters of attention for the forms taken by public action. It is not however useless to browse through its main analytical traits before proceeding to a discussion of further dimensions of the notion.

On an analytical level, the transdisciplinary viability of the notion of governance relies on a few basic conceptual distinctions. First, and implicit in the very currency of the term, is a distinction between *government* and *governance*, whereby the latter acquires autonomy as referred to the complex of activities implied in governing which insist on but ate not limited to structures and forms of action of statutory governmental authorities. Inquiry on governance, on the contrary, puts emphasis on the centrality of action orientation, in an environment characterized by a realignment of institutional rationales to a conception of an 'entrepreneurial state' as the basis for legitimation of governing action, and where non-institutional actors are an integral part of the policy-making process, if not conceived as the motors of policy-making initiative itself called to fill a void in action capacity by governmental institutions.

Governance thus represents a perspective (indeed a problematic and rather non-formalizable one) in the exercise of authority in territorial management and initiative which is no more identifiable with forms of action and with guidance-and-control functions of governmental institutions alone, and which rather structurally includes decentralized forms of action and initiative arising from the field of economic and social activities, and is defined by a broad but loose framework of interdependencies with formal governmental institutions.

The field of governance studies has been accordingly redefined, in general terms, "as concerned with the resolution of (para-)political problems (in the sense of problems of collective goal-attainment or the realisation of collective purposes) in and through specific configurations of governmental (hierarchical) and extra-governmental (non-hierarchical) institutions, organisations and practices" (Jessop 1995a, p. 317).

An interpretation which sums up this kind of approach in an exemplary manner is the conception proposed by Rhodes (1996), who sees governance activity as the *public management of networks*. The networks in question in this conception (which is more than a mere stipulative definition') are intended here as diffuse forms of social coordination: in this sense, "governance refers to self-organising, inter-organisational networks" (Rhodes 1996, p. 660) which constitute structures complementary to markets and hierarchies in realizing the allocation of resources and in exercising forms of coordination and control. Networks thus feature important elements of self-coordination and self-governing capacity, which tend to make them organizational and operational forms capable of creating their own action environment and tendentially autonomous if not resilient to external guidance by public authorities. The management of interorganizational relations thus operates in a field which extends beyond the traditional boundaries of the public, the private, and of a 'third' sector, constituting an authentic alternative for public action rather than a hybrid between given forms of regulation represented by markets and hierarchies.

The reasons for interest in such an understanding of governance (besides the risks connected with a reification of the notion of networks and with the neglect of the contextual dependency of their explanatory potential) resides in the evidence attributed to the potential dissolution of a distinction between state and civil society: "[t]he state becomes a collection of interorganisational networks made up of governmental and societal actors with no sovereign actor able to steer or regulate" (Rhodes 1996, p. 666). The challenge of government and administrative action thus becomes that of coproduction, of the pursuit of joint results of one's own and of other social actors' activity and initiative.

In this conception, reference to governance is intended not only as an *analytical* move beyond the dichotomy between alternative paradigms of the interpretation of public policy, like pluralism and corporatism. 'Modern governance' is, on the contrary, often explicitly seen (with apparent normative implications) as the emergence of a 'third way' out of crises in the effectiveness of governing activity (characterized by decreasing or suboptimal outcomes and by implementation constraints) relative to classic alternatives like the reduction to forms of 'weak' or 'lean' government (deregulation) prospectively 'hollowing out' the state, or the delegation (privatization, 'corporatization') of governing tasks in the framework of a corporatist model of society.

This perspective, with its normative implications, appears the more explicit in contributions which thematize governance from a sociocybernetic perspective, intending it as the *outcome of interactive sociopolitical forms of governing* (see: Kooiman 1993c). Besides a basic distinction between *government* and *governance*, a further distinction is proposed there between the activity of *governing,* intended as the complex of governmental interventions directed to scope, and the coevolutive outcome of this activity, i.e. *governance* as the result or sum of the effects of interventions and interactions of a sociopolitical-administrative nature which are implied in governing and which concur to its eventual perspective of effectiveness.

The notion of governing activity is redefined therefore in terms of a conceptual constellation which comprises the idea of sociopolitical governance as the interaction between government and society, and the idea of *"governability* as an expression of governance in terms of effective and legitimate adjustment of governing needs to capacities and capacities to needs" expressed by society (Kooiman 1993a, p. 43, emphasis added). If "[g]overning in contemporary society is mainly a process of coordination, steering, influencing and 'balancing'" of the interactions between public and private actors (Kooiman 1993b, p. 255), governance "can be seen as the pattern or structure that emerges in a social-political system as 'common' result or outcome of the interacting intervention efforts of all involved actors. This pattern cannot be reduced to one actor or group of actors in particular [...]. This emerging pattern forms the 'rules of the game' within a particular system or, in other words, the medium through which actors can act and try to use these rules in accordance with their own objectives and interests" (Kooiman 1993b, p. 258).

Governance in this sense is an interactional outcome as well as a system- and context-dependent outcome. The concept of governability thus appears as defined by the dual relationship between governing and governance. According to the idea of a 'duality of structure' (Giddens 1984), "in which a pattern of governance is not only the unintended outcome of social (inter)action but also the mechanism through which actors have the capability to act and govern, [...] governing and governance are subjected to a permanent process of mutual interaction. Actors who govern, or try to govern, also influence the governance structure of a subsystem. some (more powerful) actors have the possibility to rewrite some 'rules of the game' but no one has complete control. There is always some intended and unintended change, which creates maneuvering space for actors willing to change the existing pattern" (Kooiman 1993b, p. 258). In this interaction, the space is given for a possible balance between potentials and needs for order and control. *Governability* is understood as this potential equilibrium emerging from governing and governance activities.

The notion of governance hence assumes a radically interactionist stance, defined by a dual relationship between government and society, which marks a further paradigmatic shift related to a unilateral conception of relationships between governmental institutions and practices and 'civil society'. On one side, in fact, the notion of governance questions the idea that civil society represents, according to traditions of political philosophy, the field of expression *par excellence* of *bourgeois* individualism or, rather, the residual place of expression of communitarian ties, in a tendentially dualistic relationship with the sphere of the state, if not *tout court* with the public sphere. On the other side, as we shall see, governance opens up a new potential field of collective action, going so far as to render conceivable effects of "governance without government" (Rosenau and Czempiel 1992), of self-governance and self-regulated organization, in specific spheres of activity, when effective mechanisms of a regulatory type are active, even if not endowed with formal authority and legitimation.

Theories of Governance and Theories of Regulation

As has been noted, the notion of governance hold within its semantic horizon a voluntarist dimension and a normative attitude: "[i]n 'governance', ideas of leading, steering and directing are recognizable", albeit "without the primacy accorded to the sovereign state" (Le Galès 1998, pp. 494-5). *Governance* as a notion combines an acknowledgement of the emerging and eventual character of processes of political regulation with an explicit engagement in questions of *institutional design*, with a mainly mesopolitical and interorganizational orientation.

In this sense, at first glance, *governance* and *regulation* are notions which, besides their different ascendancy, seem to bear an element of paradigmatic incommensurability. On the one hand, in fact, the notion of regulation, as defined particularly by the 'French school' of regulation, presupposes an explicit denial of a *voluntarist* conception, rather viewing regulation as a 'process without a subject', in which "specific modes of regulation are always emergent, evolutionary effects of a multiplicity of actions in specific, strategically selective contexts" (Jessop 1995a, p. 329). On the other hand, regulation theories cannot refrain from integrating their vision of the social embeddedness of economic processes with the role played by organizational and strategic dimensions and by discursive production in defining the nature of the processes of institutionalization of regulatory constructs.

Significant parallels and elements of convergence among regulation theories and governance theories have in fact been underlined particularly by theorists engaged in a translation of their contributions into a sociologically oriented perspective of political economy, i.e. in an interpretive perspective of neo-Gramscian inspiration.

A first relevant trait which joins regulation theories and governance theories is to be found precisely in their anti-realism and in their attention to the constructive dimension of processes: according to this view, the constitution of their objects themselves is at stake in processes of governance/regulation. This common element is recognizable primarily in some metatheoretical similarities, concerning both structural and strategic aspects of the processes of governance and regulation: an interest in aspects of self-reproduction and self-organization of complex systems in turbulent environments, inspired by the influence of complex systems theory, and an interest in aspects of interorganizational coordination, learning, and the constitution of identity. An explicit metatheoretical kinship has been recognized precisely "in terms of their common interest in the path-dependent, constitutive relationship between modes of governance/regulation and objects of governance/regulation. Neither regulation theory nor theories of governance can be seen as teleological in character or as committed to ex post functionalist arguments: for they imply that it is in and through governance (or regulation) that the elementary objects of their attention are transformed through complex articulation into specific moments within a given mode of governance (or regulation)" (Jessop 1995a, p. 326).

While regulation theories imply that the objects of regulation do not fully preexist the process of regulation,[3] theories of governance underline on their part that the objects of governance may be recognized as such only through efforts in governing them, through governance activity itself. Thus, "the very processes of regulation or governance constitute the objects which come to be regulated or governed in and through a form of self-referential self-organisation". In this sense, "just as there is neither regulation in general nor general regulation, there is no governance in general nor general governance. Instead, there is only particular regulation and the totality of regulation, only particular governance and the totality of governance [...]. In the real world there are only definite objects of regulation that are shaped in and through definite modes of regulation; and definite objects of governance that are shaped in and through definite modes of governance" (Jessop 1995a, p. 315).

A second element which joins the two approaches is what could be defined as a common new-institutionalist stance, which understands the outcomes of governance/regulation as relatively stabilized systems of relations. The nature of these relationships descends essentially from the modes of treatment of three fundamental dimensions involved in sociopolitical processes: the coordination of activities or relationships among actors, the allocation of resources related to these activities or systems of relations, and the structuration (prevention and resolution) of conflicts.[4] In this perspective, governance and regulation are notions concurring to defining modes of institutionalization of an essentially emergent, non-intentional, constitutively pluralist nature. Again Jessop (1995a, p. 322) underlines that "[n]ew governance mechanisms, like new structural forms, emerge from a trial-and-error search process which operates through evolutionary variation, selection and retention. It is in this context that issues of strategic selectivity and strategic capacities are so crucial and that attention must be paid to the material and discursive appropriateness of proposed responses".

The constructivist, emergent and coevolutive nature of modes of social regulation implies a shift from a monistic and/or dualistic towards a pluralist interpretive perspective. Political regulation crosses the boundaries of hierarchical settings and impinges on process of restructuring of forms of interaction among institutional and social actors. The emergence of systems of governance, in this sense, may be interpreted as a process of the articulation in time and in different social sectors or spheres of different modes of regulation (Le Galès 1998), the dynamics of institutionalization of which appear to be dependent on the path taken by the process as well as on the specific and contextual nature of their integration. This however also implies the need to distinguish the mode from the object: thus, governance is not an entity 'other' than government, but rather the situationally determined principle in which forms of political regulation rearticulate themselves in specific contexts of reference, and in which forms of policy-making and institutional action rearticulate themselves in relationship to these beyond formal-hierarchical requirements.

The plural dimension of the combination of modes of regulation hence highlights the potential 'evolutive advantage' of governance regimes: i.e. on a positive level, their ability to learn and innovate in a changing environment intended in a Schumpeterian sense, and, on a normative level, the pluralism of possible combinations among 'local' modes of regulation at the disposal of policies.

Local Regulation and Territorial Governance

The theoretical alignment of theories of regulation and theories of governance also allows for a theoretical passage from governance of the economy to governance of the territory and for a reintegration into the framework of inquiry of a properly *political* issue, through a discussion of their *local* nature.

The choice of a pluralist and contextual, *territorialized* dimension of inquiry of forms of regulation and governance does not only respond to methodological reasons, to the purpose of operationalizing otherwise 'abstract' notions, however important this may be. Rather, the territorial dimension represents the level at which the sense of development policies is being reconstituted, at least as influentially as this is occurring at the level of transnational policies and of the revision of principles of sovereignty of nation-states.

Research has pointed to the semantics of the notion as a primary reason for a redefinition of problems and forms of territorial governance. Again, reference to the distinction between *government* and *governance* and to a dualism between statutory settings and actual practices of governance and regulation offers a first level of conceptual clarification. If in fact the field of *urban/local government* is defined as a field of public action structured into statutory levels of competencies and roles coincident with a definite territorial articulation, according to a classic model of sovereignty, *urban/local governance* is rather defined as an activity of regulation and decision-making of which local government is only one of the players, of the actors possibly involved (Harding and Garside 1993). The notion of local governance therefore introduces a fundamental discontinuity not only in the definition of the actors and practices of the governing of territories, but also in its *identitarian* dimension. The problem of the territorial identity of governance practices, and the redefinition itself of the geography of governance, become thus a tangible expression of the *stretching* between the global and the local and of the tension between processes of *embedding* and *dis-* and/or *reembedding* (Giddens 1990) which characterize the spatial dimension of economic development and competition and of their linkages to local social and political processes in conditions of 'late modernity'.

The relative marginalization of formal-statutory aspects of territorial government prerogatives, however, opens up new territorial forms of convergence between modes of regulation and modes of politics and policy-making. It is in particular the tradition of research on 'local' forms of regulation of development which have characterized the emergence of new forms of territorial competitiveness (as described in economic geography under notions of industrial districts, local systems of development, and the like) to point to the role of forms actually taken by public action in constituting the conditions of 'institutional robustness' or 'institutional thickness' (Amin and Thrift 1995; Savitch 1997) which enable their ability to persist and evolve. These conditions are however to be interpreted in an eminently processual and proactive dimension: in the notion of 'local action systems' (Bagnasco 1988; 1994), which redefines the peculiar form taken by politics in neo-localism, the operation of institutions and local government is inscribed in a process, the horizon of effectiveness of which resides in its interactive and pluralist dimension, in the contribution to an effective articulation and integration of different forms of regulation into concrete action situations.

The shift from an inquiry into the governance of the economy towards the inquiry into territorial governance implies therefore an understanding of the constructive and territorialized dimension of modes of regulation. "Raising the question of governance suggests an understanding of the linkage of different types of regulation in a territory in terms of political and social integration and at the same time in terms of capacity of action" (Le Galès 1998, p. 495), in a process of mutual adjustment among economic actors, social groups and institutions aiming at achieving collective objectives in an action environment of a fragmented and uncertain nature.

Reformulating the question of forms of regulation in terms of local regulation also makes it possible to reintegrate in the analysis of local development processes a properly political dimension, and to approach an analysis of the role played in them by new forms of local politics. Redefining local politics as a question of local governance, in this sense, however also implies recognizing in the new forms territorial policies are assuming a dimension which transcends issues of mere economic development or mere administrative rationality, and which, on the contrary, represent a restructuring of relations between institutions and economic and social actors which entails aspects of mobilization of interests and of abilities for action, but also aspects of collective learning and sensemaking (Le Galès 1998).

Governance and Local Development Regimes

The construction of forms of local governance is thus a response to rationales which go beyond a merely functional or organizational dimension of institutional innovation, and in which, on the contrary, the discursive-narrative dimension and the strategic dimension of public action play a relevant role (Fischer and Forester 1993).

Empirical evidence of new territorial phenomena has contributed to an acknowledgement of the complex dimension of local policies and a paradigmatic renewal in their interpretation, fostered by new approaches in the field of urban and regional policy studies, which combine lessons from political economy with interpretative frameworks from the tradition of policy analysis. The parallel between the dominance of local modes of regulation and the constitution of local forms of governance constitutes the key to reinterpreting the difficult nexus between politics and territory beyond determinist geographical identifications or clear-cut relationships of sovereignty. The constitution of 'local development regimes' may thus be interpreted as a particular mode of governance which finds its reason for territorial identification in a contextual system of interest representation, in the framework of which the relationships between interests and territories are themselves redefined.

Beyond their belonging to distinct disciplinary niches, and beyond a sometimes redundant proliferation of definitions,[5] several recent contributions seem in this perspective to bear more of an analogy in describing the most relevant traits of this policy-making environment.

Approaches based on regulation theory have in particular concentrated on the transition from a 'fordist' social order – defined by rigid forms of division of labor between the state and local governments in the social reproduction of accumulation modes – to a global competitive dimension – characterized by a progressive 'withdrawal of the state' from the role previously assumed in the framework of social welfare policies, by a growing dependence of cities and regions on decentralized forms of self-governance and self-promotion, and by evidence of new forms of performative pressure accompanying the changing relationship between institutional and economic actors at the local level (Jessop 1990; 1995b; Amin 1994). In this context, marked by the shift from settings of *government*, understood as institutionally defined systems of public action on a formal-constitutional basis, to settings of *governance*, understood as fields of activity pertaining to management and promotion of territorial development, extending to non-institutional actors, to non-constitutionally defined action rationales, and to an array of intermediate, informal, and associative forms of action, a connection is instituted between the crisis of the fordist model of accumulation, and of the role played in it by the state, and the constitution of a *local state*, grounding on the aggregation of territorially-bound *regimes* and *development coalitions* (Stoker 1990; Keating 1993; Harding and Garside 1993; Woolman and Goldsmith 1993; Stoker and Mossberger 1994; Harding 1994; 1997).

In the framework of the tradition of research on neo-corporatism, on the other hand, the revision and the questioning of systems of corporatist representation of a generalist kind, induced by the evolution of welfare regimes and of central-peripheral governmental relationships, has contributed by shifting attention towards their reconstitution in renewed forms at the local level and towards the emergence of forms of 'mesocorporatism', directed towards the constitution of abilities for action and self-promotion on a territorial basis around concerted settings featuring high levels of self-organization and low levels of institutionalization. These forms of concerted decision-making and action, developed 'in the shadow of hierarchy', and sometimes interpreted as intermediate, transitional forms between classic governmental settings and governance regimes, frequently assume the form of initiatives in 'experimental regionalism', where needs for effectiveness and innovation of public action are addressed in the framework of local neo-corporatist settings (Heinze and Voelzkow 1991; Heinze and Schmid 1997; Schmitter and Grote 1997).

Beyond differences in interpretative aims and tools, these positions stress the emergence and meaning of new forms of identity, initiative and operational ability at the urban and regional level, based on concerted agreements and explicit as well as tacit and embedded forms of negotiation, in the framework of institutional settings directed towards the constitution of concrete fields and opportunities for intervention. A key dimension of modern governance becomes "the need to create institutional arrangements which enmesh formal government structures and processes within the wider relational webs of economic and social life" (Healey 1996, p. 290). These new modes of territorial governance thus appear to be defined by a shift in the conception of public action: from a primary concern with social control to social production, and (according to: Stone 1989) from the exercise of power (intended as authority) 'over' something, to forms of power (intended as ability) 'to' accomplish something, according to objectives defined through extensive settings of alliance, exchange, and cooperation.

Thus, the basic contradiction facing local societies, set in a tension between the aims of economic attractiveness and competitiveness and of efficiency on one side, and the aims of social integration and equity on the other, becomes centered on territories and on territory-specific forms of collective action, constitutively rooted in – but also reproductive of – specific sociopolitical *milieux* and of their identity, or as 'artificial communities' (De Rita and Bonomi 1998) constituted – in analogy to 'invented communities' (Anderson 1991) – through their very enactment in a locale.

Of these expressions, innovative and creative potentials have been underlined, as have uncertainties and threats connected to their dependence on a 'risk environment' potentially conductive to forms of 'coacted consensus' based on predefined social aims, once and again dominated by quasi-natural imperatives of competitive economic performance. As underlined by its analysts, governance in fact does not only and primarily represent an (uneven) horizon of opportunities, but also an (uneven) horizon of risks, and a potential challenge to democratic representation and accountability. Governance structures, in analogy to what has been evidenced by network-theory-based interpretations (Rhodes 1996), may thus develop tendencies to autonomization and to forms of resistance to control and external regulation, contrasting with political-administrative attempts at reform; the very nature of the distribution of resources for the access to and the manipulation of regulatory apparatuses may define a structurally unequal distribution of opportunities (Painter and Goodwin 1995). Under such conditions of territorial policy-making, moreover, "a progressively more concrete interdependence is constituted between local economic processes and welfare-state processes", which may result as such in "a fragmented and potentially highly unbalanced supply of social services tied to the respective economic 'performance' of the region and dependent on the abilities, political priorities and mobilization of local political actors" (Mayer 1996, pp. 22-3). A perspective of local governance may thus give rise to significant effects of empowerment of society, but also to conflicts of representation and interests which stress our deficits in a renewal of democratic theory and practice (Mayer 1991).

Notions like those of governance and subsidiarity appear in this perspective overshadowed by dense clouds of ambiguity, situating local societies between the extremes of new potentials for empowerment and of perspectives of pure dependence.[6]

Transdisciplinary Perspectives on Territorial Governance

The universe of discourse which constitutes the reference of the notion of governance has evolved through disciplinary contributions of a highly diverse nature, which however bear a common implication in rethinking the relationships between society, economy, politics and territories, as well as their institutional dimensions. Innovative approaches to territorial governance and to institutional issues have in fact fostered significant developments in fields of research related to urban and regional studies, a cross-fertilization of which is crucial in order to link positive and normative perspectives on territorial policy-making.

The Dimension of Place-boundedness of Socioeconomic Development

The first line of research to be mentioned focuses on a renewed theoretical as well as operational interest in the role of local and regional cultures, identities, and forms of action in the perspective of socioeconomic development. Attention to aspects of *territorial embeddedness of economic processes* (Granovetter 1985; Granovetter and Swedberg 1992; Grabher 1993) has underlined the role of 'regions' as geographical contexts regulated by forms of social interdependencies not reducible to pure market variables (Dematteis 1995; Storper 1995), rather underlining their endogenous and interactive creative dimension (Sabel 1989; 1992; Liepitz 1993; Scott 1996), and stressing the potential of new directions in economic development and innovation policy, based on the enhancement of contextual forms of institutional and organizational learning.

Breaking with linear models of innovation and economic development, based on neo-classical paradigms of rationality and on traditional notions of locational and agglomerational externalities, distinctive lines of inquiry, like research on 'innovative *milieux*' (Camagni 1991; Dematteis 1995; Butzin 1996; Matthiesen 1998), extending the notion of Marshallian districts to the understanding of sociocultural determinations of local development, and the 'learning region' paradigm (Morgan 1995; Florida 1995; Hassink 1997), have attempted to combine neo-Schumpeterian notions of technical/organizational innovation with recent trends in economic geography oriented towards an analysis of aspects of the social and institutional embeddedness of economic performance. A shift has thus been particularly emphasized from an economic towards a sociological-culturalist perspective in the understanding of innovative phenomena, underlining the mutual relationships between innovation, entrepreneurial and economic structures and sociopolitical and institutional forces as the conditions for the development of growth dynamics and diffusion processes, and redefining the role of institutional forms of support and organization based on the involvement and the interaction of social actors at a regional level. The diffusion of associative, networking and cooperative paradigms in regional policies (Cooke and Morgan 1993; Amin and Thrift 1995) stresses the patterned interaction and the involvement of locality-bound actors, aiming at promoting effective modes of exchange in knowledge and information, and thus the development of 'environments of scope' based on new forms of trust and of mutual commitment, and enhancing the development of collective resources and social capital.

Respatializing Social Theory: Regions/Locales as Interactional Constructs

The reappraisal of the territorial dimension of economic processes which characterizes 'new regionalism' and recent trends in economic geography is a significant expression of a more general renewal of a spatial focus in social theory (Giddens 1984; Dickens 1990; Bagnasco 1994).

A strictly related paradigm shift is centered on a *redefinition of the notion of geographical region*. Along with the emergence of new forms of territoriality, traditional categories sustaining the notion of geographical region, based on either physical, functional, or administrative definitions, have been losing ground in terms of both their interpretative and normative adequacy, highlighting the need for more flexible interpretative frameworks, of a rather metaphoric nature, inspired by transactional and networking paradigms. 'Region' is a notion which comes to be defined as a social construct of a relational nature, as a relatively stable but coevolutively defined pattern of interactions between actors, connecting in a networking mode the places where actors of interactions are actually situated (Dematteis 1989; 1995).

According to this conception, what identifies regions becomes the flexible, emerging form in which human conducts are ordered, defined by their respective patterns of relations and activities. Regions and locales are thus seen as both mediums and outcomes of social interactions, no more as entities defined by precast identities, but as constructs, 'shared spaces' (Thrift 1983; Massey 1991, 1993; Amin and Thrift 1994) or 'theatres of interaction' (Giddens 1984; 1990), defined by intersections and articulations of relationships in the framework of actual social interactions: not as aggregates of spatial localizations, but as forms of a subdivision of time-space defined through the forms taken by concrete social practices.

'Experimental' Regionalisms and Localisms

The phenomena and the interpretations outlined bear, of course, a crucial political dimension. Its most striking expression is perhaps the political redefinition of territorial identity.

The last two decades – particularly in non-federalist contexts subject to the growing influence of multilevel systems of governance – have witnessed the emergence of a diffuse crisis of territorial representation and of a wave of regionalisms and localisms. Their origin has been indicated in the convergence of three main factors of transformation affecting local societies. First, structural change and the conversion of the *functional role of the territory* in production and accumulation processes, a phenomenon comprised in a tension between the *despatialization* and the *reterritorialization* of the economy on which much has been written. Processes of 'glocalization' impose the invention of new territorial relationships: "[g]lobalization creates a tension between the aspatial rationality of the trasnational corporation, with its multiple branches and ability to move investments around, and the spatially-bound rationality of communities which depend on these investments" (Keating 1997, p. 386). Regions and localities reemerge in this process as the levels at which regaining an effective institutional steering potential for policy-making and planning is at issue.

These new opportunities are backed by a diffuse need for *institutional restructuring* (paradigmatically exemplified by the process of European integration), and from which experiences in modernization and rationalization have emerged which have shifted from a centralist-distributive rationale to principles of decentralized (and often institutionally favored) self-organization, for the sake of contextually variable purposes of efficiency, effectiveness, or autonomy and self-determination of political-administrative practices.

Starting from the 1980s, attention to the role of the territory and to the endogenous potential of regions and localities in a perspective of development has thus fostered processes of institutional decentralization, but also an autonomous will and capacity for political initiative. It is in fact at the convergence of processes of functional restructuring, of institutional restructuring, and of *political mobilization* that the nature of regionalisms and localisms – as particularly evident in Europe – must be interpreted, fostered by both in repositioning phenomena regarding both issues of an economic and social nature (the shifting mutual relationships between development policies and welfare policies) and issues of identity, autonomy and competition.

This 'recomposition of political space' sees regions emerge as "political arenas, in which various political, social, and economic actors meet and where issues, notably to do with economic development, are debated", and as "constituting themselves as actors in national, and now European, politics, pursuing their own interests" (Keating 1997, p. 392). It is however of little use to talk in this respect of a 'reemergence' of regions: these are not natural entities, but at the same time territories, political spaces, institutions, and sociocultural identities, and as such always problematic in their identification. Regions are social constructs, systems of action in and through which – as has been noted with reference to Europe – "territory is being reinvented, as the European state restructures, collective identities are reforged, and new systems of collective action emerge in state and civil society", while "[n]ew forms of both autonomy and dependence come about" (Keating 1997, p. 395).

In light of this problematic nature, however, any perspective of reification or institutional hypostatization becomes unlikely: to the point that, "[r]ather than talking of regional autonomy as a bilateral relationship, or merely reformulating it as a trilateral or multilateral one, it is better to talk of *governmental capacity at the territorial level* [...], that is the capacity to formulate and implement a developmental and social project" (Keating 1997, p. 395, emphasis added). Regionalisms and localisms emerge thus as the outcome of processes from above and from below

(Keating and Hooghe 1996), in a perspective which makes it possible to reinterpret the scope of the ideological motor of many identitarian claims, but which at the same time imposes the acknowledgement of a new, unstable balance of powers.

From Hierarchy to Policentrism in Territorial Governance Settings

Regions and localities emerge as potentially important actors in a process which is marked by profound innovations in institutional rationales of governmental action, and which leads to a shift in attention from a primary focus on sheer competition among territories, cities, and regions, related to issues of economic development and to mobilization to attract resources and investments, to the reframing of institutional and political competition at the national and supranational level in a perspective of 'polycentric governance': territorial governance – particularly in Europe – "is witnessing increasingly unstable intergovernmental relations, with the cooperation/competition model giving way to the creation of networks and to the strengthening of intermediate level innovations" (Le Galès 1998, p. 486).

Around these phenomena a line of critical reflection is centered which aims at a revision of rationales of territorial government and of center-periphery relationships.

The main object of this revision concerns the crisis of hierarchical and comprehensive models of metropolitan-regional government. Radical transformations in economic development processes as well as in actual models of urbanization and of territorial relations, along with growing evidence of a trade-off between governance action and governmental settings, in terms of both the interpretative adequacy and of the effectiveness of public action, have put under question the formal-statutory nexus between the ends of territorial governance and the institutional forms of their achievement.

Backed by a recognition of the crisis in legitimation and effectiveness of 'metadecisional' coordination systems, and at the same time to a growing extent relying on market mechanisms, in a policy environment which has however progressively unveiled the myths of *laissez-faire* liberalism, developments in innovative practices in territorial governance are normatively as well as experimentally directing us towards an understanding of governance settings as 'institutions in action', intended as institutionally guided and facilitated forms of conjoint action oriented towards consensually defined objectives.

The Rise of Multilevel Governance Systems

The reframing of territorial policy-making settings into complex inter-organizational fields may be recognized as a dimension common to most situations witnessing structural transformations in governance conditions. Notions such as multilevel governance, although paradigmatic of the institutional experiment represented by the process of European integration, assume thus a generalizable interpretive meaning.

The reasons of interest of the notion *multilevel governance* reside in its reference to a tradition of analysis of problems of intergovernmental coordination in complex societies hardly amenable to be interpreted with consolidated models of comparative political analysis, and in its recent actualization in relationship to the development and pervasive influence of a supranational level of determination of territorial policies.[7]

The constitution of what has been termed a 'system of multilevel governance' (Marks 1992; Scharpf 1994a; Hooghe 1996), thus underlining its peculiar nature, non-reducible to usual interpretive models of the evolution of territorial government institutions, impinges decisively on the nature of authority and of competencies of central governments and on the promotion of capacities for mobilizing resources by local governments, redefining the nature of relationships between the state and the 'local state'.

Multilevel governance may be interpreted as a new experimental condition for territorial policy-making, characterized by the growing importance of supranational regulatory patterns and by the constitution of subnational, regional-local arenas of social and economic regulation and microconstitutional choice. The main feature of policy-making in multi-level governance settings is thus its 'stretching' along two dimensions, which may be defined as 'vertical' and 'horizontal', i.e. the dimension of the multiplicity of governmental and institutional settings (involved in processes which range between a supralocal, even supranational, if not 'global' level of the definition of policies aims and rationales, and local-regional levels of enactment of policies), and the dimension of the multiplicity of policy actors (broadening between the fields of the state, the market, and quasi- or non-governmental associations).

The concurrence between the introduction of progressive shifts in institutional sovereignty towards supranational regulatory systems, and the principle of subsidiarity, which entails the rooting of policy action in local initiatives and abilities, thus determines the embeddedness of territorial policy-making in multiple institutional domains and interaction arenas which blur the meaning of hierarchical settings in the development of policies. Governance of territorial issues becomes defined as a complex field of multiorganizational entities and multiplex administrative levels, through which policy inputs and policy outcomes emerge as a result of recursive flows of information and resources.

Governance and Institutional Innovation

What are the implications of the change in character of territorial policies for opportunities and initiatives in the promotion of institutional innovation, or, in other terms, for the very notion of *institutional innovation*?

The primary connotation of the complex of practices which affect territorial governance, as we have seen, is the intersection of processes which result in effects of social regulation along channels of intermediation which are no more identifiable as 'public' and which rather appear in large measure as deceptive, if not contradictory, relative to governmental rationales. The notion of governance – as far as it is normatively committed – aims at tracing of possible forms of a public function of guidance, control, direction, in an environment which impels a reframing of social components into different, innovative constellations, tending to dissolve constituted boundaries. The integration of mutually contradictory social claims however becomes more and more difficult under such conditions, and points to the dilemma between the pursuit of effectiveness of public policies and the pursuit of representativeness of their processes of formation: action capacity may in this sense contradict needs for democratic accountability of choices before a society which raises multiple, non mutually consistent, non-transparent demands. The changing features of politics thus place governance activity before a triple challenge (Le Galès 1998):

- the complexity of public action grows and the integration between differentiated universes of meaning and sense becomes more problematic;
- the environment of public policies becomes mobile, fluid, uncertain: each choice recursively sets the scene for actors responding to different statutes, and the combination of elements of concentration and decentralization in decision-making in a multilevel system redefines the autonomous abilities and the mutual constraints of each of the actors;
- the articulation of relationships between 'electoral politics' (the processes of political competition and selection) and 'substantive politics' (the politics of problem- and agenda-setting and of the building of interest and action coalitions) becomes more and more divergent.

The Changing Features of Coordination Tasks

Among the most relevant consequences of these developments is the change in meaning and sense of a classic problem of public policy-making and analysis like that of *intergovernmental coordination*.

The issue of intergovernmental coordination (as discussed in ch. 2) is classically defined, in a first instance, by the problem of pursuing effective strategies for coordinating action of different institutional actors and rationales. The task of coordination is thus understandable as a combination of a horizontal (implying the mutual coordination of decentralized units acting upon a policy on the basis of different, idealtypically sectoral rationales) and of a vertical-hierarchical dimension (defined by central-peripheral relationships, whereby decentralized units function as executives of central decisions).

The aim of coordination, however, faces increasing tendencies to the fragmentation of structures and competencies relevant for political-administrative activity, combining with a proliferation of actors endowed with relative degrees of autonomy, with the non-linear overlapping of possible problems and solutions, with a progressive blurring of competencies of political-decisional and administrative-executive order. The 'theatres of interaction' which actually structure political activity broaden the field of actors between the spheres of the public and of the private, blurring traditional categories of roles and institutional definition, reframing conditions for the insurgence and the setting of interests and conflicts.

The interplay between a dimension of multilevel governance, as we have seen, still entails a conception of coordination as the prerequisite for *active* rather than *reactive* forms of policy-making. However, the nature of this 'active' connotation of policy-making differs from that in use in the 1970s, when the notion was first formulated (Mayntz and Scharpf 1975). While policy-making aims at becoming 'active' in face of this complexity, and calls therefore for coordination as a prerequisite for effectiveness, the decentralized and differentiated character of political systems and of arenas that define the environment for policy-making activity is the source itself of new challenges for coordination, which go beyond the classical interpretation of over- or undercoordination problems, setting growing constraints and limitations to traditional solutions. In a multilevel governance environment, the very subject of policy-making becomes structurally dispersed, decentralized; the interrelation of complex policy problems becomes a function of the multiple contexts for the initiation of policies and of their multiple modes of regulation, rather than of autonomous, clear-cut, long-range policy solutions and of linear steering efforts.

The Changing Conditions for Institutional Experimentation

In a similar environment, governance signifies in the first place a governing practice in which the achievement of effective coherence of public action is no more entrusted to an élite and to a relatively homogeneous and centralized administrative function, but is the outcome of multilevel and multiactor forms of coordination, "the result of which, always uncertain, depends on the capacity of different public and private actors to define a space of commonsense, to mobilize expert forms of knowledge from different sources, and to establish forms of commitment and legitimation of decisions which may operate in the meantime in the sphere of electoral politics and of the politics of problems" (Muller and Surel 1998, pp. 96-7).

Secondly, the guiding function of formal governmental institutions is set in a fundamental tension, defined by both cognitive and legitimation aspects: as the authority and the effectiveness of their formal-regulatory, direct and sectoral forms of intervention is narrowed by growing factors of interdependence, their indirect potential of direction extends towards the ability to mobilize and activate social stocks of knowledge, resources, and relations from a broad field of potential actors of intersectoral policies. No longer are formal-institutional structures assumed to be in place which may "link the various public and private-sector forces that can influence change and make their efforts coherent" (Harding 1997, p. 293). At the same time, the vital importance for local societies and polities represented by the acquisition of negotiation and self-promotion abilities in their competitive positioning and in defining potentials for development translates into an *imperative to innovation* for governing practices.

Under these conditions, forms of innovation of policy-making that ground their legitimation on negotiation, the pursuit of agreements and the constitution of a mutual commitment of development actors on a territorial basis acquire the meaning of an experimental challenge, and impel a revision of institutional settings which regulate intergovernmental and interinstitutional relations along lines non-reducible to formal decision procedures.

In this sense, such experiences imply a more or less explicit revision of the very conception of processes of institutional innovation. On the one hand, no single collective actor of territorial governance is anymore recognizable, or formally appointed, as the 'entrepreneur' of institutional innovation; rather, the arena of territorial governance is being reframed into an *institutional field*, i.e. into a pluralist setting of practices and of their coevolutive forms of institutionalization, in which innovations are emergent and concurrent (albeit not necessarily non-conflictual) outcomes, occurring at the micro- and mesolevel of social processes in which concrete forms of action in the management of territorial resources take place; on the other hand, similarly, the issue of institutional innovation is reframed in terms of a dual outcome, emerging from new *institutional models* of relationships between actors of territorial transformations as well as from *institutions-in-action*, i.e. from the concrete evolutive outcomes of diffuse, decentralized, weakly institutionalized forms of experimentation.

This perspective thus implies the assumption that innovation and effective action are neither exclusive nor necessary outcomes of linear, rational-comprehensive approaches to problem-setting and -solving, enacted through vertical, unidirectional lines by central institutional actors, but rather the outcomes of processes of contextual interpretation, mutual identification, legitimation, and reorientation in the course of action, enacted by multiple decentralized actors in the framework of evolutive processes.

The issue of institutional innovation, in other words, calls for a re-framing in terms of an understanding of the mutual interplay between inputs of *institutional design* (as the expressions of an innovative intentionality, of a *design rationality*) and processes of *institution-building*, intended (as discussed in ch. 2) as ongoing social and interactional processes of the interpretive and negotiated definition of the nature of policy aims, enacted through contextual processes of microconstitutional choice occurring at the micro- and mesolevel of social interaction through which concrete innovative forms of action in the management of territorial resources take place.

Two further implications of the evolution of territorial policies for institutional innovation must be stressed here.

The first implication is the preeminence of their multidimensional, transsectoral and translevel connotations. Innovative approaches to the development of territorial policies, in this sense, are increasingly under-standable as 'policies of policies', i.e. as policies overarching different policy rationales as well as different policy areas. Their challenge is defined by the need to address a threefold dimension of innovation:
- a dimension of *enactment* of policies, rooting policy formulation and implementation rationales into concrete social practices;
- a dimension of *mobilization*, bundling traded and untraded interde-pendencies between relational, material, knowledge- and power-resources;

- a dimension of *collective sensemaking* and *institutional learning*, as a condition for the establishment of feedback loops and for the constitution of institutional capacities.

A further implication is a fundamental change in the rationales of governmental action and in the style of policy-making. While stressing the shortcomings of top-down, 'methodologically-constitutionalist' approaches to reform, the notion of governance is far from implying the dominance of a deregulatory policy environment; it rather poses the question of the new role of governmental institutions as 'creative mediators' between the different levels of regulation concurring to the development of multilevel policy-making. The task of the 'institutional entrepreneur' becomes defined by the assumption of an enabling, facilitating and stimulating role, setting the procedural means and the incentives for horizontal forms of self-organization and cooperation, developing 'in the shadow of hierarchy' (Scharpf 1994b).

The development of strategies for territorial governance tends accordingly to assume the features of an experimental effort, based on the mobilization of the actors' potentials in the framework of innovative inter-governmental patterns of relationships.[8] Such institutional initiatives, as forms of 'experimental regionalism', or of 'regionalization-in-action' – according to the idea of 'institutions-in-action' and of 'regionalization' as constructs and processes of a relational nature, defined by concrete patterns of interaction between actors – represent in this sense an innovation-oriented process which may be defined according to two main dimensions (as discussed in ch. 3):

- the development and experimentation of new institutional settings and rationales for intergovernmental coordination and decision making, which combine aspects of *institutional design* with an explicit aim of enhancing *institutional capacity building*;
- the development of new relationships as a condition for effective decentralized action, with a focus on the role of *consensus-building* in clustering local forces around concrete action orientations, combining agreements on representations of the place-boundedness of the actors' interests with multiple commitments of their roles towards the develop-ment of a shared heuristic.

Among the fundamental questions which must be raised in the light of these experiences, in a new-institutionalist perspective on governance and policy innovation, are therefore not so much questions of the formal consistency of planning processes to exogenous definitions of preferences, goals, objectives, and means, but questions of their ability to enhance and to sustain the insurgence of innovation in a framework defined by a constitutive plurality of modes of social regulation and by decentralized forms of self-organization.

Notes

1   Harding (1997, p. 295), for example, summarizes them idealtypically as follows:
    - "general moves away from welfare/distributive modes of governance to ones which prioritise localized supply-side measures aimed at enhancing economic performance,
    - a shift from nationally based, process-oriented governing arrangements to locally based, product-oriented ones, and
    - a change in emphasis from vertical integration, standardised rules, clear lines of authority, accountability and national equity to horizontal integration, flexibility, networking, problem solving and the realisation of economic potential through strategic competition and collaboration".

2   Rhodes (1996) has particularly underlined how the thematization of governance represents a paradigmatic shift, also in a normative sense, relative to approaches to the modernization of public action centered on intraorganizational relationships, such as *New Public Management*, in the light of reference to an environment and a style of policy-making very different from those postulated by the latter. While in fact these approaches are designed to treat problems of efficiency of in-line administrations, they seem structurally incapable of dealing with problems of effectiveness in environments characterized by a dominance of interorganizational links and by the need to negotiate shared objectives in the absence of a unitary hierarchical function of guidance and control.

3   As Jessop puts it, in the meaning proposed, "the objects of regulation do not, and cannot, pre-date regulation in their full historically constituted identity. [...] At most they could exist as a series of elements, different subsets of which could then be articulated in different ways to produce different ensembles, each with its own relative stability and unity" (Jessop 1990, p. 186).

4   Reference to a sociological matrix of new-institutionalism contributes to solving the ambiguity between the meaning of 'regulation'– as consolidated by the French school of regulation theory – and the traditional concept of regulation – intended as political-juridical regulation – expressed by the French term *règlementation*: 'regulation', as Jessop notes, should therefore be viewed as 'regularization' or 'normalization', in the sense of the general theoretical question of evolutionary economics and institutional analysis. For reasons probably connected to that ambiguity, moreover, "[w]hereas the revival of 'governance' suggestively highlights a major paradigm shift in political (and economic) analysis, the continued use of 'regulation' has partly obscured a major paradigmatic shift in economic (and political) analysis" (Jessop 1995a, p. 308).

5   To complicate the field, and to underline the limitations of these notes in this respect, we may recall the richness of theoretical-analytical contributions – sometimes competing, sometimes complementary – recently proposed by policy analysts mainly in the European context: from approaches based on network analysis (Powell 1990; Marin and Mayntz 1991; Heinelt and Smith Randall 1996) to the notion of *advocacy coalitions* (Sabatier and Jenkins Smith 1993), *cognitive coalitions* (Haas 1990) and *discourse coalitions* (Hajer 1995; 2000). For a discussion of some of these contributions see: Le Galès and Thatcher (1995), Börzel (1997), Muller and Surel (1998).

6   The relevance of these questions to an assessment of opportunities for the development of civil society offered by the new governance environment permeates the debate on possible models of societal organization originated by a revision of corporatist models (Streeck and Schmitter 1985) or rooted in idealtypes of *discursive* (Dryzek 1990) or *associative democracy* (Hirst 1994; Wright 1995; Amin and Thrift 1995; Savitch 1997; Amin and Hausner 1998).

7   Reference is, of course, mainly to the first contributions of the 'school of the Max-Planck-Institute' of Cologne and to the original definition of problems of multilevel governance given by Mayntz and Scharpf (1975; see also; Scharpf 1978).

8   Particularly in the context of a restructuring of territorial-based policies in Europe, defined by local and regional responses to the global dynamics of economic competition and by the progressive devolution of strategic policy-making assets, a wide middle ground of mixed decision-making settings based on consensus and on negotiated agreements is developing, which represents an innovative field for informal and flexible institutional experimentation. Several experiences may be recognized, in different forms and in different European contexts, which bear paradigmatic relevance in a perspective of integration between local development and supralocal policy rationales, and which have been termed either *negotiated programming, territorial pacts, regionalization* of structural policies, or have assumed the form of intergovernmental, multiparty strategic agreements for the government of urban regions in the framework of organizational settings characterized by low levels of institutionalization. For a brief introduction to similar trends in recent experiences in state sponsored growth management in the U.S., see the introduction to the case-study in ch. 4.

2 Redefining Institutional Issues: A Planning-Theoretical Review

Introduction

Recent planning-theoretical contributions to rethinking the institutional dimension of planning practice represent attempts at responding to the above outlined challenges of 'democratic governance': the problematic connection between the aim of the *effectiveness* of public policy and the changing conditions of *representativeness* and *legitimation*. Their common orientation is "to reconcile the collective spatial goals [...] with the totality of social forces in which the actual spatial developments take place" (Salet and Faludi 2000, p. 7) by facing the growing requirements for *innovation* in the institutional settings into which planning is embedded.

Identifying an alleged 'institutional turn' in planning theory is anything but an easy task, nonetheless. The main reason for this is that contributions to a renewed understanding of the institutional dimension of planning go far beyond approaches which thematize it[1] or which explicitly interpret it as an issue of 'institutional reform' or 'design'. An institutionalist turn in planning theory, if such a turn is indeed occurring, is rather the result of a pluralist but possibly convergent array of elaborations on the social and institutional embeddedness of planning.

As such, it has of course deep roots. Looking for a recognizable unitary orientation is therefore useless, if not deceptive, and all the more so if a longer view on planning theory is taken. The aim of this review is hence not that of tracing the development of a 'new paradigm' in planning theory, but rather to trace contributions which (across the lines of approaches which time and again have been distinctively defined as *communicative, argumentative, interpretive, interactive, collaborative, negotiating*, and the like) have fostered a shift in understanding aspects of institutionalization of planning and have elaborated on its consequences for normative theory and for practice.

The following review – necessarily highly selective – is subdivided into two parts, along the divide represented by a crisis in a reflection on the institutional dimension of planning centered on the planner as a *subject* – be it as an expert, a professional, a political mediator, or an advocate – and

29

on the dilemmas raised by relationships between the subject-planner, planning practices, and the institutional environment.

Overcoming a subject-centered view is seen as the condition for an understanding of planning as a set of practices constitutively belonging to an *institutional field*, and for the development of distinct approaches to the issue of institutionalization as a crucial dimension of planning processes. This finally leads to the adoption of a particular stance on the institutional dimension of interactive approaches, which is followed up in the next chapter by a discussion of consensus-building as a policy-making and planning strategy.

Towards a Pragmatic Relationship between Planning, Institutions, and Society

In the self-consciousness of the planning discipline, a nexus is recognizable between the 'discovery of social actors', i.e. between the broadening of decision-making arenas and the assumption of centrality of social interaction, and the emergence of an epistemological issue centered on a critical inquiry into planning practices.

In the distributional contest and in the conflictual environment which since the 1960s mark the decline of a welfarist myth of planning, a transformation in the social role of the planner is introduced which confronts it with the fragmented and competitive character of policies as well as with new conflicts of identity and roles.[2] Professional expertise becomes involved in a progressive politicization of administrative activity, in a process of ongoing interactive definition of policies implying the systematic recourse to negotiating abilities (Rabinovitz 1989). The involvement in administrative rationales increasingly recognized as 'political' marks a new phase in the auto-analysis of planners as actors of institutionalized practices, which detaches their ethical aspirations from Davidoff's (1965) judiciary metaphor, but also implies a crisis in a univocal definition of their identity. Attention to issues of social interaction becomes central in the reformulation of relationships between substantive forms of knowledge (such as the expert's) and action orientations. A significant reflex is a progressive shift from inquiring into planning practice from within a disciplinary perspective to questioning its social and institutional embeddedness, which delegitimizes a distinction between a *normative* and a *positive* view of planning processes (between a 'theory *in* planning' and a 'theory *of* planning': e.g. Faludi 1973), with significant consequences for the understanding of professional behavior and strategies.

A paradigm shift in social planning from problem-solving approaches to the challenge of multiple interaction processes introduces a focus on the managerial and organizational components of the planner's ability (Friend and Hunter 1970). Interorganizational policy-making and implementation are progressively recognized as domains of the planner's competence (Kelman *et al.* 1980). The widening of the field of planning activity, the changing rules of representation in planning practices highlight new and inherently non-technical functions of the planner, such as that of a mediator, raising interest for procedures of dispute resolution and to dialogic forms of conflict management and developing commitments to new communicative, even social-pedagogical abilities (Fischer and Ury 1981; Raiffa 1982; Sullivan 1984; Susskind and Ozawa 1984).

Throughout these influences and experiences, planners are confronted with the dispersion and pluralism of social subjects and of their claims, and, along with this, with the biases, the imbalances of power, the potentials for manipulation as well as for integrative outcomes entailed in practices of social interaction. The deontological understanding of the planner's role and expertise progressively shifts away from a vision of planning as embedded in a self-regulating model of pluralist society, ruled by principles of partisan mutual adjustment, and producing consensus as a byproduct of rational decision-making procedures, and rather acknowledges the need to explicilty face conflict and to build consensus as a condition for effective and just action.

Beyond the Gap between Knowledge and Action:
Inquiring into Planning Practice

Behind the most relevant contributions to rethinking the conditions of the social and institutional embeddedness of planning in the 1980s is an appeal towards 'bridging the gap' between theory and practice through a reflection on practice oriented by principles of critical pragmatism. A major role in this orientation is taken by a reassessment of the function that the disciplinary abilities of the planner may assume in realizing the constructive mediation between the self-interested intentionality of the actors and the technical, procedural, normative as well as symbolic-cognitive aspects of social rationality at play in pursuing purposive action within planning arenas. The new agenda for planning theory, according to this attitude, is defined in exemplarily broad terms:

> It must be centrally about practice and be descriptive and predictive as well as normative, and it must be grounded in empirical research on the practice

and context of planning. Such research should be qualitative and storytelling rather than primarily quantitative and hypothesis testing. Its goal must be in part to develop a new metaphor or imagery for planning practice. It should draw heavily on phenomenological thought and critical theory rather than positivist analysis. (De Neufville 1983, p. 1)

Interpretations of such a shift from a linear nexus between knowledge and decision-making to a circular relationship between knowledge and action, mediated by conditions of institutional embeddedness, may be found in two main line of reasoning: in a renewed but minority reflection on the organizational conditions of planning,[3] and in a much more influential commitment to a critical assessment of the planner's social identity and expertise, based on an hermeneutics of the behavioral dimension of 'everyday' planning practice (Forester 1989; Forester and Krumholz 1990; Healey 1992; Krumholz and Clavel 1994; Forester 1999).

A critical reassessment of participation and of the democratic nature of planning processes represents a first important matrix for such an auto-analysis of planning practice. The framing of issues of effectiveness in planning in terms of both policy rationality and social equity and the assumption of its dependence on the equitable representation of partisan interests, on public support in the treatment of complex issues, on a role in social education to decision-making rather than in the mere provision of resources or opportunities for access, constitutes the inspiration for the growing attention to the non-technical dimension of the planner's expertise. The redefinition of the 'politics of expertise' (Benveniste 1977) results in the key interpretive meaning conferred to the specificity of action situations and to the instrumental, communicative and argumentative mediation realized by the planner in public contexts. The inquiry into the planner's abilities in organizing and framing interactive processes and in enhancing and realizing social communication turns in this perspective into a strategy for social change, into a new 'political economy' of planning (as in Forester)[4] or even into a radical ideal (as in Friedmann).[5]

The empirical perspective which frames this attention to the planner's 'everyday practice' bears to a certain extent radical features, as far indeed as to justify an explicit reference to methodologies of ethnographic research (Forester 1993a; Rein and Schön 1993). The reason for interest in it goes beyond the more or less explicit emancipatory perspective it pursues, however. The focus on argumentative processes bears broader implications in terms of rethinking the fortunes of practices of participatory and deliberative democracy and the potential of planning as an institutionalized practice in fostering their progress.

In these terms, at least potentially, the focus on planning as communicative and argumentative practice bears implications in terms of a reflection on both its 'constitutional' and 'constitutive' dimensions.

The critical epistemology of political decision-making, on the one hand, broadens the understanding of planning as belonging to a pluralist domain of practices. Planning processes thus becomes akin to giving up the postulation of consensus (Christensen 1985) in favor of its construction within situated agreements (Susskind and Cruikshank 1987; Bryson and Crosby 1992) and of the recourse of methodologies of social inquiry defined by abductive inferences (Blanco 1994) stressing the argumentative dimension of problem-setting and social inquiry.

Attention to the extradisciplinary, non-technical components of the communicative and symbolic-cognitive universe displayed through interaction in planning practice moves therefore beyond a mere attitude to the deconstruction of bias and power structures (De Neufville and Barton 1987; Healey 1993b), and develops towards their assumption as interactive resources for a constructivist approach towards new forms to collective action.

Narratives, styles of writing, ways and means of listening and reading, and different levels of iconic representation which are mobilized by and through planning practices are, on the other hand, recognized as the features of a collective process of argumentation endowed with an emancipatory potential as well as with a constitutive dimension (Mandelbaum 1990; 1991; Fischer and Forester 1993; Innes 1990; 1992a; 1995b; 1996; Healey 1993a), based on which the features of the embeddedness of planning in the context of broad democratic processes should be analyzed.

The emancipatory and social-ethical perspective on equity and effectiveness accordingly turns from the struggle of a planner-subject against the constraints of the institutional rationales which comprise planning practice towards an attention to the complex interplay of factors which define its dimension of institutionalization in a combination of factors of a *constitutional* nature (i.e. to a dimension defined by and amenable to inputs of *institutional design*) and of factors of a *constitutive* and *coevolutive* nature (i.e. to a dimension of *institution-building*).

Before addressing the features of this turn, however, further contributions towards a detachment from a subject-centered attitude in planning theory must be defined. The most relevant influence in this perspective has been exerted by two lines of research: by contributions from social psychology and from the sociology of organizations to the analysis of the dynamics of *social learning*, and by contributions from policy analysis and implementation studies to an understanding of the dimension of policy-making as a *symbolic-cognitive process.*

Social Reflexivity and Social Learning

The 'Reflective' Dimension: Planning as Professional Practice

Issues like the strategy of attention and the communicative competence of the expert have become central since the 1970s to a line of research, backed by growing resort to methods of sociological inquiry in a perspective of disciplinary auto-analysis, pursuing a critical epistemology of planning practice under the influence of the paradigm of social learning (Michael 1973).

Despite its rooting in an important tradition of organizational studies, the most readily and extensively influential contribution is marked by an essentially deontological conception of action, which is symptomatic of prevailingly subject-oriented approaches to an epistemology of planning practice. The 'reflective practitioner' theorized by Schön (1983; 1987) represents in fact a convergence of contributions from research on organizational learning (e.g. Argyris and Schön 1974; 1978) on a perspective centered on the professional. The turbulence of decisional and operational contexts and their constitutive condition of uncertainty are acknowledged as conditions for a dialogic and interactive linkage between knowledge and action, reframing the issue of the effectiveness of professional practice in terms of a circular relationship between technical expertise and problem-setting, between theory and empirical experience: their mediation is realized through forms of inquiry and learning as systematic forms of interaction, defining both their frame of cognitive reference and their conditions of legitimacy.

Despite moving from a criticism of traditional models of professional practice of epistemological more than political inspiration (as in the case of Forester or Friedmann), Schön's contribution nonetheless focuses its pragmatic commitment to action on action situations which remain basically defined by a quite specific set of relationships (i.e. the dual relationship between professional and client), which strongly limits its generalizations in terms of a strategy of action-learning as formulated in organizational studies. A major consequence of this attitude is a weak thematization of the complexity of institutional practices and of the diversification of demands raised to expertise which is experienced in planning as social practice.[6]

The 'Reflective' Dimension: Planning as a Practice within an Institutional Field

A renewal in the meaning and influence of this line of research, as well as a relevant shift in its domain of pertinence, have been fostered by consolidated attention to the collective meaning of communicative and argumentative practices, to the institutionalization of negotiating practices, and to the involvement of planning in broader settings of public deliberation.

The practice-oriented attitude of inquiry into processes of collective framing of policy issues developed in the framework of the 'communicative-interactive' turn in planning theory (Healey 1993a; Innes 1995) has contributed in turning the focus of the social learning paradigm towards aspects of the social embeddedness of planning processes and towards a perspective of institutional reflexivity. A relevant influence in rethinking the possible contribution of the institutional design of planning processes in enhancing forms of learning has been offered by a conceptualization of their embeddedness in *interorganizational fields*[7] (DiMaggio and Powell 1983) as defined by the intersection and interdependence of different action- and organization-rationales (Nathan and Mitroff 1991; Gray and Wood 1991a; 1991b). This conceptual alignment has also favored a broader reception within planning theory of contributions from a rather minority approach in the social sciences, *action research,* and particularly from its 'strong' version pursuing an interpretation of negotiated order and evaluation as the basis for "a systematic community learning process for the collaborative review, improvement and development of policies, programmes and practices" (Marshall and Peters 1985, p. 273; Whyte 1991).

A paradigmatic attempt at merging approaches to policy design within socially-constructed institutional contexts with the tradition of social learning is Schön and Rein's concept of *framing* as a condition for social-institutional reflexivity (Rein and Schön 1993; Schön and Rein 1994; Rein and Laws 2000).

According to this reflexive model of action-learning, addressing the symbolic-cognitive components of behavior which defined the autonomous strategic identity of the involved actors not as a reason for confrontation, but as the issues at stake in collective processes, and, as such, as the *locus* of interactions, is the condition for a reflection on collective theories-in-use and for learning. Preferences and modes of attribution and evaluation are treated according to the assumption that behaviors of multiple actors may translate into action strategies which make it possible to illustrate how value schemes and modes of evaluation have formed, and hence to act towards forms of mutual inquiry and monitoring. It is through this dynamics, which does not escape conflict and which rather encourages dealing overtly with embarrassment and threats, that defensive routines and barriers to learning may be minimized, theories-in-use redefined, and new modes of conjoint action eventually formed.

The principle which rules such settings of collective action, as defined with reference to analytical categories by Simon, is that of a *design rationality* understood as a form of intentional rationality of conjoint action, as an alternative to a *design causality* understood as a form of action grounded on the postulate of a causal link between intentionality and action in everyday life, which is seen rather as the reason for a mismatch between the real reasons of action and their 'rational reconstruction', and for the constitution of forms of *pattern causality* systemically inhibiting *deutero-learning* (Bateson 1972) and the transformation of dominant governing values.[8]

Schön and Rein's effort in the desubjectivation of design functions, through the multiplication of the subjects involved and their incorporation in a relatively stable policy object, seen as the 'memory' of the cumulative moves of actors and as the condition for turning disjointed, chaotic and conflictual policy dialectics into a learning experience, is apparent; it is, however, still at odds in conceptualizing the embeddedness of practice in a multidimensional and evolutive institutional field (Baum 1995). In this sense, a perspective centered on the unfolding of theories-in-use according to the model of *double-loop action-learning* (Argyris 1993) meets the challenge of its hermeneutical condition. The model of *triple-loop action-learning* is an attempt to make the dimension of institutionalization of cognitive processes explicit (fig. 2.1). According to its definition,

> In triple-loop action-learning, there is a change in the embedded tradition system within which the governing values of a behavior can be nested. That is, the social tradition is both criticized and treated as a partner in mutual action-learning. The tradition system is valued and respected but also is considered to have potential negative biases that can be reformed and transformed just as individuals and the governing values that individuals accept can reform and transform. (Nielsen 1993, p. 118)

In this extension of the double-loop action-learning model, the social and institutional dimension of the frames which underlie collective action are made explicit, as well as the potential of social traditions incorporated in institutional settings of action, intended as abstract operators of actions oriented to the innovation of practices. The actors are thus enabled to monitor changes in discrete actions, in guiding values, as well as in the traditional forms of action into which such actions and values are embedded.

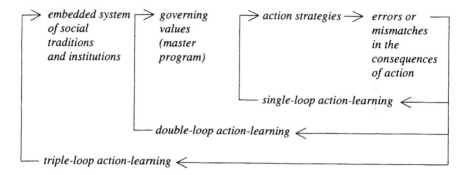

Figure 2.1 Conjoined scheme of the three models of *action-learning*
Source: adapted from: Nielsen (1993, p. 120)

While the background to such a conception lies primarily in the methodologies and approaches of ethnographic research, its assumption as a policy strategy is connoted in an experimental sense. This points to the shortcomings of an assumption of reflexive processes limited to a bilateral conception of relationships between 'planning' and 'society', which is clearly indebted to an institutional hypostatization of the professional-client link convincingly criticized by planning-theorists like Mandelbaum (1985) and by Crosta (1990); it furthermore points to the need of addressing social learning as an issue pervading the broader field of practices which define the features of the social-institutional embeddedness of planning.

Policy Analysis and Implementation Research

A significant factor of convergence towards an understanding of processes of collective framing and sensemaking as a relevant dimension of institutionalization entailed in planning processes may be traced back to the developments and mutual influences of two important lines of research:
- research on issues of *intergovernmental coordination*, addressing the crisis of functionalist paradigms of policy coordination and developing ment new orientations towards policy design and institutional design;

- research on policy processes as unfolding in arenas defined as *multiactor contexts of interaction*, and the revision of relationships between processes of policy formulation and policy implementation emerging from *implementation studies*.

Limits and Dilemmas of Interorganizational Coordination

The key role of interorganizational coordination has increasingly been reframed since the 1970s in the light of a critical appraisal of a hierarchically centered normative perspective and the pluralistic, non-hierarchical and decentralized multiactor environment which emerges from a positive view of decision-making. Shortcomings of *goal-oriented* perspectives of coordination are identified in the influence of relevance criteria built on methodological-individualist assumptions. Their descriptive and normative adequacy is questioned in the light of environments in which the pluralism of factors affecting decision-making relates not only to the strategic identification of resources for implementation, according to a linear means-ends relationship, but to the very definition of the contents of policies. In the framework of intergovernmental relationships defined by formal institutional arenas, a conceptual distinction is thus proposed between the different inferences involved in coordination efforts (Thompson 1967). Standardized coordination practices are seen as possibly adequate in the case of shared objectives and of tasks of a repetitive nature: where a direct unidirectional input-output relationship is recognizable between functional parts of a system, coordination strategies based on hierarchical conformity may prove to be effective. Where this relationship takes a bilateral form, however, the need for strategies is envisaged which allow for practices of mutual adjustment among the actors. In settings where reciprocity is implied, i.e. where outputs of one side are inputs for the other one and *vice versa*, like in multilevel governance systems, coordination tasks entail therefore a dimension of intertemporal comparison of preferences among the actors involved.

An instrumental perspective of coordination thus radically changes in nature when decision-making issues are at stake which do not only concern the sharing of means and resources for action, but also the sharing itself of the sense and of the conditions for self-identification with policy choices and relevant policy actions: "[p]olicy formation and policy implementation are inevitably the result of interactions among a plurality of separate actors with separate interests, goals, and strategies" (Scharpf 1978, p. 347). Precisely with reference to such conditions, this conclusion seemed to lead to a dilemma: that of a discrepancy between the policy's own relevance criteria and the policy's objectives of equilibrium and coordination.[9] The issue of interorganizational coordination thus increasingly blurred distinctions between normative and positive assumptions, entering the strategic and symbolic-cognitive domains of policy-making.[10]

From Coordination to Policy Design

In conditions of policy development characterized by high complexity and rapid change, the need for effectiveness induces therefore the need for a shift to 'constructivist' forms of coordination, addressing a "systematic and explicit combination of prescriptive and empirical approaches" (Scharpf 1978, p. 349). Entering an experimental dimension of policy-making, however, cannot fail to alter its meaning, objectives and means. Addressing the implications of interaction processes affects the very nature of the goal-orientation of coordination tasks.

A critical assumption of this is recognizable in approaches related to a 'cognitive approach' to the analysis of decision-making processes. An ascendancy of approaches to policy development through consensus-building can be identified in a tradition of policy analysis which has pursued an alternative to both rational-comprehensive, hyperinstitutio-nalized, and incrementalist, hypoinstitutionalized policy-making models, focussing on the possibility of 'open' institutional settings of policy-making which might proactively address the challenge of social pluralism and the need for the construction of agreements in a perspective of effective policy action. The origin of such a tradition is usually identified in what has been called *policy design movement*, i.e. in a line of theoretical contributions which, as a reaction to the tradition of policy science, pursued the ambition of addressing "the entire policy stream" (DeLeon 1988, p. 2, cit. in: Schön and Rein 1994, p. 84), with the aim "to re-embed policy making in the social and political environments from which earlier researchers had abstracted it, compensate for earlier writers' neglect of implementation, and restore problem setting to its rightful place alongside problem solving" (Schön and Rein 1994, p. 84).

This line of thinking was highly influenced by the revision of a dualism between policy formulation and policy implementation, accounting for the constitutive intersection between practices of problem-setting and practices of mutual adjustment and monitoring among the actors of problem-solving and implementation, as was emerging from a growing tradition of implementation studies.[11] Acknowledging the basic feature of implementation as the "prosecution of politics with other means"

(Bardach 1977) implied a shift from a linear conception of coordination tasks towards a proactive attitude, based on a normative approach to societal transformation through political planning which recognized a reality of structurally unequal distribution of knowledge and power resources.

As constitutively embedded in a *metapolicy* environment, policy processes are thus recognized to be facing new contradictions and dilemmas. Consensus-building was increasingly acknowledged as a condition for innovation, but the need for consensus was seen to rise exponentially with the aim of innovation.

Under such conditions, a design-centered orientation to social change was seen to face the threat constituted by the tendency to institutional inertia, i.e. to forms of a "complexity-based conservatism" (Scharpf 1973, p. 68) inherent in pluralist-democratic societies.[12]

Policy-making in this conception stands before a dual challenge. On the one hand, it is called upon to define institutional means and procedures which may render growing transaction costs 'socially compatible' and results amenable of control (Bardach 1977). On the other hand, an orientation to effectiveness implies a transformation of conditions for implementation from the very definition of setting of policy formulation: building consensus and public involvement in the policy development process bears the meaning of a strategy for the transformation of the structure of actors, relationships and resources involved in implementation.

A perspective of *policy design* therefore entails a dimension of *institutional design,* the need for a *metadesign* perspective going beyond coordination and rather representing a 'coercion to institutional innovation' (Scharpf 1973) through policy actions aimed at an alteration of the distribution of power structures and resources.

Interpretations of the possible solutions to these dilemmas have clustered around two main coordination strategies, the former based on the transformation of decision-making structures, the latter on the transformation of the actors involved. On the one hand, a coordination strategy may ground on a reduction of complexity of decision-making: complex correlations and inter-dependencies of problems may be solved and broken down to simple decisions, thus reducing the need for consensus and increasing potentials for innovation through the constitution of a pluralist structure of relatively independent subsystems which multiply occasions for semi-autonomous initiative and action. This perspective however stresses problems of coordination towards goals: decentralization of decision-making may induce fragmentation and synergy may become dubious, contributing to the elusion of formal-hierarchical control. A perspective of effectiveness is thus defined by a tension between the aim of a central definition and control of objectives and the local and decentralized nature of action and knowledge involved.[13] While highlighting the role of informal and non-hierarchical strategies of coordination (Chisholm 1989), a shift in perspective towards institutionalization into weakened hierarchical models, as a response to alternative tendencies to over- and undercoordination, runs into the threat of specific decision-making deadlocks, of 'joint decision traps'.[14]

On the other hand, a coordination strategy may ground on the reduction of a patterned complexity of decision-making arenas: the political system may pursue the involvement and commitment of consolidated interests through forms of selection of the relevant actors, realizing a form of institutionalization of interest representation (as typically in neo-corporatist models or in strategic models based on a 'stakeholder approach'). This however induces forms of hypostatization which entail a threat of incorporating into the policy process elements of resistance to innovation. Favoring policy innovation thus becomes possible only through the emergence of crises, of external shocks, or rather through the emergence of new interests and new actors in the competition: policy change thus assumes the meaning of a change in the arenas and in the system of relationships of the actors involved and, accordingly, of the modes of assumption of representations, expectations and forms of mobilization in the process of problem-setting.

From the Design to the Social Construction of Policies

Throughout these strategies, the need for policy innovation tends to shift the meaning of policy effectiveness from a predefined goal orientation to the process of collective framing of goals. It is in this sense that normative orientations to policy design are increasingly being challenged, and progressively affected, by positive studies of decision-making and policy processes later to become identified as a 'cognitive-constructivist' approach (Dryzek 1982; Majone 1989; March and Olsen 1983; 1984; 1989; Dunn 1993).

Questioning a linear conception of policy-making intended as a problem-solving activity or as an exercise of power, this tradition of policy analysis, in diverse but recognizable ways, has centered its attention on the dimension of the social and interactive construction of meaning and sense inherent in policy processes, redefining its analytical and interpretive tools. Policy is thus understood in the first instance as an irreducible duality between *means* and *outcomes*, as a collective process amenable of representation as a form of puzzle.[15] The dual nature of policies thus highlighted, as a point of departure (as an *input*) and as a result (as an *outcome*) of decision-making processes, redefines the understanding of the intertemporal dimension of policy-making. Policies do not appear anymore as mainly the sequential outcome of discrete processes of decision-making, but as a coevolutive outcome of tendentially iterative and ongoing decision-making processes, in the framework of which policies are never decided once and for all, but rather endlessly defined and redefined through processes of mutual adjustment (Lindblom 1965). Criteria of policy evaluation are thus reframed according to a relativization of a 'logic of consequentiality' as well as of a 'primacy of outcomes'.[16]

This conception bears significant implications, in particular, in terms of their contribution to a radical reframing of relationships between policy analysis and institutional analysis.

First, the acknowledgment of the cognitive, behavioral and inter-temporal dimensions of policy-making processes implies, along with a redefinition of the notion of policy, a revision of the conception of power, breaking with a 'distributional' paradigm: "the question as usually put – who has power over policy – is badly stated along all its dimensions. Policy is not a constant but a variable, and power is not a finite sum but expandable on all sides" (Heclo and Wildavsky 1974, p. xiv). Shifting from a distributional conception of power to aspects of its behavioral, symbolic-cognitive and cultural conditioning (Lukes 1974)[17] opens to an understanding of its constructive, relational, strategic dimensions (Crozier and Friedberg 1978) and of its constitutive duality (Giddens 1984).

Second, the dimension of uncertainty and ambiguity which is introduced underlines the *constitutive* nature of policy-making. Contexts and processes of policy-making appear as the contingent and situationally determined environment for the construction of relationships of sense between possible problems and possible solutions, breaking with assumptions of consequentiality and time consistency between problem-setting and problem-solving (Cohen, March and Olsen 1972; Kingdon 1984). A *coevolutive* rather than *sequential* conception of processes is thus implied. A policy is accordingly redefined as a concrete course of actions, questioning the existence of a linear-causal nexus between preferences, forms of knowledge and power resources of the actors involved, and stressing the process-dependent nature of relationships established between these factors in the process of agenda-setting within complex *metapolicy* environments (Barret and Fudge 1981; Majone 1989).[18]

Third, the essence of policy-making becomes identifiable in its nature as a legitimation mechanism, in the role it plays in attributing sense to what happens, in the institutionalization of meaning (March and Olsen 1984; 1989). Addressing politics and policy-making as "an interpretation of life" (March and Olsen 1984, p. 741) thus shifts attention to their symbolic-cognitive dimension towards a very different heuristic attitude than the classic understanding of politics as the 'mobilization of bias' (Schattschneider 1960).[19]

Fourth, these theoretical perspectives affect normative views on the issues of policy reform. Questioning attitudes of 'methodological constitutionalist', they rather stress an understanding of policy-making settings as 'institutions-in-action', which highlights the 'empirical challenge' facing attempts at institutional reform as well as at the renewal of institutional analysis (March and Olsen 1984; 1989).

Accordingly, such approaches radically question the effectiveness of measures of institutional design in matters of 'democratic governance' (March and Olsen 1995). Demand and supply of solutions to issues of policy reform are not seen as linearly emerging from within the framework of parameters defined by the problems which are the object of policies: they are not given, but have rather to be framed and defined in a constructive way through the transformation of the very parameters which define problems. This position hence requires the abandoning of rationalistic strategies of coordination aimed at a reduction of complexity, in favor of a conscious assumption of the complexity of the social environment and of the task of enhancing factors of innovation and change which may determine *ex post*, as an outcome of concrete processes, effects of coordination and synergy between policies, interests and strategies which are not amenable to be designed *ex ante*. In other terms, the issue of policy effectiveness calls for a shift from a dominant focus on the regulatory function of public policy to attention to the constitutive dimension of its processes (Lowi 1972)[20] as a condition for an *anticipatory* rather than *reactive* policy-making style (Richardson 1982).

A constitutive assumption of policy-making implies (from a positive point of view) the rejection of a predefinition of relevance criteria for policy issues as a function of power relations and organizational factors. Similarly, an anticipatory style of policy-making implies (from a normative point of view) addressing the embeddedness of factors for strategic coordination and implementation within processes of a situated, context-bound and coevolutive character, according to a necessarily 'weak' form of intentionality.

Along this conceptual itinerary, the tradition of positive policy studies thus nurtures a revision of normative rationales, which draws political theory once again towards an idea of policy-making as a collective experience coming from the tradition of American pragmatism, redefining its possible institutional identity as that of an exercise in social inquiry conducted through patterned activities of collective *framing* (Schön and Rein 1994), *probing* (Lindblom 1990) and *sensemaking* (Weick 1995).

Policy-Making and Planning as Argumentative Practices

An important outcome of the convergent influence from these lines of research is a significant shift in focus in attention to the 'everyday dimension' of planning practice. The 'argumentative turn', while deeply intermeshed with a Foresterian attitude of self-examination on the part of planners, is in fact a decisive step towards a desubjectivized assumption of the institutional embeddedness of planning within the broader environment of policy-making processes. The focus shifts towards the dimension of *institutionalization* through which the actual constellation of forms of action and knowledge displayed in interactive processes shapes the trajectories of action. The inspiration is an understanding of policy-making as discourse: "public policy is made of language. Whether in written or oral form, argument is central in all stages of the policy process" (Majone 1989, p. 1). A policy may thus be analyzed as an intersection of tropes (as e.g. metaphors and metonyms) unfolding in the form of a narrative, of a 'storyline' (Throgmorton 1993). The interpretive burden put on the notion of argumentation is, accordingly, quite high:

> The institutionally disciplined rhetorics of policy and planning influence problem selection as well as problem analysis, organizational identity as well as administrative strategy, and public access as well as public understanding. (Fischer and Forester 1993, p. 2)

On the one hand, this line of research is the expression of an interpretive stance which relates it to the influence of such thinkers as Foucault and Bourdieu and to attitudes of radical deconstructionism. On the other hand, attention to discourse[21] bears the seeds for a new critical pragmatism in normative orientations. The focus is again on the social production of knowledge and on its effective linkage to action through the mediation of argumentative processes.

Reinterpreting the issue of the interactive dimension as a key for bridging different cognitive spheres towards a 'usable knowledge' and for policy effectiveness (Lindblom and Cohen 1975), a 'transactive model of argumentation' argues that:

processes of knowledge production and use are symbolic or communicative actions involving two or more parties who reciprocally affect the acceptance and rejection of knowledge claims through argument and persuasion. Thus, knowledge is not exchanged, translated, or transferred: it is transacted by negotiating the truth, relevance, and cogency of knowledge claims. (Dunn 1993, pp. 265-6)

Policy reform may accordingly be reinterpreted as a "process of rational argumentation", whereas 'argumentation' is intended as a transactional process, as "a metaphor whose roots lie in the everyday social interaction of policy makers, scientists, and citizens at large" (Dunn 1993, p. 256). Its connotations are thus those of a coevolutive process, where the codetermination of rules and contents, of processes and policy aims (i.e., of procedural and substantive rationality) proceeds through the constitution of concrete action situations, defined by their participation in a "conversation with their situation" (Rein and Schön 1993, p. 163) and by an environment shaped by the mutual influence of forms of communicative *and* strategic rationality (Rusconi 1992), contributing to redefining preferences and identities of the actors and their positions and roles in the policy-making arena.

Along this line of reasoning, two significant consequences emerge for a 'new planning direction' (Healey 1993a). Building on the interactive-communicative dimension of planning reframes the understanding of relationships between *arguing* and *bargaining* (Elster 1991), and underlines their belonging to a continuum of interactive practices through which social identities and preferences are continuously renegotiated (Douglas 1992), allowing for an assumption of negotiated approaches in planning processes which generalizes and, to a certain extent, transcends the lesson of models of alternative dispute resolution. Furthermore, conceiving planning as a critical form of public discourse makes it possible to think of proactive ways of 'inventing democratic processes' which may enable the constitution of reflexive abilities:

there are many possible democracies. Learning and listening and respectful argumentation are not enough. We need to develop skills in translation, in constructive critique, and in collective invention and respectful action to be able to realize the potential of a planning understood as collectively and intersubjectively addressing how to act in respect of common concerns about urban and regional environments. We need to rework the store of techniques and practices evolved within the planning field to identify their potential *within* a new communicative, dialogue-based form of planning. (Healey 1993a, pp. 248-9)

Far more interesting than 'voluntarist' normative versions of this ideal of 'discursive democracy' (e.g. Dryzek 1990) are nonetheless three directions of research related to the central assumption of an argumentative dimension of planning. All of them also stress the difficulties and experimental challenges in turning an institutionalist analysis of argumentative and discursive processes into a normative perspective of institutional change.

On the one hand, the multiple layers which constitute policy discourses are assumed as the key challenge for a renewed interpretation of interactive strategic planning, moving beyond rationalist as well as technocratic strategic choice models inspired by corporate planning.

Exemplary of this line of reasoning is the assumption of a multidimensional view of social interactions as a reference for 'designing' contexts for the collective framing of policy issues proposed by Bryson and Crosby (1992). Relating to the threefold notion of power proposed by Lukes (1974) and to the assumption of a pluralist societal condition of 'shared powers', the aim of a strategic steering of public choices is reframed as an experience in the 'staging' of the meaningful dimensions of strategic action in settings which allow their intersubjective transparency and critical appraisal among the actors involved. The attempt is to combine patterns of social interaction which are given in the established formal and material constitution of social practices (such as the idealtypes of *forums*, *arenas,* and *courts*) into complex settings for democratically playing out the underlying principles of preference formation (the constitution of principles of *signification*, of *domination*, and of *legitimation*) throughout the relevant stages of the policy process (argumentation and the creation of meaning, agenda-setting, policy development and implementation, and conflict resolution and enforcement).

This idealtypical representation of a stakeholder-based model of policy-making is a remarkable attempt at making sense of the continuum between strategic and symbolic-cognitive determinants, while pursuing a normative orientation in the design and use of patterned forms of interaction. As has been noted, and as is typical of stakeholder-approaches, its assumptions on mutual learning however have "less to say on how power relations in interaction work out, and little appreciation of the potential communicative difficulties which face many stakeholders in getting involved with governance processes dominated by well-educated elites" (Healey 1996, p. 263).

Research on policy discourses and 'discourse coalitions', on the other hand, has tentatively moved towards a possible 'normative use' of references to discourse formation, addressing it as a proactive means, as

Healey (1996, p. 277) puts it, "in the process of strategy development, shifting the 'storyline' of policy debate from one account to another".

Such attempts represent a crucial move in pointing to the discursive dimension as a crucial institutional factor and in pointing to the key relevance of institutionalization processes for an analysis of policy effectiveness and power. A possible normative turn of discourse analysis must accordingly move from the challenge "to find ways of combining the analysis of the discursive production of reality with the analysis of the (extradiscursive) social practices from which social constructs emerge and in which the actors that make these statements engage", as the conditions for discursive institutionalization (Hajer 1993, p. 45). The possible contribution of discourse analysis thus defined to a normative perspective lies precisely in its ability to underscore the concrete social mechanisms through which discourses are constructed and institutionalized.[22] This means moving towards an understanding of the constructive dimension of discourse, which is – for instance – hardly attainable by inquiries into the forms of social inference as proposed by approaches embedded in a tradition of social learning based on the philosophical tradition of American pragmatism (e.g. Blanco 1994).

A normative aspiration towards change from a perspective of discourse analysis may be summed up, accordingly, as a commitment to the constitution of 'reflexive discursive practices'. Discursive *reflexivity* is intended here as:

> a relational notion that should be seen as a quality of discursive practices in which actors engage. Such practices are reflexive if they allow for the monitoring and assessment of the effect of certain social and cognitive systems of classification and categorization on our perception of reality [...]. The reflexivity of actors is thus related to the extent to which they are able to mobilize and participate in practices that allow for the recognition of the limits to their own knowledge-base. (Hajer 1995, p. 40)

Clearly, as this definition suggests, a dimension of 'discursive reflexivity' implies institutional conditions. The building of 'reflexive institutional arrangements', however, cannot be intended as a mere 'external' framework of conditions, nor their provision as an exogenous process: as a framing factor, institutional conditions are always intimately involved in discourse formation. A discourse on 'reflexive institutional arrangements' is thus itself only conceivable, hermeneutically, as an emergent result of a 'reflexive discursive practice'. And such a perspective, once again, underlines the need for an experimental attitude which may combine – 'reflexively'– elements of institutional design with elements of institution-building, and the need for such a perspective to focus on the concrete social mechanisms through which processes of institutionalization actually develop.

According to a such an 'experimental' interpretation of normative ideals, a different understanding thus finally emerges of the role played by the representations which are displayed, i.e. produced as well as 'received' in planning practices, constituting an important foundation for a plea for a renewal of the "comprehensive planning ideal" (Innes 1995b). Beyond the positive and social-critical stance taken by many contributions of discourse analysis which have marked the heyday of the 'argumentative turn', the embeddedness of interactive-communicative practices into an experimental institutional framework of planning is thus increasingly seen as a condition for proactively addressing (and critically scrutinizing) processes of *framing* and the role of images and representations involved, and for possibly constituting a 'shared heuristic' (Innes and Booher 2000).[23]

Towards an 'Institutional Turn' in Planning Theory?

As Healey (1996, p. 28) argues, concluding her own account of the planning-theoretical field, "[a]ll these traditions, as they have evolved, provide pointers to the development of institutionalist analysis and communicative approaches". But is a corresponding 'turn' recognizable today in planning theory?[24]

Pleading for an 'institutional turn', after so many others, is not the aim of this book. It is however not pointless to question its existence. A particular institutional focus may in fact be found today in several different strands of planning theory. A simple but useful distinction has been proposed between two 'styles of institutional thought' in planning theory which may be adopted in the first instance as an introduction to the issue:

> The first concentrates on the issues regarding the *legitimization and embedding of planning and policy*. Some speak of the normative function that institutional approaches have in this regard. [...] The second dominant application of institutional thought is aimed at the *sustainable embedding* of certain normative targets via 'institutionalization'. Institutions are here viewed as instruments to be used by planning subjects to secure the implementation of certain principles. Therefore this view stresses the functional aspect, and includes the 'institutional designers'. (Salet and Faludi 2000, pp. 7-8)

As the authors of this distinction note, these two approaches – "aimed at legitimization and implementation respectively" (ibidem) – are not necessarily mutually exclusive, and are possibly complementary. They express, however, quite distinct heuristic strategies, and imply distinct sets of normative assumptions. Let us therefore briefly address the question of how a distinction may be useful.

It may be first noticed that, at first glance, a definition of 'institutions' is hardly what distinguished these approaches. A dominant *koinè* derived from sociological institutionalism is rather recognizable, assuming institutions in general as *patterns of social rules*, focussing on the continuum between their formal and informal determinants.[25] Both 'styles of institutional thought' in planning theory rather stress aspects of *embeddedness* through processes of *institutionalization*. Moreover, it would be misleading to identify a cause of distinction (as we shall see below) in an alleged 'positive' versus 'normative' orientation, although a distinctive bias is recognizable in their dominant theoretical inspiration.

Much more relevant in distinguishing such approaches are two further aspects: the kind of assumption and normative operationalization which is made of a common central aspect, the dynamics of institutionalization and institutional change; and the relative weight and combination of influences from the broad (and highly diverse) field of new-institutionalist studies. The former aspect may help in introducing a distinction: in doing so, we may again provisionally build on a general, elementary definition of two possible ways in the construction of institutions (i.e. in institutionalization).

According to this definition, institutions may be seen as "created and changed by human action either through *evolutionary processes of mutual adaptation* or through *purposive design*" (Scharpf 1997, p. 41, emphasis added). The former definition of institutional change stresses the *unintentional, emergent, path-dependent dimension* of institutionalization processes: we may call this (in a rather intransitive sense) the dimension of *institution-building*. The latter conversely stresses the *intentional dimension* of institutional change: we may call this (in a transitive sense) the dimension of *institutional design*. In the following, planning-theoretical positions are distinguished according to their dominant orientation to the notion of *institution-building* or tasks of *institutional design*. This will lead to evaluate the viability of this distinction in a normative perspective.

Planning Theory between Institutional Design and Institution-building

An orientation to institutional design represents the most readily recognizable set of contributions to an 'institutionalist turn' in planning theory.[26] These are in fact among the few contributions which thematize institutional issues in a recognizable and single-dimensional way. Addressing questions of institutional design is assumed here as the main way to account for the institutional embeddedness of planning practice.

One of the earliest approaches in this line of reasoning is Bolan's proposal of the notion of institutional design as a means for facing the challenges of planning. Bolan directs his broad and eclectic review of new-institutionalist contributions towards the basic observation of the meaning of the dynamics of institutionalization: "[w]e need to become more explicitly concerned with the fact that our plans are only partly involved with the engineering of artifacts and equally concerned with coping with the dialectical vibrancy of socially coordinated behavior and relationships". The 'dynamics of institutionalization' must therefore be seen "as an important part of the agenda for planning theory and practice" (Bolan 1991, p. 31).

The instrumental viewpoint of Bolan's assumption of the dimension of social and institutional embeddedness of planning practice becomes clear however in his further elaborations on the notion, despite a move from a formal-constitutionalist to an interactive dimension of institutions. Institutional design is a means which responds to the assumption that "institutional arrangements matter considerably for urban planners" (Bolan 2000, p. 36), while assuming that "a lack of adequate institutional arrangements is painfully clear" in the new governance environment (p. 25). Governance represents thus primarily a task in the design of new institutions, and this is seen as 'a critical task facing the planning profession". As such, institutional design is conceived by Bolan as the necessary complement of "advances made in communicative action theory" (p. 37) and, similarly, focuses on social interaction as the material and as the 'challenge' for institutional design, addressing the behavioral foundations of interactive practices and their guidance through the introduction of a framework of incentives rather than through a traditional focus on regulation and enforcement.

A similar argumentative role to that attributed by Bolan to the "social interactional fabric underlying institutional design" is taken by inter-organizational coordination in Alexander: interorganizational coordination is understood as:

> a form of social structure, that is, a set of rules and norms that enables and constrains action, and which is itself continually reenacted. Mediating between structure and action is the actor's knowledge of their society. The perception of their mutual interdependence is what motivates people in

different organizations to interact and coordinate their actions. (Alexander 2000, p. 161)

Thus, interorganizational coordination may be seen as the 'architecture of institutional design', and institutional design as its 'structural aspect', as "the creation or transformation of institutions – systems of rules or norms, organizations and interorganizational networks – involved in a common undertaking" (Alexander 2000, p. 160).

Alexander's idea of institutional design is in line with previous reflections on the multifaceted dimensions of rationality at play in policy-making and planning processes and with his inquiries on the dynamics of interorganizational fields in shaping the trajectories of policy formulation and implementation. As in his discussion of 'coordinative planning' inspired by notions from new institutional economics, however, where the notion of transaction costs cannot account for the symbolic-cognitive components of strategic behavior (Alexander 1992; 1994), planning and institutional design appear to be assumed basically as institutional tools for collective choice intervening in a field of collective choices, the problems and dynamics of which are seen as an *external* problem. Even if stressing the interactive and transactive dimension of institutional design and rejecting its view as "scientific institutional engineering" in favor of an assumption as "a kind of art or craft, based on what can be deduced from experience" and "on pragmatic intuition when applying the lessons of experience as a criterion of 'goodness of fit'" (Alexander 2000, pp. 167, 174), such an approach to institutional issues can hardly contribute to an understanding of institutional embeddedness as the dimension where meanings and the precontractual conditions for agreements are constituted.

While helpful in attaining a more precise definition and understanding of what institutional design may be,[27] such contributions highlight, in our view, a major difficulty of planning theory in dealing with new-institutionalist thinking. The elaboration of the features and role of institutional design proposed there appears to be still fully embedded within a paradigm of theory *in* planning, which is however clearly at odds in thematizing planning as part of an *institutional field*, as a set of practices itself subject to processes of institutionalization. Curiously enough, a consequence of this irreducible subjectivist and instrumentalist view of institutional design as a tool for strategic planning is the difficulty in thematizing the consequences of a critique of causal-linear models in terms of mutual learning, i.e. in terms which comprise the behavior of the 'institutional entrepreneurs' themselves in the 'fabric of social interactions' involved in planning processes. The question of the subject of a 'design' activity, a question, as we have seen, at the center of over a decade of planners' auto-analysis, is apparently disregarded in favor of a problematization which is based on the assumption of an abstract institutional subject. Interaction, bounded rationality and preference formation are seen as the tools for an enlightened 'institutional entrepreneur' whose subjectivity and identity and, accordingly, whose own preferences are not questioned.

In these contributions, in sum, the idea of institutional design as a planner's task falls well short of stepping out of a subject-centered perspective on planning (Salet 2000). The question to be addressed in overcoming such a perspective is therefore not so much that of a convincing thematization of how planning may 'make use' of tools of institutional design as part of its strategy. What is at stake is rather a new understanding of the domain of planning as a field of practices defined by a coevolutive interplay of factors of institutional design and institution-building and, accordingly, a new understanding of its possibilities, strategies and tools as defined by its position in this interplay.

While approaches to institutional design appear to be easily identifiable, but less amenable to an adequate conceptualization of the link between institutional design and institution-building, further approaches emerge from an eclectic field of contributions which differ significantly in the way they – more or less explicitly – address institutional issues. They range from a thematization of informal aspects of institutionalization (such as *discourse* formation: Hajer 1993; 1995) to a relativization of formal efforts in policy reform[28] and institutional change[29] down to an explicit approach to aspects of institution-building which assumes related notions – like *social capital* (Coleman 1990; Putnam 1993), *institutional capacity* (March and Olsen 1995) or a redefined conception of *collective action* (Ostrom 1990) – as the key to a reassessment of the notion of institutional embeddedness and of its implications for planning.

In the framework of this short review, our interest is directed mainly to contributions which pursue an explicit normative commitment towards planning, and which have accordingly undertaken a conscious effort in an operational, albeit non-reductionist, merging of the two dimensions of the institutional embeddedness of planning into a viable framework for action. Significantly, such efforts in operationalization bear the signs of an *experimental* attitude.

A contribution of outstanding importance in merging these dimensions is represented by two theoretical works in progress since the early 1990s, which, in some sense ironically, have recently been converging around the – in our view misleading – label of *collaborative planning:*[30] more profoundly, they share a pragmatic commitment to planning as a coevolutive field of social action which, moreover, represents a compelling pragmatist response to the drift towards a radical solution to the planner's subjectivist dilemma outlined in Friedmann's *opus magnum*.

Significantly, Innes's position moves from an 'epochal' statement of the need and possibility for a new experimental (and experiential) commitment to planning,[31] shifting away from an 'ethnographic' focus on the planner's everyday practice as a struggle for justice and emancipation, which still dominates the focus on the role of the planner as a 'deliberative practitioner' (Forester 1999), and rather stressing the incremental and path-dependent evolution of social settings of planning towards the definition and achievement of collective goals. The microanalytical dimension of communicative action and interactive processes thus gains a new significance as it is normatively linked to the macrostructural dimension of institutional embeddedness and to its potential trajectories of change.

Breaking the taboo of comprehensive aspirations of planning and conferring on them a new meaning is a task addressed by Innes in the light of a new phenomenological consciousness of the constructive character of processes through which the plan's action orientation is achieved. Practices of consensus-building – in a sense which is, in our view, closely akin to the meaning attributed by Mandelbaum (1990) to 'reading plans' – result in both a transvaluation of the object of planning and in a reidentification of the subjects of planning. The comprehensive character of planning in the framework of communicative-interactive approaches, more than in the case of the rhetorical-persuasive function of visioning in most experiences of marketing-oriented strategic planning, becomes both the medium and the outcome of a process of collective learning and identity formation, built around the shared representations of problems and objectives:

> consensus building does focus on specifics, and yet its concepts are general and designed to anticipate the future. Consensus building groups that become deeply enmeshed in understanding and dialogue usually seek comprehensive solutions, aided by scenario building. Perhaps most important, these groups can develop new 'public interest' conceptions of themselves as being part of a regional system or a community through this kind of imagining process. (Innes and Booher 1999, p. 21)

What must be highlighted in such positions is the radical overcoming of an instrumental or goal-oriented assumption of interaction processes and of consensual decision-making procedures, as well as of the representational activities which are subject to institutionalization in the framework of planning: the multiple and diverse actors of planning processes are not only bound to and constrained by institutionalized contexts, but also embedded into them and, as such, actors of their coevolution. Implicitly, this changes the ethical-constitutional connotations of planning as a social practice, definitely detaching them from a subjectivist stance.

In this sense, as Innes (1995a, emphasis added) proclaims, "planning *is* institutional design". It may be seen as ironic, however, that the acknowledgement of the innovative potentials conferred to planning by (rather than despite of) its being an institutionally embedded practice is labeled here as 'design'.[32] This is all the more clear in the light of the contribution this line of reasoning offers to an operational framing into practice of indirect aims of policy-making and planning (like the constitution of *social, intellectual* and *political capital* or, in more general terms, the issue of building *institutional capacity*: Innes *et al.* 1994) which rather stress the incremental, emergent, path-dependent and paraintentional features of processes which have been previously defined as *institution-building*. While probably the signal of a conceptual struggle, and of a more general theoretical difficulty in dealing with (new-)institutionalist notions, this identification is nonetheless a significant hallmark of this approach: it expresses a strong claim for an experimental, transformative attitude, attuned to address, in an evolutive perspective of social change, a mutual fertilization between the purposive rationale of institutional inputs and the social intelligence of interactive processes.

The crucial contribution of Healey's view on 'collaborative planning' is a move towards a micro-macro analytical linkage which, while adhering to similar assumptions on the meaning of structuration processes and of the relationships between structure and agency in thinking about trajectories of social change, introduces a decisive boundary-spanning move. The object of institutional issues moves towards a consideration of the *inter-institutional* and *interorganizational field* which comprises planning practices and defines their patterns of institutionalization. The object of an institutionalist focus is the multiple domains of knowledge and action involved in territorial transformation and their specific processes of institutionalization. Understanding their intersection in concrete territorial contexts is thus crucial in making the "links between economic, environmental and social dimensions of issues as they interrelate in places" (Healey 1996, p. 69), in a relational-constructive as well as path-dependent understanding of place-boundedness. Not secondarily, this attitude offers a theoretical framework for a cross-fertilization between different disciplinary fields concerned with the dynamics of territorial development (e.g.

regional economics and geography) traditionally rather alien to planning-theoretical thinking.

The merging of these contributions underpins an understanding of the institutional dimension as a social-constructive process, as the building of an *institutional capacity of places*. Healey's syncretic effort in understanding the institutional embeddedness of planning practice is thus probably the most advanced step beyond the subjectivist stance which has characterized much planning theory as an existentialist struggle with a constraining assumption of the institutionalization of the planner's practice. It is however, most notably, still the expression of strong normative, if experimental, aspirations.[33] The 'institutional capacity of places' is a constructive challenge requiring efforts in 'systemic institutional design':

> systemic institutional design is important because it carries substantial power to frame the specific instances of governance activity. These systemic designs are not autonomous, isolated from the wider relations of governance. They are enmeshed in networks of relations which contribute to their articulation and realisation [...]. The task of institutional design at the level of appropriate institutional frameworks which could have the capacity to encourage collaborative, inclusionary consensus-building is therefore the design of appropriate systems or regimes. It involves the transformation and re-making of 'abstract systems' [...] which will perform useful shaping, structuring and framing work, rather than inhibiting the development of collaborative consensus-building. (Healey 1996, p. 287)

The experimental combination of 'systemic institutional design' within situated contexts of institutional capacity is thus seen as a collective endeavor, the achievements of which are only conceivable as involving all the subjects contributing to the identity of a place. The idea of a practice of 'collaborative consensus-building' becomes the metaphor for a continuum of practices linking the prospects for purposive action to the enhancement of the social and institutional capital on which abilities to act ground.

In the following chapter, the idea of consensus-building is assumed as a paradigmatic expression of a development in planning theory which recognizes the necessarily experimental and experiential dimension of its normative ambitions within conditions of institutional embeddedness. Consensus-building approaches are thus assumed as the most mature efforts in proactively addressing the duality of planning as an *institutionalized practice* and as a *factor of institutionalization* through a combination of institutional design and institution-building. Their discussion will however step back from planning-theoretical arguments, addressing them from a perspective of political theory and policy analysis.

Notes

1   Among the earliest contributions (some of which are discussed below) explicitly addressing the institutional dimension of planning as a new theoretical challenge are: Mandelbaum (1985), Bolan (1991), Innes (1992), Alexander (1992), Healey (1996).

2   This critical itinerary is exemplarily mirrored by Webber's reflections. After the critical detachment from ideals of 'scientific planning' and from a centrality of instrumental knowledge in the substantive definition of policy objectives (Rittel and Webber 1973), the acknowledgement that "there are no scientifically or technically correct answers, only politically appropriate ones" (Webber 1978, p. 157) turns Webber towards an ideal of 'permissive planning' (Webber 1969) based on the acknowledgement of differences entailed in the plurality of contributions to policy choices and in the irreducible character of their value orientations (Webber 1978). Planning is thus redefined according to its responsibility in enhancing pluralist expression and exchange among societal differences, as a 'cognitive style' and rational technique for democratic evaluation of choices, mediated by its multidisciplinary knowledge base.

As a form of procedural and informational intelligence, centralized (institutionalized) planning does not abandon, in Webber's view, its comprehensive ambitions: these however are reframed according to the view of an institutional role of planning in setting the rules for the expression and promotion of social pluralism.

This conception entails a criticism of formal-institutional hypostatization of forms of participation in the planning processes: the aim of planning is rather an idea of effectiveness targeted to the aim "to insure that all parties can champion their favored ends with the best available intelligence and analysis, that they are free to transact their affairs and to argue their causes, and that the political system remains informed and open to all contenders" (Webber 1983, p. 99). The role of planners is accordingly seen as that of 'expeditors of pluralistic political processes' (Webber 1983, p. 97).

While not going far in contributing to the analysis of concrete planning practices, Webber's conception of the institutional embeddedness of planning practice and his idea of the planner's role move in the direction of a desubjectivization, which stands very much in contrast with the existentialist centrality of the planner which still is the moving point of analysis, for instance, in the work of Forester and Friedmann.

3   A few contributions in the early 1980s stress the former aspect in a vaguely new-institutionalist perspective. The aim is to understand organizational factors as political-strategic conditions for achieving ends, with a focus however on cognitive aspects of the planner's embeddedness in an interorganizational environment and on the need to overcome their understanding as exogenous constraints (Baum 1980a; 1980b; 1983). Baum's critique of the profession's ideal of effectiveness is thus radical: it addresses a conception based on the representation of a conflict between planners' own expectations and the organizational realities, which responds to a cognitive map seen as reflecting elements of 'professional socialization' typical of the tradition of most planning schools, and which determines a peculiar forms of myopia, focussing on distant ends and near techniques without addressing the mediating role played between them by organizations. The consequence is a misunderstanding of the need to correlate interests, alternatives, actors and constituencies in concrete organizational contexts. The need to break with rationalistic expectations based on the idea of a planner mastering problems as an *entrepreneurial problem-solver* points to the dimension of organizational learning entailed in processes of problem formulation.

A general rethinking of the roles taken by planning as part of interorganizational fields is thus advocated. Similarly, Alexander's idea of the strategic *design* of courses of action makes the link between the 'rational' and the 'creative' components of effective decision-making, pointing to the dialectics of processes of organizational learning. Policy implications thus point to the design of rationalized procedures for constructive and creative exercises and for heuristic research on policy alternatives, which combine the institutionalization of rational strategies with an openness to non-rational processes and to the 'paradoxes of creativity' (Alexander 1982).

4 Forester moves from a rejection of an idea of planning rationality as a cognitive technique in favor of the idea of a 'cognitive style' similar to Webber's. His originality consist in turning this critique since the 1970s towards an idea of critical and pragmatic rationality aimed at a combination of the normative, empirical and interpretive components of the planner's expertise into a new understanding of political action (Forester 1980). Forester's itinerary is paradigmatic of a theoretical attitude which, having recognized the delusion of institutionalized planning practice as a means for social-welfare policies, points to the everyday interaction with power-endowed actors as the main possibility for a commitment of the planner to a balance of power relationships (Forester 1989). Unlike Friedmann, however, Forester pragmatically keeps to a disciplinary perspective of theoretical inquiry, centered on the institutionalized role of the planner as an expert and a professional. Based on an inquiry into the planner's practices framed by reference to Habermasian discourse-ethics and to the notion of communicative rationality, Forester methodically refrains from addressing structural factors of power focussing on the inherent potential of emancipation and empowerment of the practices of social interaction involved in planning processes. Having acknowledged the basically pragmatic and communicative character of planning action, its irreducibility to an autonomous cognitive sphere, and rather its full embeddedness into social action, Forester is mainly interested in defining criteria for what we might call a 'discursive ethics' of disciplinary action: besides and beyond technical and organizational abilities, the focus on the planner's expertise is broadened towards the nature of communicative acts involved in its social practice and towards the strategies available for an emancipatory, progressive use of power resources implied in information, argumentation and social participation (Forester 1980; 1982; 1989). This confers on the planner as an agent of social communication a fundamentally political role, directed to the achievement of substantive democratic values through interactive-communicative processes, summing up in this the responsibility of belonging to the 'public domain'. Equity planning is thus connoted in an essentially pragmatic way, according to a critical pragmatism centered on issues of constitutional and distributive conflicts, and aimed at principles of commutative justice (Forester and Krumholz 1988; 1990). In a perspective of pluralist democracy set into contexts of multiple interactions, the power inherent in the use of information and communication introduces, however, a shift in the democratic function attributed to planning: it turns it into an instrument for a 'political economy of attention and argumentation' (Forester 1989) as part of a broader process of "political alphabetization" (Forester and Krumholz 1988, p. 82), of the reconstruction of citizenship.

The planner regains his/her center precisely in this function as a mediator: no more (or only) a bureaucrat, the planner's specific identity, gained in the turn from a *provider* to an *enabler,* is in any event precisely defined within the boundaries of the form of rationality determined by his/her institutional position.

Forester is thus interested in institutional-political contingencies and in the potential and vulnerability of concrete social dialogue rather than in abstract principles, like an alleged 'ideal speech situation' (Forester 1993b). It is thus difficult to overestimate the meaning of his key question, i.e. what could happen in social practice "if social interaction were understood neither as resource exchange (microeconomics) nor as incessant strategizing (the war of all against all), but rather as a practical matter of making sense together in a politically complex world" (Forester 1993b, p. ix). In his perspective on communicative practice, and even in his contribution to the 'argumentative turn', however, a central focus remains constant on the subjectivity of the planner. This attitude is clearly indebted to the legacy of *advocacy planning*: accordingly, Forester defines the practice of *progressive planning* he proposes as a refinement of the idea of advocacy based on a mature understanding of the belonging to concrete systems of action (Forester 1982; 1994).

5   Contrary to Webber and Forester, Friedmann explicitly thematizes the hermeneutic circularity which defines the planner in his metaphor of a 'Janus-faced' social role: 'in terms of social space, radical planners occupy a position tangential to radical practice at precisely the point where practice intersects theory' (Friedmann 1987, p. 392): a position which projects planning practice in a radical perspective of social reform.

Since the 1970s, Friedmann has centered his reflection on the need for planning to face the turbulent, entropical features of the social environment, based on a critical reconsideration of models of policy development (Friedmann and Abonyi 1976). The mediation between knowledge and action becomes the essence, the very meaning of planning practice, and is interpreted as the exercise in the social epistemology of practice, in line with a tradition of social learning. Its connection to the issue of a reconstitution of the social system bears however radical implications for planning involvement. The conditions for social innovation are seen in practices of social learning, which enable the emergence of new opportunities for collective action through the development of group relationships. The centrality of communicative acts in mediating theory and practice, as in Forester and Schön, stresses the role of the planner as an agent of transactive relations in contexts defined by reflective behaviors and mutual learning, mediated by a repertory of abilities defined by formal as well as experiential codes (Friedmann 1973). Sensibly more complex, however, is Friedmann's thematization of the planner's role as an agent of change, and in particular of the planner's position in the institutional universe, as his/her implication in radical social practice is assumed as more organic. Since "in radical planning, the relevant knowledge, embedded as it is in a transformative theory, is always and necessarily contextual", and "[t]his contextualization is a profoundly social process in which those who stand in the frontline of action [...] make a decisive contribution" (Friedmann 1987, p. 394), in the arenas of radical action planners face the contradiction of being subject *and* object of experimentation of their own social role.

The contradiction of this subjective dilemma is significantly solved in a non-subjectivist perspective: i.e. in an evolution of the idea of a transactive foundation of planning rooted in the tradition of social learning (Friedmann 1973) towards the a self-centered perspective of territorial development, which merges the dialectics of knowledge and action into a strategic direction of social mobilization and of recovery of 'human territoriality' (Friedmann 1987; 1990), inspired by the culturalist tradition of American regionalism (Friedmann and Weaver 1979), which ties the action of the planner to a coevolutive process aimed at the cooperative construction of a political-economic unity rooted in the social determinants of locales (Friedmann 1992).

6   It has been accordingly observed that the approach proposed by Schön (1983; 1987) to the question of the nexus between knowledge and action, i.e. between the production and the use of knowledge, is beset by the limits of a model which tries to solve the dichotomy between relevant knowledge and profession "in the terms in which it is, if not misplaced, less relevant – i.e. the formal detachment of the actors" (Crosta 1990, p. 147). Friedmann's critique of much of the tradition of social learning points similarly against an ideology of 'learning to learn' which disregards implications concerning power relationships and cannot therefore critically account for issues like the heterodirect and centralist character of most measures in a 'politics of knowledge' and its influence in framing action (Friedmann 1987). It would, however, be misleading, in our view, to underestimate the inherent and potential political meaning of Schön's epistemology of practice, as is made clear by the substantial influence it has exerted on critical planning theory.

7   For the notion of *field*, see ch. 11, note 4.

8   In Schön and Rein's expression, these self-limited forms of experience, based, as would be said in terms of sociology of science, on a detachment between the 'context of discovery' and the 'context of justification', progressively constitute a form of *twilight discourse* characterized by a predisposition to "slow learning and speedy forgetting" (Schön and Rein 1994, p. 72).

9   Scharpf (1978, p. 348) expressed this in the following terms: "the plurality of individual goals and strategies actually involved in public policy processes, while sufficient for the empirical explanation of policy outcomes, cannot serve as a prescriptive frame of reference for policy evaluation. The aggregate of partisan goals will not be hierarchically integrated, and they are likely to be contradictory. Hence, no strategy could be designed that would fully realize the aggregate of individual goals, or whose actual performance could be evaluated in terms of such a goal set. Thus, empirically identifiable goals, whether those of one individual or those of all participants, appear to be equally irrelevant for the purpose of prescriptive policy analysis and of policy evaluation".

10   The following discussion of contributions to planning theory from the sociology of organizations and from policy analysis and implementation research is intended as complementary to the discussion presented below in ch. 11 with particular reference to the development of new-institutionalist schools of thought and of inquiry.

11   A significant shift is recognizable in implementation studies oriented to an inquiry into strategic behaviors in multiactor environments from an initial focus on a 'functionalist' assumption of policy effectiveness confronted with the limits of authoritative implementation settings (as e.g. in: Pressman and Wildavsky 1984, orig. ed. 1973; Mayntz 1977; Bardach 1977) to a focus on effectiveness tied to the interpretive dimension entailed in implementation processes (Majone and Wildavsky 1979; Barret and Fudge 1981; Browne and Wildavsky 1983a; 1983b).

12   This aspect is well mirrored in this representation of the dilemmas of planning in coping with the need for social innovation: "[t]he first dilemma derives from the fact that innovations to be developed in the long term rarely are compatible with presently prevailing orientations and interests [...] and that an exclusive involvement of interest groups acting in a comparable long-term perspective would entail disparities in political influence. Interest groups less capable of articulation and organization would be systematically underrepresented. If innovation however has to direct against dominant interests and orientations, this poses a serious issue of power. Hence – this

the second dilemma – the need for consensus becomes greater the more innovative policy approaches are. [...] The inertia of the *status quo* reveals its power, since change becomes possible only on the basis of a redistribution of benefits and costs" (Häussermann and Siebel 1994, pp. 55-6). Democratic-pluralist systems, as noted by Scharpf, may therefore stick to a form of 'conservatism by excess of complexity', reproduced by the same networks through which the democratic representation of interests is realized: as power structures built on consolidated clusters of interests, these networks may become factors for 'institutional deadlocks'. In other terms, resistance to innovation may become institutionalized, embedded in an institutional system which reproduces consolidated interests.

13   This dilemma is typical of what has been defined as 'coordination in the shadow of hierarchy' (Scharpf 1994b; 1997), as the interorganizational version of the principal-agent problem defined by intraorganizational relationships in the model of the firm (for a discussion of implications in public policy, see: Moe 1984).

14   In its classic definition (Mayntz and Scharpf 1975), overcoordination typically affects hierarchical systems, whereas centralization hinders the access and effective contribution of decentralized, 'local' knowledge, while involving complex bargaining (i.e. efforts in 'positive coordination') between the parties, thus causing inefficiency and ineffectiveness as well as barriers to change and innovation. Undercoordination may conversely be seen as a typical outcome of shared-power systems, where the need for an active consensus on joint choices either induces political compromises or is bypassed through embedded forms of negotiation (i.e. through means of 'negative coordination') which, again, tend to push contradictions and problems out of the system and to avoid institutional rearrangements, thus limiting the potential for innovation. In both situations, coordination settings relying on consensus among the actors as a basis for joint action tend to suboptimal outcomes in terms of problem-solving capacity, whereas the rationale for getting to agreements among the parties involved in the policy-making process is based on given interests and definitions of the situation by the actors, i.e. on a *bargaining* decision style, leading to the establishment of 'joint decision traps'. For a further discussion, see ch. 3.

15   In Wildavsky's famous definition, "policy is evermore its own cause": "[p]roblems are defined by hypothetical solutions; the problem's formulation and the proposed solution are part of the same hypothesis in which thought and action are fused" (Wildavsky 1979, pp. 62, 83). Policy is thus "a process as well as a product", a term which "is used to refer to a process of decision-making and also to the product of that process" (Wildavsky 1979, p. 387). Heclo (1974) speaks of the production of policies as a form of learning which entails the constitution of forms of knowledge as well as processes of decision-making: "[p]olitics finds its sources not only in power but also in uncertainty – men collectively wondering what to do. Finding feasible courses of action includes, but is more than, locating which way the vectors of political pressure are pushing. [...] Policy-making is a form of collective puzzlement on society's behalf; it entails both deciding and knowing" (Heclo 1974, p. 305).

16   This aspect was already present in Rittel and Webber's idea of the 'wicked' character of policy-making: "[t]he information needed to understand the problem depends upon one's idea for solving it. [...] The problem can't be defined until the solution has been found" (Rittel and Webber 1973, cited in: Wildavsky 1979, p. 58).

17   A revision of the understanding of power in the social sciences was notably introduced by debates around urban policy issues, starting from a critique of élite theories of oligarchies (Hunter 1953) and of 'reputational' approaches to polyarchical

patterns (Dahl 1961; 1971; Polsby 1980) by Bachrach and Baratz (1962; 1970) and further by Lukes (1974), including more recent contributions (e.g. Stone 1989).

18 This also questions ed a distinction of phases of policy-making which conceptually separates the stages of decision-making and of the formulation of policies from those of implementation: a distinction which, as has been noted, "tends to attribute heuristic value to categories of a juridical type", running into "the risk of restating the empirical validity of a distinction between politics and administration" (Dente 1990, p. 16).

19 A perspective is thus introduced which questions the meaningfulness of a traditional repertory of analysis "when ample processes of the redefinition of the meaning itself of public policy are at stake": "[i]n public policy, in fact, the cases are rare in which facts speak for themselves, in which problems have the power of imposing themselves on the attention of policy-makers due to some kind of instrinsic power [...]. This means that the traditional mode of framing the study of a public policy – asking which objective causes have determined the crisis of old solutions, which context variables have intervened, how the relationships between the main actors have been defined – is just one of the many possible ways of understanding how things happened. And possibly not the most reliable, when wide-ranging processes of redefinition of the meaning itself of public policy are at stake" (Regonini 1993, p. 363).

20 The notion of *constitutive policies* was introduced by Lowi (1972) and is used by Dente (1990, p. 18) to refer to policies "which transform the organizational and procedural modalities of the development of public activities". New-institutionalist approaches in policy analysis and sociology have extended the implications of this concept to the symbolic-cognitive and strategic dimensions of policy-making.

21 Hajer (1995, p. 44), building – among others – on Foucault, defines *discourse* as "a specific ensemble of ideas, concepts and categorisations that are produced, reproduced, and transformed in a particular set of practices and through which meaning is given to physical and social realities".

22 "If a discourse is successful – that is to say, if many people use it to conceptualize the world – it will solidify into an institution, sometimes as organizational practices, sometimes as traditional ways of reasoning" (Hajer 1993, p. 46).

23 The notion of *planning doctrine* proposed by Faludi (Faludi and Van der Walk 1994; Alexander and Faludi 1996) is perhaps the most interesting attempt (even if initially rather on an analytical-descriptive level) at connecting the role of territorial representations to a specific context of institutional relationships and of social and professional planning practices. Attention to representations has assumed a growing importance in the analysis of institutionalization processes (e.g. Faludi 1996; Zonnefeld 2000). Of course, this aspect is also important in approaches to the analysis of discourse and of the building of *discourse coalitions* (Hajer 1993; 1995). A new-institutionalist approach to the role of images which inquires into their 'constitutional' and proactive dimension has been proposed by Neuman (1996). It must be noted that the social-psychological and philosophical foundations of these approaches, even in their possible normative turn, confers on them a clearly alternative attitude towards social-constructivist processes compared to the conception of 'visioning' in use in corporate-inspired models of strategic planning.

24 The question of an alleged 'institutional turn' in planning theory has been for instance raised as the subject of a roundtable during the plenary session of the XIII Aesop Congress in Bergen (Norway) in July 1999.

25 Three examples of the scope of institutional analysis implied in this definition, randomly taken from planning literature: Young (1994, p. 3, cit. in: Low 1977, p. 94)

who echoes that "[i]nstitutions are sets of rules of the game or codes of conduct that serve to define social practices, assign roles to the participants in these practices, and guide the interactions among occupants of these roles"; Salet (2000) who, building on Mayntz and Scharpf (1995), speaks of three levels of institutional analysis as centered on *social rules and systems of beliefs, formal rules of regimes,* and *institutional reflection in p*ractice; or Bolan (2000) who, building on the matrix (not a 'map'!) of types of power as forms of strategic games proposed by Low (1997, p. 91), considers the institutional dimension of interactions as the intersection between *structural* and *performative* aspects of games with the *public* and *tacit* dimensions of action.

26   Significantly enough, 'institutional design' is reemerging as a central issue in strategic planning just at a time of raising 'institutionalist awareness', whilst it has in fact always been its dominant aspect, in rationalist or technocratic strategic-choice models extending to approaches to the strategic steering of shared-power systems as in Bryson and Crosby (1992). The most interesting attempts in readdressing approaches to strategic planning in a new-institutionalist perspective are represented by: Healey *et al.* (1997) and, more recently, by contributions to the Amsterdam symposium collected in: Salet and Faludi (2000).

27   Alexander (2000, p. 165) defines institutional design as "the devising and realization of rules, procedures, and organizational structures that enable and constrain behavior to fit held values, achieve desired objectives, accomplish set purposes or execute given tasks", implying "a view of institutional design which is pervasive and manifest at all levels of social deliberation and action, including legislation, policy making, planning and program design and implementation".

28   A similar attitude is expressed by positions from administrative science inspired by a 'cognitive' and interactionist approach to policy analysis. An example are recent elaborations by Dente (1998) on the need for parsimony in institutional design: the 'construction' of new institutions is seen as a possibly useful means for the design of public policies in cases where the aim of reframing conditions for problem-setting and of restructuring decision-making arenas is at stake in facing 'wicked' policy issues.

29   Thus, for instance, in the imagery of 'glocalization', a transformative perspective of institution-building implies that "even more important [than institutional design is] to understand the way in which global changes are mediated at every level below the global and how they become interpreted by people, spread through networks, and finally embedded in tacit assumptions and practices" (Low 1997, p. 108).

30   This is, in other terms, a good reason for looking at planning-theoretical labels with an *ironic* attitude, the more so being ourselves involved in reproducing them.

31   Reference of course is to Innes's plea for an experimental and practice-oriented 'communicative-interactive' paradigm in planning theory and for a corresponding reappraisal of the 'comprehensive planning ideal' through a critical confrontation with the critique by Altshuler (1965a; 1965b), contained in two well known and closely related papers from the mid-1990s (Innes 1995b; 1996).

32   Alexander (2000, p. 159) makes a significantly different comment, noting that "some planning is not institutional design", as in his definition, rather than stressing the existence of a dimension of institutional change potentially inherent to planning practice which is *not reducible* to intentionality, i.e. to purposive design inputs.

33   Healey (1996, pp. 68-71) expresses this in her threefold set of normative criteria for judging public policies from an institutionalist viewpoint: we may summarize them as a criterion of *goal effectiveness*, a criterion of *situational learning*, and as a criterion of *inclusionary involvement and networking* of stakeholders.

3 Perspectives on Consensus-Building in Planning

Introduction

In a recent paper A.O. Hirschman (1994), discussing social paradigms centered on the resurgence of a call for common and shared values, pleads for an understanding of *conflict* as a *constructive* and *constitutive element of social relations*, as a source of its strength and ability to innovate. According to this view, the possibility of integration within modern societies, contrary to commonsense in social thinking, derives from their ability to develop across conflicts, and to constitute capabilities in dealing with the insurgence and variety of conflicts. In modern societies, according to Hirschman, conflicts might just as well be sources for building social capital as they might be factors for the dissolution of social bonds. Their definition, however, either as opportunities or threats, as 'constructive' or 'destructive' conflicts, is itself subject to change as the rationale of politics is changing; as such, their very nature in this respect is hard to assess, and their identification is only possible through experiencing them.

In discussing the meaning of approaches to consensus-building in policy-making and planning, we would like to start from this apparent paradox and, similarly, plead for a conception of consensus-building which understands 'consensus' as the 'other side of conflict', as a possible dimension indissolubly tied to the everyday reality of conflict.

Consensus-building, in this sense, stands for a constructive attitude in which 'consensus' (however situationally defined) neither constitutes a precondition for action nor a means for bypassing conflicts, but a possible outcome of a deliberate confrontation with conflicts through the enactment of concrete action situations. As far as consensus is understood as the other side of conflict, and conflict as a resource, consensus-building may be understood as *a strategy for constructively dealing with conflicts*.

The awareness of the integrative and generative potential of creative approaches to dealing with conflict is a major lesson of experiences in alternative dispute resolution, a primary source of inspiration for consensus-building approaches in planning. The contextual reference for our discussion of the meaning of consensus-building strategies in this

63

chapter, however, is not primarily given by situations of social conflict which constitute the typical field of application of mediated forms of negotiation. It rather relates, in more general terms, to the fragmentation, differentiation, complexity, and increasingly multiactor connotation of multilevel governance environments. As we have pointed out, policy-making in these conditions reaches beyond an activity of the mere construction of solutions: it becomes an activity through which actors address the very construction of policy problems and the constitution of conditions for jointly addressing possible policy solutions. Policy-making processes thus bear an important social-constructivist and discursive dimension, a dimension of sensemaking. Conflict and consensus, accordingly, are not only or mainly to be seen as ways and means through which policy solutions are defined, i.e. for *problem-solving*, but as part of the 'arts and crafts' of defining policy problems, i.e. of *problem-setting*.

The emergence of approaches to consensus-building is hence strictly related to factors which increasingly affect the understanding and the culture of public policy-making: the changing features of tasks of intergovernmental and interorganizational coordination, and the social-constructive and sensemaking dimension of politics.

Consensus-building assumes in this context the meaning of an alternative approach to policy-making, the foremost feature of which is the multiplex, flexible, and tendentially non-hierarchical definition of policy arenas and issues, challenging institutional rationales and settings and bringing to the fore need for institutional experimentation and innovation.

As such, consensus-building will be regarded in the following as an issue to be inquired into rather than as a solution amenable to model-like reductions and to reproduction. In this sense, the meaning we attribute to consensus-building and to concerted and negotiated policy-making as notions useful for understanding innovative practices in territorial governance differs conceptually from the meaning it bears within formalized approaches (particularly within certain models of strategic planning and within the literature on negotiation), as well as from a view of it as merely a decision-making procedure.

Rather than as a formalized technique for settling controversies and taking decisions among defined sets of preferences, consensus-building will be regarded as a creative, institutionally supported and ongoing practice of adjustment and positioning among policy actors towards the definition of shared preferences, and will be explored in its role in innovating institutional approaches of territorial policy-making and, to a certain extent, in challenging their formalization: i.e., in stressing their dimension of institutionalization as a critical issue and as a challenge.

Consensus-Building and Territorial Governance: A Positive View

As we have noted in introducing a discussion of the analytical and normative challenges of governance, developments in the field of public policies seem to highlight the growing importance of – as well as the 'practical need' for – approaches to decision-making based on consensual procedures and unanimity rules, and this "even in situations where formal decision rules would permit, or even require, either unilateral/hierarchical or majority decisions" (Scharpf 1989, p. 155).

The dominance of formal rules of decision-making of a hierarchical type seems thus, from a positive point of view, to be fading even in contexts characterized by the highest level of formalization, traditionally identified with procedures of democratic deliberation in the public sector. This tendency, in the light of empirical evidence, entails a revision of the dichotomy between the public and the private sphere, as well as a progressive blurring of oppositions between arenas of command-and-control and arenas of trading and bargaining, between hierarchy and the market, both in the field of politics and of socioeconomic processes.

In the wake of the restructuring of territorial policies defined by local and regional responses to the global dynamics of economic competition and by the progressive devolution of strategic policy-making assets, a wide middle ground of mixed policy-making settings based on consensus and on negotiated agreements is developing into a new field of institutional experimentation. Several experiences may be recognized, in different forms and contexts, which bear paradigmatic relevance in a perspective of integration between local development aims and supralocal policy rationales, assuming the form of intergovernmental, multiparty strategic agreements for the government of urban regions, in the framework of organizational settings characterized by low levels of institutionalization.

Common to these experiences in the reform of the institutional rationale of territorial policy-making is the role of consensus-based forms of decision-making in facilitating and mediating the enhancement of self-promoted and self-committed, territory-based forms of competitive cooperation; institutional conditions of progressive decentralization and devolution of powers are thus assumed as opportunities for concrete, integrated and targeted paths of development defined by locally shared agreements towards action.

Institutions of territorial governance are thus reframed as complex *interorganizational fields* of relations between actors and forms of decision and action, connected in a coevolutive dimension in which innovation is an emergent and concurrent (albeit not necessarily non-conflictual) outcome, occurring at the micro- and mesolevel of social processes in which concrete forms of action in the management of territorial resources take place.

It is unmistakable that these phenomena, especially in the context of strongly territorialized policies, share a renewed dominance of aspects of politics which recall its dimension of collective sensemaking, shifting attention from the exercise of power to the social construction of meaning. Conversely, it cannot be denied that such practices are developing within a general crisis of representative democracy, against the background of processes which often stress rules of legitimate democratic expression and, consequently, point to our actual deficits in democratic theory and practice.

The experimentation of consensus-building in territorial policy-making develops therefore in a contradictory horizon. The aim of our arguments is to focus on some of the relevant questions which may be raised through a critical scrutiny of their diffusion as a means for innovating institutional policy-making rationales.

Towards the Social Construction of Policies: Consensus-Building as a Strategy for Facing 'Wicked' Policy Problems

As has been pointed out in previous chapters, a major reason for the increasing demand for institutional experimentation in the changing governance environment is a different understanding of the challenges of intergovernmental and interorganizational coordination.

The issue of *coordination* is defined, in classical terms, by the problem of pursuing effective strategies for focussing the actions of different institutional actors and organizational rationales. The aim of coordination defined as such, however, faces tendencies towards an increasing fragmentation of structures and competencies relevant for political-administrative activity, combined with a proliferation of actors endowed with relative degrees of autonomy, with a non-linear overlapping of possible problems and solutions, and with a progressive blurring of tasks of a political-decisional and administrative-executive order. The 'theatres of interaction' (Giddens 1984) which actually structure political activity broaden the field of actors between the spheres of the public and of the private, blurring traditional categories of roles and institutional definition, reframing conditions for the insurgence and the definition of interests and the settling of conflicts.

Coordination in this sense bears the features of a *wicked* policy problem, of a rather multidimensional nature and, as such, is hardly amenable of model-like reductions. The level of vertical coordination among institutional levels of territorial government is backed by a level of horizontal coordination, among the multiple actors who concur to an effective implementation of policies, which decisively imposes a shift to the domain of the arenas pertaining to the policy process, away from rigid, institutionally based forms of predefinition.

The 'discovery of social actors' in the framework of a renewed paradigm of action-oriented coordination confronts the institutional design of policy processes with the paradox of the ineffectiveness of mandated and regulatory forms of coordination, and with the need for experimental self-governing forms of coordination, in a framework of progressive relativization of formal or linear consistency requirements.

The idea of intergovernmental coordination which is shaped by this attitude towards the 'governance of fragmentation' tends therefore towards the definition of institutional forms which may be capable of enabling constructive interactions among multiple actors as well as enabling the transformation of zero-sum games into mutual-gain situations. In such a form, processes in policy development and in the definition of conditions for effective implementation become involved in the reconstruction of the settings of actors and of their relational framework around concrete action situations. As such, these relational settings tend to assume the meaning of coevolutive forms of 'institutions-in-action'.

Coordination of action therefore tendentially assumes the features of both a practice in *institution-building* and of an issue of *collective action*. The establishment of 'coordination' as an effective form of action acquires the meaning of the ongoing constitution of conditions for a shared orientation of multiple strategies and action rationales, i.e. of the primary conditions for collective action: the building of settings which allow the constitution of credible commitments towards action, of formal as well as informal means for the enforcement of agreements, and of stable forms of self-monitoring.

Practices in the building of consensus, in this perspective, may be interpreted both as a strategic perspective and as a necessary means for realizing conditions of *intersubjective communication* and *symbolic-cognitive mediation* which are indispensable for the constitution of forms of agreement towards action and for their assessment and coevolution throughout time. Approaches to the resolution of dilemmas of collective action through the proactive involvement of structured forms of interaction thus bear the features of an activity for the *collective framing* of policy issues, committed to the argumentative interplay of representations, which highlights the *political* dimension of the descriptive frames through which planning inputs make their way into public discourse, making their nature explicit as that of a political stake.

The conditions for such a political dimension to become explicit, however, entail the non-coercive structuring of participation, the pursuit of integrative aims through open settings of argumentation and negotiation, and the constitution of forms of self-organization, self-monitoring and mutual commitment, which calls for practices in the anticipation, acknowledgement, internalization, and public treatment of conflicts.

The processes of institution-building through which consensus-building thus conceived is pursued become, circularly, both opportunities for and outcomes of the constitution of conditions for a strategic realignment and for a reidentification and mutual recognition of involved actors. Their conditions, intended as both premises and reasons for effectiveness, may thus be seen in the constitution of a form of *capital* (Gruber 1993; Innes *et al.* 1994), forged through interaction and constituted by the merging of three components defined as follows:
- the constitution of *social capital,* defined by shared norms, rules, and networks of behavior and communication, based on the building of mutual acknowledgement and trust;
- the constitution of *intellectual capital,* defined by shared stocks of knowledge and information and by opportunities for learning;
- the constitution of *political capital,* defined by the gain in procedural and substantive abilities in reaching agreements and in initiating shared strategies and action orientations.[1]

Practices of policy development based on consensus-building thus take the shape of "mixed systems of shared power and joint deliberation" (Innes 1991, p. 2), based on the loosely structured conduct of group processes and interactive procedures and progressively shifting from their assumption as mere consultative tools to the constitution of collective deliberative practices (Forester 1999). Within their argumentative settings, the 'substantive' functions of negotiation and exchange merge into an experience of *collective sensemaking,* of the *collective framing* of rationales and reference criteria for policy choices.[2]

The Contradictions of Consensus-Building:
Limits of Decision-Centered Consensual Approaches

The experimentation with consensus-building strategies in policy-making develops within contradictory horizons, where reference to 'consensus' may become a means for an inquiry into the pluralism of social claims, as well as an expression of the dominance of a constellation of social positions and powers. While we attribute a paradigmatic meaning to consensual approaches to policy making in reconsidering the dimension of institutionalization, we nonetheless need to address an understanding of their internal contradictions.

These are highlighted in an exemplary manner by institutionalized *decision-centered* approaches to the building of consensus. Typically, in the context of non-hierarchical forms of decision-making and policy development, such contradictions emerge in terms of a trade-off between the benefits of opening decision-making processes with regard to the formulation of policy options, and the escalating complexity of the task environment in terms of legitimation, acknowledgement, and appropriation of responsibilities and abilities for implementation.

Openness and flexibility in adopting collective options is accordingly often limited to a short-term perspective, i.e. to the phase of development of policy options itself, without extending further to the 'constraining' phase of action, traditionally conceived as subsequent to decision: a contradiction which may become a threat if the actors' commitment is exhausted in this short timeframe, limiting interactions to a dimension of symbolic politics and, eventually, to the symbolic restructuring of a precast field of powers.

Idealtypically, such contradictions may be represented in the form of a series of dualisms. The most classic is that between *top-down* and *bottom-up* strategies. In the former, development of options at a higher institutional level, and their approval in principle, is followed by modes of ratification by assembly-like bodies representing formal institutions or relevant acknowledged interests. Opportunities for innovating institutional arenas range between a merely formal transformation of decision rules) and a selective opening towards the field of potentially relevant actors (according e.g. to a 'stakeholder approach'). The advantages of this kind of policy orientation reside in its (potential) political efficiency: consensus is relatively easily and rapidly achieved, and is relatively targeted towards implementation; moreover, it is achieved through limited amounts of negotiation and characterized by the low probability of the emergence of vetoes. Its limits, however, are inherent to the short-term connotation of its efficiency, i.e. to the eventuality that the easy achievement of consensus in policy development may not be backed by the abilities and commitments actually needed in the 'real world' of policy implementation. Top-down approaches are moreover tendentially exclusionary, or at least highly selective, and this may result in a low level of openness towards innovation.

By contrast, bottom-up strategies provide for greater long-term effectiveness by an orientation which explicitly tends to the enhancement of diffuse learning and innovation processes. In principle, however, this attitude highlights problems related to the costs and complexity of processes for the achievement of an active consensus and for its policy effectiveness: these consequences are further connected with multiple problems of an organizational nature in a strict sense, which, by the externalities they imply, may themselves exert selective effects on societal pluralism (e.g. affecting the constitution of abilities, of channels of communication, and of minimum warranties for effective access and participation to decision-making arenas).

The opposition between *administrative* and *political* strategies of consensus-building highlights similar problems. Administrative efficiency essentially short-termed, based on the reduction of costs and organizational inputs through reliance upon routines and existing relational patterns, as in the case of vertical and horizontal intra- and intergovernmental ties of a formal-institutional kind; even in this case, however, the rooting and the accountability of political will is modest, and neglectful of the need to back an internal *ex ante* kind of consensus with the *ex post* construction of external forms of consensus. A vicarious strategy typically developing in the framework of prevailingly administrative rationales is therefore based on the constitution of forms of anticipatory political consensus, mainly relying on leadership and on a reduction of the complexity typical of inclusionary political strategies, which however implies a dependence on key actors and interests and on a basically exclusionary model of participation.

A dualism between *exclusionary* and *inclusionary* tendencies is therefore always present in these competing strategies. While limits in terms of effectiveness and innovation of exclusionary models of policy-making are apparent, based on conditions of access to policy processes defined through administrative selection of groups, prevailingly of a small size and representing macrogroupings of societal interests, contradictions may affect even the opposite models, prevailingly based on cross-sectional representations of diversified social interests and groups. Problems in realizing mobilization, access, and facilitation, at the origin of frequent 'paradoxes of participation', may result in ineffectiveness and manipulation even in the framework of inclusionary strategies. Thus what could be termed the 'fallacy of pure inclusiveness' may take on different shapes: it may be tied to questions of representativeness and of the strategic dimension of representation (e.g. putting social parties on the same level of representativeness and importance in an administrative way may turn out to be as mystificatory and ineffective as excluding them); or it may be tied to matters of resources and information flows and to their influence on motivation (e.g. the balance of costs of information and communication needed to effectively participate in an inclusionary and consensual context of deliberation may turn out to be unaffordable and inconvenient for restricted and minority groups compared to the expected benefits and to the cost of alternative strategies of an oppositional and conflicting kind); or it may result, even more subtly, in the disciplinatory effects induced by consensual rules (e.g. the openness and voluntary character of participation in consensual settings may confront representatives of dissenting groups with the paradox of losing legitimacy and ground, with effects possibly more disruptive than in environments ruled by a formal respect for minoritarian positions). Their outcomes, accordingly, may result similarly in a merely symbolic form of inclusiveness.

It may be of interest to review these contradictions in a more formalized perspective of decision-making, building on the analysis of policy outcomes proposed by Scharpf (1989) in relation to the nature of *decision rules* adopted. Governance activity is interpreted in this perspective as concerned with two fundamental aspects of the regulation of conditions for the exercise of organized power: *boundary rules*, aiming at the definition of collective identities, intended as the reference units for the evaluation of policies, as well as of areas of intervention pertaining to governance action; and *decision rules*, which define who is entitled to participate in specific decisions, and which may be summarized into the idealtypes of *hierarchy*, *majority* and *unanimity*.

The constitution of collective identities clearly represents a decisive factor for the social cohesion of choices and for their potential orientation towards criteria of a supraindividual order. *Boundary rules* thus also constitute a decisive factor for the constitution of a shared sense of action: "any of these collective identities may, at one time or another, become the effective referent for the comparison of alternative courses of action. Whenever that is true, individual action can only be explained and predicted by reference to the utility of the relevant collective unit whose membership is circumscribed by boundary rules" (Scharpf 1989, p. 153). At the same time, however, institutionalized boundary rules both unite and divide along the lines of the different identities they define. And this division entails a distinction between alternative modes of agreement and choice.

The quality of policy choices according to their reference to alternative decision rules is analyzed through three further dimensions and through their respective normative criteria: the *interpersonal* dimension and criterion of *inclusiveness*; the *intertemporal* dimension and the criteria of *stability*; and the *substantive* dimension and criteria of *allocative efficiency* (or *optimality*). Contradictions in the outcomes of consensual approaches to decision-making, already highlighted with regard to problems of the constitution of collective identities which build the frame of reference of decisions, become all the more apparent in the face of the effects on collective choice of the intersection with criteria so defined.

Consistency with criteria of *interpersonal inclusiveness* tends to emphasize questions of mutual recognition and membership and to favor decisions corresponding to interests which may be referred to such aspects: the consequence, however, in the framework of inclusive settings subject to unanimity rules, is the potential constitution of strong veto powers. Consensus in these conditions tends to turn from an opportunity into a problem: compared to hierarchy or hegemony, "consensual decision rules permit each party to defend the existing pattern of distribution" (Scharpf 1989, p. 157). Its paradoxical consequences are a lowered capacity of disposition of distributive patterns of policies and a tendency to sub-optimality and conservatism.

Consistency with criteria of the *intertemporal stability* of policy choices in concurrence with unanimity decision rules raises similar paradoxical requirements: "freed from (some of) the pressure of party competition and more secure in their expectations of continuing participation, decision-makers are less compelled to maximize short-term advantages. But given the high transaction costs associated with Unanimity, effective policy choices will often depend on complexity-reducing and conflict-avoiding redefinitions of the problem at hand [...] – and limiting discussion to incremental changes and their short-term consequences is surely one of the most common techniques for reducing complexity" (Scharpf 1989, p. 157). Thus, conditions for the intertemporal consistency of policies are conversely attached to a tendency to conservatism in policy formulation.

Consistency with substantive criteria of the *allocative efficiency* of policy choices, as defined by the author, pushes these contradictions to the extremes, as their correlation to features of inclusiveness of decision-making settings retroacts on the definition of their objectives: in contexts defined by ongoing interactions, in fact, the alternatives given to unanimous forms of decision are either 'disagreement' and, accordingly, the prosecution along preexisting dominant patterns of decision-making and distribution (i.e. the protection of established interests), or 'agreement' and, again, a form of 'political conservatism' based on the reproduction of traditional political positions and rents.

Innovative Potentials of Consensus-Building: Combining Consensual Decision Rules and Decision Styles

This idealtypical review of decision-centered approaches to consensus-building clarifies why an identification of consensual practices with decision-making procedures proves to result in reductionist assumptions. A *decision-centered* perspective on consensus-building is, in fact, a reductionist perspective since the emergence and coevolution of social practices and their innovative potential cannot be reduced into *discrete* decisional procedures, and rather always entail an *ongoing* rather than a discrete dimension of *sensemaking* and *collective framing*.

This furthermore clarifies why real experiences in consensus-building are never coincident with pure unanimity rules, and why unanimity rules must be intended in an idealtypical sense: reality rather presents a wide field of mixed forms of decision-making, and it is precisely their composition which makes for their original and innovative character. Our concern in pointing to this aspect, however, is mainly to draw attention to two aspects which stand in reciprocal tension in the framework of consensual practices of policy-making: the *representation of problems* and the *inclusiveness of settings* relative to the identities and volitions of their potential actors. Boundary rules in use, in an evolutionary perspective of policies, bear in this sense the meaning of an element defining as well as itself being defined by the representation which is given of policy aims.

The meaning of what have been termed *boundary rules* (the criteria for the access and the inclusiveness of a context for the development of policy choices) may be understood in this duality only by intersecting the nature of *decision rules* with the nature of what Scharpf calls *decision styles*. It is in fact only through the analysis of this intersection that a further dimension of inquiry becomes possible, leading to an understanding of the question of the *constitutive* role of representations in policy-making, and of the fact that "propositions regarding rules and styles of decision-making need to be connected to propositions about commonalties and differences of belief systems or 'cognitive maps'" (Scharpf 1989, p. 166).

The interpretive exercise of an intersection between *decision rules* and *decision styles* is of great interest. Decision styles may be defined as three fundamental models of the coevolution of interests and preferences (i.e. of the development of 'volitions': Lindblom 1990) through decision-centered interaction processes:
- an *individualistic* model, aiming at an utilitarian maximization of benefits;
- a *competitive* model, aiming at a relative maximization of benefits;

- a *cooperative* model, aiming at the solidaristic transformation of benefits and at the development of integrative solutions.[3]

The comparative application of these models to different typologies of mixed-motivation games highlights the importance of a decision *style* oriented towards integrative and cooperative solutions in overcoming the contradictions of consensual decision *rules* taken as such. 'Optimal' collective outcomes are in fact given in each solution in which an integrative orientation prevails in policy choices: "institutional arrangements make a difference if, and to the extent that, individuals who would otherwise pursue different or conflicting strategies need to be coordinated or constrained. By the same token, however, if solidaristic goals and common cognitive orientations can be generated and maintained among participants, decision rules, and institutional arrangements generally, have much less of an influence on policy choices" (Scharpf 1989, pp. 167-8).

Thus, decision styles bear a double potential for restructuring decision-making contexts. They can affect decision-making directly through a transformation in the perception of what is at stake, and this potential is dependent on the innovative input connected to the sharing of knowledge. But they also bear the potential to affect decision-making indirectly by questioning and reframing its embeddedness in an institutional setting, and this opens to a perspective of *institutional change*.

In overcoming the threats of consensual decision rules, and particularly the constitution of deadlocks which may derive from the parallel pursuit of agreements on conflicting interests, the integration of other factors into a cooperative and integrative decision style may certainly be assumed. Reaching effective agreements may, for example, depend on the ability to address decisions sequentially, to separate decisions on collective benefits from decisions on their distributive aspects: thus, "the parties might collaborate in successful 'productivity coalitions' without forcing one side or the other to generally accept an inferior distributive outcome" (Scharpf 1989, p. 171). One of the tasks of interaction processes, according to this conception, would be to develop shared relevance criteria on distributive aspects, e.g. the definition of compensation mechanisms. In other terms, partial outcomes of interactions might bear a fundamental meaning for implementation. This however is nothing but an aspect of the incorporation of conditions for implementation into policy development, and of the basic impossibility of their distinction: it implies that what is at stake in a confrontation on (to stay with our example) distributive matters are the very relevance criteria the sharing of which may contribute to solving it (Majone and Wildavsky 1979; Barret and Fudge 1981).

According to this formulation, also the limits of *substantialist* conceptions of the conditions for effective agreements become clear. In fact, if their outcome is conceivable as a form of *contract*, their *pre-contractual* conditions have to be developed and displayed through interaction. Thus, a theoretical discussion on consensus-building approaches cannot neglect the *symbolic-cognitive dimension* of the possibility of agreement and cooperation or, in other terms, the 'power of common orientations' which makes for their fundamentally *conventional* character.

The coevolutive character of the intersection between forms of the structuration of interaction and forms of the constitution of social identities and preferences, which has been analyzed idealtypically on the basis of the intersection of consensus-based decision rules and styles, highlights the inherent limits of decision-centered approaches to consensus-building, suggesting the need for a reinterpretation of the generative and learning dynamic of consensus-building practices in the broader framework of their institutional environment, beyond a mere decision-centered attitude.

Our arguments thus lead us to reflect on the opportunity of overcoming the dilemmas facing consensual decision-making strategies in a perspective of institutional innovation oriented not only towards inputs of *institutional design* (e.g. through the experimentation of consensual *decision rules*), but also towards the development of self-organized and self-policing forms of collective action and networking (in a path-dependent perspective allowing for the involvement and coevolution of consensual *decision styles*).

An interesting interpretation of this direction is offered by actor-centered approaches which, proceeding from the shortcomings of hierarchical models in the coordination of collective action, have interpreted the possibility of the emergence of effective forms of negotiated coordination within non-hierarchical structures, being constituted 'in the shadow' of hierarchical settings (Chisholm 1989; Ostrom 1990; Powell 1990; Mayntz 1993; Scharpf 1994b). On the one hand, these approaches respond to a positive perspective on the 'political economy of institutions', and to the realist institutionalist stance connected to the aim of conciliating requirements of hierarchical control and of organizational autonomy: in fact, in a 'real world' perspective, as we have noted, consensual approaches to policy-making are usually set in a field of tensions between vertical coordination forms of a hierarchical type and horizontal self-coordination forms of a negotiated type. On the other hand, these approaches acknowledge the constructivist and coevolutive dimensions connected to the development of symbolic-cognitive frames in consensual interactional settings, and the need to address these dimensions beyond the framework of institutionalized models of decision-making.

A possible condition for overcoming the dichotomy between hierarchical and non-hierarchical policy-making settings and for the emergence and sustainment of patterns of networking and self-organization, intended as emerging, loosely-coupled structures of relationships based on informal or negotiated forms of coordination, may be seen in the combination of two distinct, structurally *embedded* forms of negotiation and consensual practices, which may be defined respectively as *positive* and *negative coordination.*

Negative coordination acquires importance and potentials for effectiveness, whereas conditions of intersection and overlapping of cooperative networks and patterns of factual interdependence are given, i.e. in conditions which would require highly complex multilateral forms of positive coordination and where formal orders of a non-hierarchical type would not effectively protect collective action from the consequences of individual choices. In their combination, these mechanisms thus extend the potential scope of coordinated action beyond what may be expected from hierarchical or from negotiated forms of coordination taken singularly.

These forms of coordination express their potentials particularly in their ability to develop non-hierarchical environments. Self-organized networks of cooperative relations must in fact be intended as *emergent* structures, i.e. not as connected to intentional objectives or actual outcomes in a deterministic way, but rather through relationships of a constitutive and coevolutive nature.

Conclusion: Consensual Approaches between Institution-Building and Institutional Design

Discussing the idealtypical features of consensual policy-making approaches has led us to question the capacity of the *institutional design* of consensus-building processes in favoring forms of institutional learning and innovation. Acknowledging the symbolic-cognitive dimension involved in reaching integrative agreements and cooperation entails a critique of a mere procedural, decision-centered focus of consensual approaches. In the terms adopted in the our discussion, this has been expressed as the need to combine the choice of *boundary rules* (defining the entitlement and conditions for access to public decision arenas), of *decision rules* (defining criteria for legitimation of participants and the nature of the agreement on the basis of which decisions are taken), and of *decision styles* (defining modalities for the assumption and treatment of contrasting interests and issues) in terms oriented towards integrative solutions based on self-commitment and self-monitoring among the actors and to the development of conditions for durable self-governing and self-policing forms of action. Aspects of the *institutional design* of consensus-building processes have been hence put in strict relationship with the need for a commitment to processes of *institution-building,* allowing for an iterative, coevolutive and path-dependent dimension of interaction and for the development of practices of collective *framing* (Schön and Rein 1994), *probing* (Lindblom 1990) and *sensemaking* (Weick 1995).

Inquiring into the interplay between *institutional design* and *institution-building* should thus be intended as part of an incremental strategy based on a critical assumption of phenomena of *institutionalization* and *institutional change.*

By this, however, we also turn again to the problematic nature of any reference to 'consensus' as a way towards understanding non-hierarchical and cooperative forms of action. While analytically useful in understanding alternative combinations of *decision rules* and *styles,* the notion as such apparently loses meaning in understanding the actual development and trajectories of their combination and coevolution. Let us conclude on this – as we are going to address the meaning of 'consensus' and of consensus-building practices in a case-study – before getting back to the lessons which may be drawn, in a planning-theoretical perspective, from a commitment to such an elusive concept.

Notes

1   It is important to note, as these authors suggest, that *intellectual, social,* and *political capital* are not only to be seen as preconditions, but, even more interestingly, as important emergent 'by-products' of consensus-building processes, which are constituted besides their possible achievements in terms of problem-solving, i.e. beyond their goal-orientation (Gruber 1993; Innes *et al.* 1994). The issue of social capital is thus strictly connected to issues of institution-building and institutional change. For a further discussion of this topic, see: Gualini (forthc.).

2   Interpretive contributions on the innovative potential of consensus-building approaches for planning processes are still rare; reference here is mainly to the work of: Innes (1991; 1992a; 1992b; 1996), Gruber (1993), Innes *et al.* (1994).

3   Scharpf's original definitions are, respectively, *bargaining, confrontation,* and *problem-solving*; in our view, however, the latter in particular harbors some ambiguity with reference to the critical discussion previously presented about 'problem-solving' approaches; the choice has been therefore made to reinterpret it along with the other terms according to their discursive definitions.

PART II
AN EXPERIENCE IN INSTITUTIONAL INNOVATION: THE CROSS-ACCEPTANCE PROCESS IN NEW JERSEY

4 Introduction:
A New Planning Style

The New Jersey Experience in the 'Third Wave' of State-Sponsored Growth Management Programs in the United States

Growth management in the US, and particularly the initiatives being developed in the framework of what has been called the 'third wave' of state-sponsored growth management programs, offers a paradigmatic example of a planning effort whose main challenge is represented by the ability to reconcile different systems of preferences, beliefs, and values (DeGrove 1984; 1986; 1993; Stein 1992; Porter 1992). In this sense, intergovernmental coordination represents a major issue in growth management programs.

The evolution in the most recent generation of state-sponsored growth management programs in the U.S. is defined by a progressive diversification between the idealtypical dualism of policy orientations characterized basically as either *top-down* (as in the case of the legislation enacted since 1972 in Florida) or *bottom-up*; developments in institutional approaches have given rise to a broad typology of growth management programs (fig. 4.1) which, in the course of their political evolution, have evidenced a progressive shift in the definition of intergovernmental relationships from a preemptive-regulatory model towards models of conjoint and collaborative planning (Bollens 1992).

To the latter category belong programs characterized by an essentially locally-based level of implementation of supralocally defined goals and objectives, through planning actions of a prevailingly voluntary nature, mindful of the sensitive balance of power and autonomy between jurisdictional levels, and subject to systems of incentives as primary instruments towards consistency with supralocal objectives.

Developments towards consensus-building approaches to the issue of policy coordination at the state level are rather recent, however, and limited to a few innovative cases, whereas a shift towards collaborative rather than mandatory attitudes is recognizable in experiences which remain basically oriented by regulatory rationales. As part of this tendency, a single planning experience, namely the process introduced by growth management legislation in New Jersey, has taken a radical path in developing a consensus-building approach to state planning. In the following, its rather unique character will be briefly outlined as an introduction to the case study.

1. *state-dominant*		*regional-local cooperative*	*state-local negotiated*
2. *top-down bureaucratic* > *bottom up*		*laissez-faire, quasi-judicial*	*collaborative, consensus-building*
3. *preemptive- -regulatory*	*conjoint planning*	*cooperative planning*	
Hawaii ('61) Vermont ('70) Florida ('72) California (coast, '72) North Carolina (coast, '73) New Jersey (coast, '74)	Oregon ('73) New Jersey (Pinelands, '79) Hawaii ('78) Florida ('85) Maine ('88) Rhode Island ('88)	Vermont ('88) Georgia ('89) Washington ('90)	New Jersey ('86)

Figure 4.1 Typology of approaches to intergovernmental relationships in state-sponsored growth management programs in the U.S.
Source: adapted from: 1. Gale (1992), 2. Innes (1992a), 3. Bollens (1992)

State Planning in New Jersey: the Generative Framework of Institutional Innovation

The experience of the process that led to the adoption of the *New Jersey State Development and Redevelopment Plan* (henceforth SDRP) in 1992 represents a *unicum* in the context of growing practices of state-sponsored intergovernmental growth management programs in the US.[1]

Among these, the program launched in 1986 by the state of New Jersey emerges by dint of several of its formal features, such as its statutory setting and the role attributed to the plan in the policy-formulation process: it is, in fact, the only state-sponsored program built around a comprehensive large-scale planning document of a strategic character, elaborated at the state level and meant to be operated in a direct relationship between state and local powers, without the formal mediation of intermediate planning documents, but attributing a key role in the planning process to the political, technical, and organizational abilities of intermediate levels of administration.

The start of the experimentation with new forms of intergovernmental relationships in the development of territorial policies was, in the first place, a response to an institutional crisis, determined by a normative shock: it was the sanction of unconstitutionality of local planning practices and the call for a new 'constitutional pact' at the inter-jurisdictional level expressed by the judicial power through the famous series of Mt. Laurel decisions that pressed the New Jersey polity towards a new commitment to regional planning. The emergence of a political shift in intergovernmental relationships was based, however, on broader reasons common to most experiences in state-sponsored growth management, related to the perception of a lack of effectiveness of public policy in the governance of territorial development processes. The origins of this new demand for policy effectiveness lay in the growth of a diffuse consciousness on issues concerning the environment and the demand for quality and identification in local living conditions, as well as in the outcomes of a progressive devolution of competencies and tasks in the provision of resources between state governments and federal administration, advanced through different areas of regulation, and particularly in the field of environmental and infrastructure policy. In a situation of progressive erosion of public finances as well as of significant disparities in administrative performance throughout territorial contexts, conditions for the effectiveness of the framework of intergovernmental relationships were put under stress both at the horizontal level (i.e. at the level of programming of lines of action and of resources among state agencies) and at the vertical level (i.e. at the level of relationships between state policies, regional autonomies, and local governments).

State Planning in New Jersey: a Consensus-Building Approach

The experimental path taken by New Jersey towards a solution of the institutional crisis marked by Mt. Laurel and towards greater effectiveness in managing territorial development bears quite peculiar features. The *State Planning Act* of 1986, which provides for the draft of a new state growth management plan, is a comprehensive document based on the identification of six major objectives of development of the state, but free of prescriptive inputs concerning the means of implementation and rather generic about the relationship between the contents of the plan and the policy-making styles and activities of state agencies. The framework of political debate in which the act is developed stresses its importance as a regulatory means for state budgeting. The nexus between budgeting and implementation of the plan is not made at all explicit, however; it rather acts at an argumentative level, through the consideration attributed to the issue of an efficient allocation of infrastructure investments and to the assessment of infrastructure needs at both the state and the local level in the framework of the fiscal crisis of the state.

The image of the plan which emerges as an expression of political will is therefore rather that of a 'powerless' plan. Moreover, the relationship which its development process establishes between powers in the state's political arena is a new one. The concept guiding the process is in fact the outcome of a mediation between interests worked out by a state commission framed in inclusive representative terms, as common in American politics; however, by the same token, the act consequently provides for bypassing the legislative oversight of the outcomes of the process. The statute of the planning process in fact assigns the development of the plan to a newly established state planning office, mandating its adoption to the state commission subsequent to the reaching of the broadest possible consensus on its contents. 'Consensus' is thus understood as the outcome of a procedure, *cross-acceptance,* pursuing a concerted agreement on substantial planning issues, and intended as "a process designed to elicit the greatest degree of public participation in order to encourage the development of a consensus among the many, sometimes competing, interests in the State".[2] It is conceived as a construct, as a negotiated outcome, which entails an interplay of communicative and strategic forms of rationality, displayed through concrete situations of interaction.

The primary feature of the New Jersey growth management program may be termed a *state-local negotiated* approach, based on the rejection of a *juridical model* of intergovernmental relationships. The aim of pursuing consistency among planning levels, and, through this, an effective steering role for the state plan, is rather attributed to voluntary procedures of inter-institutional cooperation, framed into a negotiating process, without provision of formal means of coercion to conformance or revision of plans. Opportunities and challenges for defining and implementing the plan's choices are thus placed, to an extraordinarily broad measure (Gale 1992), on horizontal forms of intergovernmental coordination, on the cooperation between the public and the private sector, and on the constitution of informal modes of legitimation and commitment.

Cross-acceptance is formally intended as a recursive procedure for the adoption of the plan, articulated into phases, to be repeated at any triennial deadline for the update of the plan (fig. 4.2). Significantly, the drafting of the state plan has itself taken the connotation of an iterative process, accompanying the development of the *cross-acceptance* experience with three (six including drafts) subsequent versions: from the preliminary stage of 1988, through the interim stage of 1991, summarizing the outcomes of the first phase of negotiations, to the final adoption of the SDRP in 1992.

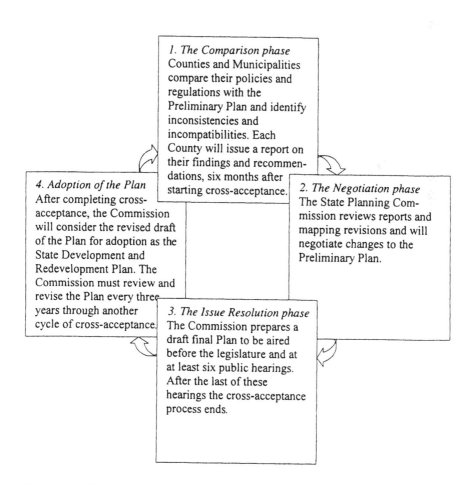

Figure 4.2 Phases of the *cross-acceptance* process
Source: SPC (1988a, p. 27)

However notable the formal-statutory features of the plan are, they only constitute the framework for the real innovation of the New Jersey experience, which lies in the way the planning process is conceived. Its most interesting feature, which will be at the core of our case-study, is the aperture to forms of internal self-definition and self-organization of the processes, i.e. to coevolutive outcomes, through the autonomy granted to the actors throughout the consensus-building process and through the backing of formal deliberative processes with informal arenas for collective argumentation.

The original feature of the New Jersey experience is thus based on the adherence to a "new planning style" (Neuman 1992) as a means for achieving a more effective governing capacity. Its policy approach by-passes references to top-down and bottom-up approaches in favor of a process-based, back-and-forth, recursive and iterative conception of policy development, based on a proactive assumption of interaction processes. In choosing the way of *power-sharing* (Innes 1992b; Bryson and Crosby 1992; Bryson and Einsweiler 1992), the process for plan development introduced in New Jersey acts on the embeddedness of the planning process in a metapolicy environment (Majone 1989) as a condition for effectiveness. This has consequences for two perspectives of institutional innovation: overcoming traditional alternatives in dealing with the mismatch between local and supralocal policies (i.e. mandatory conservative approaches, essentially environmentally focused and based on regulatory instruments, and adaptive approaches, economically focused and based on incentives: Bollens, 1992), in favor of a joint approach to policy development; and a revision of a metapractical order of the traditional instrumental-sequential conception of relationships between the formulation and the implementation of objectives of territorial policy, in favor of a coevolutive conception of policy development.

Notes

1   Among the most important discussions of the New Jersey approach to growth management, are the following: DeGrove and Miness (1992), Epling (1992), Innes (1992a), Bollens (1992), Gale (1992), Buchsbaum (1993c), Burchell (1993), Luberoff (1993), Luberoff and Altshuler (1998).

2   *State Planning Act*, N.J.S.A. 52:18A-202.1.

5 Genealogy of the Cross-Acceptance Process

A Crisis in the Institutional System of Localities: Mt. Laurel

> At first glance the idea that New Jersey might be in the vanguard of a movement towards state growth management seems absurd. As a bastion of 'home rule' the state has throughout the post-war period allowed its suburbs to grow in a rapid, sprawling fashion. This pattern was strongly encouraged by the development of the state's extensive highway system, which opened up previously rural areas for development. As a result of these and other forces, between 1950 and 1985 more than one half of all farmland in the state was converted to residential and commercial uses. During the same period, the state's six largest cities lost about 13 percent of their population and one quarter of their jobs. This pattern accelerated in the mid-1980s, when the total amount of office space in the state grew by about 50 percent and, by 1988, totaling some 99 million square feet, equaled the total amount of space in downtown Chicago and Los Angeles. Almost all the new space was in suburban locales. (Luberoff 1993, p. 3)

In 1975, at the conclusion of a series of proceedings started two years before with a lawsuit brought by the local section of the NAACP, the Supreme Court of New Jersey issued a decision[1] on Mount Laurel Township, an affluent suburb in Burlington County, near the Philadelphia metropolitan area, declaring its zoning ordinance unconstitutional.

The series of decisions known as the Mt. Laurel case was soon to become one of the most important juridical interventions in a line of disciplinary evolution frequently and strongly marked by a confrontation with constitutional principles of the public order. Mt. Laurel thus entered the history of judiciary opinions which, since *Euclid v. Amber*, have marked the interpretation of relationships between planning practices and the universe of individual and collective rights (Cullingworth 1993).

In the constitution of the normative apparatus of local planning, decisions of New Jersey courts have often assumed an avant-garde position in the national context. With its historically 'shifting involvement' between alternatively conservative or progressive interpretations, the judiciary has expressed an innovative function towards the executive and the legislative

87

powers in interpreting the evolution of social practices and their interdependence with governmental action.[2] The courts of New Jersey, and particularly the Supreme Court, have thus assumed in post-war years a primary role in defining public policies, gaining a reputation as one of the most aggressive and independent bodies in the country and sharing political center stage with the executive and legislative powers: sometimes to the discomfort of both (Pittenger 1986), but often also acquiring vast consent, by contributing to limiting forms of discretional behavior on the side of political and governmental actors and by consolidating a form of state identity and, above all, a form of power and a source of prestige which have induced in the system a peculiar mode of construction of political consensus and leadership. As an outcome, a traditional and peculiar consensual relationship has been established over time between the orientations of the superior court and of the executive.

This, however, had occurred with a significant exception – that of land-use and housing policies; and never so evidently as in the Mt. Laurel case, faced since the very beginning with overt aversion by two governors from opposite parties, B. Byrne first and T. Kean later, both very much keen of tying their political fortunes to state welfare policies.

All this, above all, had happened traditionally within a precise and not only formal balance of powers, sanctioned by the constitutional principles of the State of New Jersey. The new constitution of 1947 explicitly affirms the exclusive sovereignty of the state. The local and regional level are not entrusted with an inherent right to the exercise of governmental prerogatives: local authority is exercised upon delegation by the state through enabling legislation.

However, the constitution acknowledges *home rule*, i.e. the tradition of local decisional and administrative autonomy and self-government rooted in the state, and explicitly calls for a 'liberal' interpretation of the statutes concerning the delegation of powers to municipalities and counties. In this sense, the judiciary has traditionally assumed a delicate mediating function in the constitution of this balance of powers, in actual fact by upholding the principle that local governments are 'creatures of the state' (i.e. emanations of its power) only in case of an explicit assumption of initiative or of a call for exclusive competencies by the legislative, with the support of a strong commitment by the executive (Connors and Durham 1993).

Precisely this conception was addressed by the Mt. Laurel decision as it pointed to its contradictions, rooted in the institutionalization practices related to the exercise of *home rule*: it did so, however, under very different conditions, highlighting rather a crisis in this balance of powers, and assuming the constitutional principle as a foundation of radical critique of that balance:

> The basis for the constitutional obligation is simple: the State controls the use of land, all of the land. In exercising that control it cannot favor rich over poor. It cannot legislatively set aside dilapidated housing in urban ghettos for the poor and decent housing elsewhere for everyone else. The government that controls this land represents everyone. While the State may not have the ability to eliminate poverty, it cannot use that condition as the basis for imposing further disadvantages.[3]

This critical assumption of the constitutional mandate, this unilateral, politically 'inactual' call for an ethical conception of the state, laid the foundations for a radical crisis of the institutional and political system. But what was the political and social environment in which a sentence passed on one out of the 567 of the state's municipalities could assume such a paradigmatic meaning as to cause a fifteen-year judicial confrontation and a serious institutional conflict, with traumatic consequences for the constitutional identity of a state and significant echoes[4] even at the national level? And what made it the origin of a renewal of state policy-making much more extensive than could have been induced in previous decades by explicit governmental intentionality?

In its epochal 1975 decision, the Supreme Court acknowledged that, through the exclusionary use of municipal zoning ordinances, the community of Mt. Laurel was violating the 'equal protection' clause contained in Art. 1, Par. 1 of the State Constitution.[5] In doing so, it appealed to a principle of constitutional solidarity, which turned into the request for localities to contribute to the satisfaction of a need and of a social welfare right (in particular the right to access to affordable housing) in the measure of their 'fair share': the Court indicated the assumption of this 'fair share', in still rather vague terms, as an obligation ranging between communitarian sensibility or obligations and political programming, to be addressed in the framework of a regional domain of relationships.

Mt. Laurel I, as it became known, was only the first of a series of decisions of the highest and subsequently of the lower courts of the state to radically question the ethical-constitutional and thus even political-institutional basis of an established model of territorial development and of its normative foundations.[6] The principle of belonging to a supralocal and cooperative sphere of governmental practices, to which the decision made reference by calling for a regional approach to housing needs, raised principles of coordination and programming of social policies which directly impinged on the segregational and unequal effects in the distribution of resources and opportunities grounded in the actual forms of local sovereignty of which *home rule* had become a symbol.

The state's development patterns in fact delineated a geography of segregation and social inequality even more underlined at the threshold of the 1980s by the mismatch between the lessening amount and effectiveness of public resources and the growing potential of the state's economy. Mt. Laurel and zoning, as shown by persistent questions of constitutionality concerning a school funding system largely based on local resources and on the respective positional potentials of localities,[7] were not the only issues at stake in the crisis of a territorially fragmented and structurally unequal societal model, perpetuated by its institutional hypostatization.

From the point of view of a state whose supreme court, with the *Euclid v. Amber* decision, had laid down fifty years earlier the foundations of the progressive myth of planning, hence, "[i]t may be ironic, but it should not be surprising, that land use planning, for all of its reformist or radical connotations, has played basically a conservative role in the development of American residential communities. For every Mount Laurel breakthrough there have been countervailing decisions in the pattern of Belle Terre and Eastlake" (Simmons 1977, p. xvi). As has been noted, in fact, since the suburban turn of urbanization in east-coast states, "[l]ocal control over private land has withstood with incredible resilience the centripetal political forces of the last generation" (Babcock 1966, p. 19).[8] Disciplinary tools, in light of territorial governance structures, have thus become the means for radical internal differentiation of opportunities for social and economic development:

> To a greater degree than in other societies, urbanization in the US has separated people along economic and social lines. Nowhere in urban America is the heterogeneity encompassed by a metropolis reproduced in its local jurisdictions and neighborhoods. (Danielson 1976, p. 3)

At the threshold of the 1980s, however, this same resistance marked the stabilization of a peculiar societal bloc, the centrality of which has been underlined in constituting a new image and a new style of policy-making in the state: *suburban politics* (Salmore and Salmore 1993). Its governance tools were set under the sign of exclusion:[9]

> In separating themselves politically from the city, the residents of the newer neighborhoods superimposed a fragmented political system on the spatially

differentiated population of the metropolis. The proliferation of suburban jurisdictions, each with independent control over access to residential, educational, and recreational opportunities within its borders, greatly reinforced the social, economic, ethnic, and racial differences among rural neighborhoods. [...] Given the spatial differentiation of residences and jobs in the metropolis, neither the need for services nor the ability to pay for them is evenly distributed among local governments whose activities are financed primarily by property tax and other locally based taxes. As a result, the metropolitan political economy is characterized by great variations in the costs and benefits of public services to residents of different communities. In the words of the Advisory Committee on Intergovernmental Relations, "political splintering along income and racial lines is akin to giving each rich, middle class, and poor neighborhood the power to tax, spend, and zone. Such decentralization of power can and does play hob with the goal of social justice". [...] suburban exclusion affects the ability of a large and growing number of Americans to live and work in those parts of the spreading metropolis where the most land is available for housing and where the most of the nation's economic growth is occurring. (Danielson 1976, pp. 17-8, 26)

It was therefore the whole domain of practices and relations which had historically clustered around the notion of *home rule* that was put at stake by the court's appeal to a constitutional perspective of reform:

The strong ethical mandates to extend the protection of the law equally to each person and to attend to the general welfare threatened the order of a field of communities in which public and private domains were deeply intertwined. The expansive application of 'equal protection' would require either a catastrophic disengagement of government from private realms or an enormous expansion of the capacity of the state to proscribe and, if necessary, redress governmental-supported inequality. The obscure bargain that sustained the image that there was a 'private' domain where none existed would be undone.

In a similar way, a patterned, substantive conception of the 'general welfare' would require an agency empowered to define politically privileged claims and to oversee and (if necessary) initiate public actions. It would ensure the priority of basic human needs when they could not be satisfied without reducing the political salience of ordinary wants or violating claims authoritatively, based upon rights or deserts. A list of needs that were easily met might be accommodated without difficulty within a pluralistic democratic polity. If, however, the needs challenged deeply embedded practices – if they included, for example, 'decent housing', open space, participation in a socially diverse community, equal educational opportunity, and self-respect – then the 'general welfare agency' would be a demanding (even a threatening) political master. (Mandelbaum 2000, pp. 202-3)

Hence it can be said that the policy-making issues raised by the Mt. Laurel decision, and soon to emerge in the course of its aftermath, in their radicality manifested a crisis in the institutional system of localities.

This crisis was supported by some basic, strongly rooted dichotomic features of the political culture of the state, "a magnificently diverse state of great extremes", "something of a jigsaw puzzle", made of pieces which hardly seem to possibly "ever fit together into a coherent whole" (Zukin 1986, p. 3). The first, as we have seen, opposed state authority to the political-cultural tradition of *home rule*, grounding its local interests on an alleged sense of communitarian identification and on a defense of its decisional autonomy, in constant tension with supralocal regulatory policies. The second almost equally opposed the north and the south of the state, along lines approximately coincident with the borders of the central region's sprawling development, and marking radical differences in political culture and representation. The third is, accordingly, the harsh opposition between urban and suburban societies, cultures, ways of life.

Those dichotomies, as far as planning issues are concerned, were further related to the peculiar features of the state's recent territorial development or, in other terms, to the complex reality confronting the aim of identifying any 'regional' identity as a reference for political action.

New Jersey, widely regarded as the 'urban state' *par excellence*, is in fact the most densely populated and urbanized in the country,[10] but according to an extremely accentuated sprawl pattern. Growth in post-war years, even in periods of maximum development, has privileged suburban housing models featuring high land consumption rates, which now turn into the stigma of a highly fragmented society. In the vast 'urban field' constituted by the urbanized region centered on the corridor between New York City and Philadelphia, Route 1, and on the New Jersey Turnpike, only Newark and Jersey City count more than 200,000 inhabitants, while cities like Trenton, Elizabeth and Paterson, the only ones over 100,000, have a population little larger than many suburban centers, like Cherry Hill near Camden, or Hamilton near Trenton, which are often their strong political rivals in the quest for resources and external appeal. Older towns like Camden, Trenton and Newark, once important industrial centers, are at the highest levels of distress in the country, in striking contrast with many affluent suburbs with centers of international reputation, like Princeton, as well as with the artificial urban life pattern of the east-coast entertainment metropolis, Atlantic City.

Development patterns of the state moreover are internally highly differentiated, exhibiting strong disparities.[11] Not only imbalances internal to urbanized areas emerge, but also between these and an extensive non-urbanized territory with high environmental as well as economic potentials, and between respective social profiles and polities.

In this context, the reality of governmental and planning practices constitutes another paradox, which may be explained by the nature of its local entities. Enabling legislation of the state in planning matters, according to disciplinary parameters, is 'advanced' and calls for a comprehensive and sectorally highly detailed standard of plans at the municipal level, furthermore entrusting the counties with functions of supervision, coordination with regard to specific issues, and powers for drafting comprehensive planning tools for addressing their tasks and responsibilities. The effective authority of counties in planning matters, however, is traditionally quite modest, as well as highly dependent on local political initiative, and thus quite variable throughout the state.

This aspect is tied to further structural features of territorial government. Unlike other states, New Jersey has no unincorporated land, traditionally set under the jurisdiction of the counties: all 567 municipalities are incorporated and coincident with their own territorial jurisdiction, covering together the whole of the state's territory. At the same time, the diversity of organizational forms of territorial government institutions is relatively high, contributing to a picture of institutional fragmentation (Wood 1961; Danielson and Doig 1982). In this situation, counties suffer on their part from a great disparity in institutional profiles and from a modest level of competencies, mutually contributing to their marginality in constituting political leadership and to the prevalence of municipalist patterns of constituency formation (State of New Jersey County and Municipal Government Study Commission 1985; 1986).

Contradictions internal to this socioeconomic and territorial model of development are not only expressed in constitutional terms, however. After the era of the 'new suburban frontier' and the constitution of a new regime of localities, permeated by a familyist kind of communitarianism, only recently has New Jersey been constituting a sense of identity as a state, as well as a distinct kind of political leadership. Reflexes of this process may be seen at the formal level of national politics, as well as at the more informal level represented by the emergence of development coalitions and initiatives framed into a regional domain of cooperation, another aspect which separates the outcomes of different political cultures along territorial lines.

As far as planning experience is concerned, New Jersey is again a sort of paradox: at the forefront of the country in many respects, as the first state to adopt a unitary zoning ordinance model, and as the promoter of internationally known projects like the Garden State Parkway and, more recently, the establishment of the Hackensack Meadowlands Development Corporation, of the Pinelands Reserve, and of one of the most rigorous coastal protection programs in the country. In fact, at the time of Mt. Laurel I, the state features a regime of protection covering about one third of its territory, combined with pioneering experiments in compensatory planning tools for the safeguard of rural areas (Kolesar 1992; Lenz 1985; MSM Regional Council n.d.), as well as three special regional authorities which, as in the case of concerted decision-making practices promoted by the Pinelands Commission, have experimented with innovative approaches to intergovernmental coordination (Moore 1986; Robichaud and Russel 1988). However, diffuse reality is much more prosaic: although extensively planned, the state's territory is almost exclusively dominated, alternatively, by local or by sectoral rationales, and almost exclusively without measures in the structuration of intergovernmental relationships and in supralocal integration and cooperation. Regional planning experiences remain limited to special areas, and do not offer political tools for solving the problems of most of the state's territory.

The pattern of development of urban society is thus defined mainly by an accentuated localist frame, focused on conciliating the pursuit of maximum fiscal benefits with the safeguard of communitarian identity in affluent suburbs, and on emergency measures backed by inadequate resources for a perspective of endogenous development in declining urban areas (Burchell and Listokin 1985): a frame, however, which highlights rising contradictions in the changing political conditions of the state.

For this reason, what will be defined as "the most complex American state plan ever developed" may be seen as the symbol for "the complexities and continuities of contemporary policy-making in New Jersey. Cross-acceptance embodies an intricate balance between the old localist culture and emerging state identity and authority" (Salmore and Salmore 1993, p. 300): a balance gained through a difficult and innovative path of reform.

From Mt. Laurel I to Mt. Laurel II

> Not since the school crisis of mid 1970s has the Supreme Court issued as clear and strong a call for action from the Executive and Legislative branches as it has in its forceful and unequivocal Mt. Laurel II decision. (G.R. Stockman, in: New Jersey State Legislature 1983, p. 1)

For about ten years, despite the acknowledged relevance of the decision and the growth of a host of studies on 'fair share housing allocation', almost nothing happened in local zoning practices and in the state's political orientations towards an assumption of the regional approach to housing needs advocated by Mt. Laurel I. One major reason for this inertia is certainly to be found in the nature of the decision itself:

> The Mt. Laurel I decision remains a luminous declaration of principle. It left many details unresolved, however. Did its holdings apply to all communities or just developing towns in the path of inevitable growth? What kind of zoning changes were required? Did towns have to meet a fair share quota or just zone a reasonable portion of the community for multifamily housing? Did they have to do more than zone to foster affordable housing? (Buchsbaum 1993b, p. 13)

In the opinion of many sympathetic observers, Mt. Laurel I expressed a great potential for reform, without however offering answers to the questions it raised. If in fact "[j]udicial policy making can present an attractive alternative to the fractious push-pull of mutual adjustment between elected officials, bureaucrats, and the public" (Anglin 1994, p. 437), being in principle entitled to act in the 'public interests' beyond the compromising constraints which bind public policy-making, this does not free it from the need for clear action directives and to implementation constraints. While marking the origin of an explicit and radical confrontation of powers, the decision did not provide guidance for changing attitudes at the local level, allowing for substantial elusion.

A great deal of responsibility for this, however, resided in its reception at the level of the state polity and policy-making system. While the absence of a definition of issues crucial for public intervention (the conception of a regional level of needs and the consequent allocation of fair share quotas to municipalities) exacerbated the importance but in practice weakened the effectiveness of the courts' intervention, the whole issue of affordable housing and its potential for conflict with consolidated systems of local powers and *home rule* led to a systematic elusion even at a superior level of political representation. Apparently, if the aim of the Supreme Court was an appeal to central governmental authority and competencies and to a prompt mobilization of debate and political initiative, the result was on the contrary, at least initially, a substantial inertia.[12]

It is in this contradictory situation that the Mt. Laurel litigation evolved, highlighting the necessary dialectics between 'ethical mandates' and 'the Virtue of Prudence' (Mandelbaum, 2000).[13]

The compromise which seemed to emerge between the judiciary branch and local political powers did not result in a smoothing over of contradictions, however. The lack of any solution of issues concerning the fair share principle and the introduction of a builder's remedy produced a wave of litigations at lower courts, forcing the Supreme Court to get back into the game in 1980 with six new cases, all of them centered on the problematic definition of criteria for locally satisfying and regionally affordable housing needs. In this situation of uncertainty, as well as of absence of political initiative, by taking over the Mt. Laurel case, the court renewed its intervention, leading to the disruptive decision known as Mt. Laurel II.

Starting with the actual setting of proceedings, the court adopted an operating style similar to that of a legislative commission. The case was introduced by a series of hearings to which organized social representatives and coalitions of interests were called (municipalities, developers, civil rights associations and others):

> The 'argument' itself was as extraordinary as its setting. Because virtually all lawyers involved were intimately familiar with the actual process of housing development, their arguments had the feel of testimony at a legislative hearing, rather than appellate advocacy. The members of the court in turn slipped readily into the role of legislators, peppering the speakers with well-informed questions to elicit facts (seldom law) about housing economics and the mechanics of land development. The court heard arguments and deliberated for 27 months. (Payne 1983, quoted in: Anglin 1994, p. 439)

The judges' approach thus transformed the setting (to quote Bryson and Crosby 1992) into a *forum* and an *arena* as well as into a *court*.

Mt. Laurel II,[14] passed in June 1983, pointed strongly to the resistance of local powers to the mandate of Mt. Laurel I. In the famous words of Chief Justice R.N. Wilentz, the author of the Court's opinion:

> The doctrine has become famous. The Mt. Laurel case itself threatens to become infamous. After all this time, ten years after the trial court's initial order invalidating its zoning ordinance, Mt. Laurel remains afflicted with a blatantly exclusionary ordinance. Papered over with studies, rationalized by hired experts, the ordinance at its core is true to nothing but Mt. Laurel's determination to exclude the poor. Mt. Laurel is not alone: we believe that there is widespread noncompliance with the constitutional mandate of our original opinion in this case. (Mt. Laurel II)

At the same time, the decision represented a strong accusation of the political system and a strong declaration of intent towards a new, direct and proactive level of initiative:

We have learned from experience [...] that unless a strong judicial hand is used, Mount Laurel will not result in housing, but in paper, process, witnesses, trials and appeals. (Mt. Laurel II)

In its Mt. Laurel II decision, the Supreme Court took over most of the principles which had informed the preceding decision, and which had been hollowed out by a substantial vacuum in political response. However, in contrast to the previous decision, in this case the assumption of responsibility by the judiciary was "direct and decisive" (Anglin 1994, p. 439). And it is the rupture introduced in the balance of powers by this assumption of initiative, rather than the ethical issue *per se*, which turned the constitutional issue into a cause of political and institutional crisis.

The decision generalized in the first place the application of local obligations to the satisfaction of regional housing needs in terms of quotas, extending it to all the communities of the state. At the same time, the concept of 'regionalization' called for a definition of requisites and highlighted the need for support from a knowledge base and through a programming approach capable of specifying relationships between housing issues and the evolution of development patterns and socio-economic conditions throughout the state. The decision thus became an appeal to the political-constitutional function of planning. And, again, this appeal emerged from evidence of a lack of commitment:

The lessons of history are clear, even if rarely learned. One of those lessons is that unplanned growth has its price: natural resources are destroyed, open spaces are despoiled, agricultural land is rendered forever unproductive, and people settle without regard to the enormous cost of the public facilities needed to support them. Cities decay, established infrastructure deteriorates for lack of funds; and taxpayers shudder under a financial burden of public expenditures resulting in part from uncontrolled migration to anywhere anyone wants to settle, roads leading to places they never should be – a pattern of total neglect of sensible conservation of resources, funds, prior public investment, and just plain common sense. These costs in New Jersey, the most highly urbanized state in the nation, are staggering and our limited ability to support them has become acute. More than money is involved, for natural and man-made physical resources are irreversibly damaged. Statewide comprehensive planning is no longer simply desirable, it is a necessity, recognized by both the federal and state governments. (Mt. Laurel II)

The court's intervention pursued three basic principles. First, a principle of 'affirmative action', which translates into the allocation to municipalities of an obligation towards active policies, through the recourse to fiscal and normative tools oriented towards the safeguard of low-income destination of housing realized according to regional distributive criteria, extending to the elaboration of incentives and compensatory measures, which was later to be introduced by the *Fair Housing Act*.[15] Second, a 'remedial' principle, descending from the consideration of a rooted condition of inequality, which connected judicial threats towards local governments with the private threat represented by the interests of the building community: the court stated the principle according to which each developer succeeding in a lawsuit against an exclusionary municipality had the right (in the absence of proof by the municipality of local inadequacy for further development) to have its property's zoning changed according to parameters granting the construction of four free market housing units for each affordable housing unit. The 'builder's remedy' thus became the key to the constitution of a cross-sectional coalition of interests against exclusionary zoning practices.

It was, however, the third principle that threw the institutional system of local governments into a state of shock and uncertainty: the direct connection established by the court between its judicial function and the supralocal planning authority of the state.

The court in fact established a special committee of three judges delegated to unitarily assess all litigations related to affordable housing issues, the so-called 'Mt. Laurel litigations', which are rising throughout the state. The committee is entrusted with tasks of directing the revision of local zoning ordinances, of suspending initiatives pending revisions of the decision, of overruling any normative element of local policy deemed exclusionary, and of approving specific measures of incentive for low- and moderate-income housing. Above all for solving the issue of defining local obligations related to estimated regional housing needs, originated by the troubled problem of assessing regional growth potentials, the court makes reference to the *State Development Guide Plan* (SDGP) adopted by the Division of State and Regional Planning in 1980[16] as the only tool available. The support of the SDGP enabled the judges to define with relative clarity specific regions as subject to growth rather than to containment and conservation. In growth regions, the court distinguished between *present needs*, including existing housing plagued by decay and overcrowding, and *prospective needs*, an attempt at defining needs based on prospective growth of families eligible for low- and moderate-income housing in a period of six years.

With this move, in the view of the political system, the court enacted a kind of *hybris*, an explicit rupture of the balance of vertical inter-governmental relationships grounded on the *home rule* principle: it took a political tool of the executive, which basically lay buried in the archives, and adopted its interpretative categories of development as criteria for assuming judicial decisions and for imposing normative sanctions to local communities. It did so, however, by explicitly acknowledging the exercise of a remedial initiative. The text of the decision thus expresses a clear appeal to the institutional reappropriation of tools for territorial policy-making which could help in overcoming the crisis signaled on a constitutional basis: and it suggested, with a significant anticipation of the future *State Planning Act*, that "in order for it to remain a viable remedial standard, [...] the SDGP should be revised no later than January 1, 1985 (and in absence of proof of a more appropriate period, every three years thereafter)" (Mt. Laurel II).

After Mt. Laurel II: Institutional Shock and Radical Uncertainty

The fact is [...] that the court's ruling forced elected officials to confront the issue that they had avoided for five years because of their collective perception that their constituents would be against inclusionary housing. (Anglin 1994, pp. 440-1)

The consequences of the judicial initiative taken by the Supreme Court with the Mt. Laurel II decision on the system of local powers and interests were radical. Almost none of the 567 municipalities of the state was in compliance in 1983 with requisites for set-aside ordinances and, in particular, for 10% reserves each for low- and moderate-income housing. No community, moreover, following the smoothing of judgement criteria introduced in 1977 by the *Madison Township* decision, had pursued an attempt to clearly define its fair share. "Thus, the Mt. Laurel II decision, in effect, declared almost all suburban zoning ordinances in New Jersey unconstitutional. In well over 100 communities in the state, the promise of a builder's remedy induced a multiplicity of suits to take advantage of New Jersey's mid-1980s housing boom" (Buchsbaum, 1993b, p. 14). The courts' adoption of parameters set by the SDGP made many municipalities subject to fair share housing allocation. In fact, in the immediate aftermath of Mt. Laurel II, between 1983 and 1985, over 70 communities were sued for exclusionary zoning based on the SDGP (Martha Lamar Ass. 1988).

The response to Mt. Laurel II, however, may again be traced back to judiciary initiative: consistently with the affirmative stance of the decision, the judiciary in fact actively pursued its mandate through the committee, and did so on the basis of the only assessing tool available, the SDGP.

Political uncertainty hence could not have been more radical. The state's political-administrative system was challenged on a most sensitive issue, that of *home rule*, i.e. of a complex, hardly alterable equilibrium of consent and political leadership: and this on the basis of the 'improper' – albeit temporary – appropriation of an administrative instrument, even if demised, of its own creation by one of the superior powers of the state.

Despite the explicit mandate of Mt. Laurel II, however, it took over two years for the state to start a policy-making process according to the urgency determined by the institutional crisis induced in the system. The first reactions of the state's polity acknowledge this crisis, but in purely defensive terms. The most influent pressure came from representatives of suburban municipalities, which raised phantoms of a loss of local identity in face of the threat of the builder's remedy and of an allegedly uncontrolled growth, threatening to appeal against the decision's provisions by means of constitutional amendments.

The official reaction of the executive was largely in line with the 'localist' and conservative imprinting of the uproar against Mt. Laurel II. Its strategy seemingly aimed at withdrawing the foundation for the courts' judicial activity by overtly delegitimizing the state plan of 1980. A few weeks after the decision, the Kean administration decided the demise of the state planning office, which was fulfilled in 1984 with the definite abolishment of the Division of State and Regional Planning in the Department of Community Affairs, advocated by its own head, J.P. Renna. The official motivation for this act referred to cuts in the agency's budget introduced by the state's financial act in early 1983. The administration in fact faced a severe fiscal crisis, and withdrawal of federal funding directly affected planning: in 1982, the Reagan administration had cancelled the *Urban Planning Assistance Program* which supported local authorities with federal funds under the supervision of state governments (Roper *et al.* 1986). Shrinking resources thus represented an objective factor which weakened an important tool for strategic coordination between state and local land use, undermining the function of the department.

However, behind this move stood an ambivalent attitude. While Kean, who a few years later ended his mandate praising its own 'politics of inclusion' (Kean 1988), shouted against the court's alleged 'judicial overactivism', calling the decision an attempt in 'communist subversion', consciousness of the need for a different policy approach was rising within the executive. Its first expression was a form of *tabula rasa*: and its main advocate was C.W. Edwards, chief counsel of the governor who, while dismantling the planning division, was to become a most vigorous supporter of a new course of action in state planning.[17]

The initiative of elaborating a new planning strategy, however, did not stem from the statutory system of political deliberation. In fact, the legislature did not take any serious initiative as a response to Mt. Laurel II, while the executive stuck to a defensive public rhetoric.[18] The path to a reform of power relations and to a renewal of the role of state planning apparently required creative inputs and negotiating abilities of a different nature. They were to emerge from a much more diverse cross-sectional representation of cultural attitudes and interests, convening around mid-1983 to form the first 'institutional entrepreneur' of the new planning process, the Ad Hoc Committee on the reform of state planning.

From the Demise to the Rebirth of State Planning

> The coordination and intergovernmental relations functions seem more important than the land use and regional planning aspects of the current planning exercise. The situation results from two facts: first, much of New Jersey's development and land use has already been determined into the foreseeable future; and second, the role of state government has so dramatically expanded in the past two decades that a new governmental relationship must be forged. (Bierbaum and Nowicki 1991, p. 31)

The establishment of a committee of wise men was accompanied by a growth in public concern at the lack of initiative to contrast the effects of the court's decision. The reasons for this concern were tangible, and consequences of the situation soon to arise. Among the first voices calling for a solution of the crisis were those of local planning officials, in an obvious appeal to the autonomy of their action; in the meantime, however, various representatives of interests and categories, suchas legal consultants, conflict resolution agencies, county governments, planners' federations, stressed the risks of inaction, of a confrontation among state powers, as well as the costs of diffuse litigation and of piecemeal programming.[19] Furthermore, public debate underlined the amplitude of development issues which the conflict on exclusionary zoning itself threatened to overshadow: the 'wicked' character of many local land use decisions, as well as the partiality of equity issues that may be addressed in a judicial way. Thus, the substance of the Mt. Laurel decision was progressively extended to concerns such as the decline of urban conditions and the ability of local governments to adequately respond to welfare requirements, calling for supralocal cooperation and for coordination between state authorities.[20] Mainly in highly developed counties, moreover, policy communities were emerging which advocated supralocal development strategies, backed by significant sections of entrepreneurship in a concern for the decline of locational and economic assets due to the lack of a viable long-term planning framework.

The Ad Hoc Committee constituted for over three years an arena for mediating between quite differing views on the problem.[21] Its activity was certainly, albeit not exclusively, the outcome of political mediation. It became, however, an opportunity for developing an integrative initiative.[22]

Three positions dominated the committee's discussions. The first, mainly backed by environmental groups and regional initiatives such as the New Jersey Conservation Fund and the Regional Plan Association, advocated delegation for the elaboration of a state plan to an independent agency to mediate between the interests of state and local governmental bodies according to a regulatory and comprehensive planning ideal based on local compliance obligations and on a strict linkage between local planning and budgeting policy. The proposal soon appeared to be politically unviable.

A second position, backed by the committee's chairman, was inspired by more political-administrative realism and advocated the sacrifice of comprehensive disciplinary ideals in favor of a strict integration with resource allocation through the budgeting process: in this view, the plan basically was to become a tool for directing infrastructure investments, which established financial constraints to local land use choices as a primary means for implementation. Even in this version, however, which interpreted regulatory powers of the state plan on local authorities in weak and indirect terms, the idea of tying planning to budgeting appeared to be politically intractable.

The third model proposed, which turned out to be the most influential in leading to an agreement, was based on the complete demise of direct mandatory powers and on the construction of a political commitment throughout all levels of intergovernmental relations. The drafting of the plan was to be entrusted to an autonomous deliberative body, placed under the control of the executive, and its adoption and implementation subordinated to the pursuit of consensus and of voluntary adjustments to proposed development choices by local governments. The image of a 'powerless plan' which thus emerged, very much to the discontent of comprehensive planning advocates, imposed itself as the only one capable of achieving consensus from groups which are politically determinant for any attempt at public reform. Many of their most influential representatives admittedly made their consent to state planning dependent on the weakness of formal powers and on the image of a plan which might have remained 'like law on the books' without an explicit will of the executive, thus leaving well-known decision-making mechanisms untouched.[23] However, these positions proved to be affected by a peculiar underestimation of the scope of the process involved. The proposal appeared as the only feasible compromise, as the expression of a lowest common political denominator: but, while the main players in the committee tied their consent either to the safeguard of *home rule*, or to a generic appeal to the consultation of local authorities,[24] or to a resigned realism,[25] an innovative policy-making attitude was to make its way through the new planning process.[26]

The committee elaborated a first draft of the *State Planning Bill* at the beginning of 1984. Gaining bipartisan legislative support, under sponsorship of Senator Stockman, turned out to be relatively easy: further and politically more direct concerns were however raised by prominent – and mainly Republican – assemblymen. The bill in fact did not directly address the conflictual elements of Mt. Laurel II which most concerned local authorities, like the judicial committee's power to overrule local plans. The bill was therefore frozen for a long time, during which the Kean administration entrusted the same Ad Hoc Committee with the task of drafting a bill proposal addressing the issue of local contributions to regional affordable housing needs, called the *Fair Housing Bill*.[27] Assembly debates on the *State Planning Bill*, presented at the Senate in 1985,[28] explicitly subordinated its approval to the adoption of the *Fair Housing Act*, which occurred in July of the same year.[29]

The adoption of the new act established a formal obligation of the legislature towards the *State Planning Bill*: the linkage between the two was in fact evident, and was confirmed by the explicit references in it to the function of state planning contained in the latter.

But conversion of the proposal into law was not yet assured. The political system in fact found itself in a typical preelectoral mood of expectation. In November 1985 Governor Kean was, however, easily reelected, and his party regained a large parliamentary majority after over ten years of Democrat dominance. Influential members of the committee took over the campaign in favor of the law and, in an end-of-legislation session during which "[m]ost of the reps didn't know what they were voting on",[30] without any substantial opposition, the law was finally approved. Governor Kean signed it and made it effective in 1986.[31]

With the *Act*, state planning activity was overhauled from its roots, and was delegated to two new agencies, the State Planning Commission (SPC) and the Office of State Planning (OSP), respectively with deliberative and technical-operational tasks, established in the Department of the Treasury.

The term 'cross-acceptance' thus entered the formal vocabulary of New Jersey politics. Let us however have a look – before addressing the features of the process and the style of operation and the nature of working relationships developed through it by the planning agency, the OSP – at the composition and the identity of one of the main collective actors and 'institutional entrepreneurs' of the New Jersey state planning process.

The State Planning Commission

The composition and the identity of the SPC stand in direct relationship to the features of the process that led to adoption of the *State Planning Act*.

The SPC was established as a commission in (and not as a division of) the Department of the Treasury, a feature which – compared to the former Division of State and Regional Planning of the Department of Community Affairs – signals both its autonomy and its interdepartmental role.

The SPC was meant by the original version of the *State Planning Bill* to be the representative body of the legislature, of the executive, of local governments, of the planning community and of the general public. As a matter of fact, the approved *Act* was in this respect the result of an amendment required by governor Kean in order to underline its *executive* function. A related amendment concerned the reduction of its members from 21 to 17, to the expense of representatives of the legislature. The commission was thus basically composed by *cabinet members*, municipal and county officials, and *public members*.

Membership of the SPC mirrored the balance of powers within the executive branch of the state. Seven out of 17 members (i.e. cabinet members and executive representatives) were in fact directly appointed by the governor. The governor's influence was expressed by the presence in the commission of the state treasurer and of four cabinet members, while access to meetings and proceedings was granted to any member of the cabinet. The chairman is also appointed by the governor, and supported by a vice-chairman chosen by the commission itself among public members or local representatives. The appointment of local representatives is again by the governor upon approval of the legislature; among them, the presence is explicitly required of a representative of urban areas as well as consultations with the NJ State League of Municipalities, the NJ Conference of Majors, the NJ Association of Counties, and the NJ Federation of Planning Officials.

Among the departments of the executive, five have been always represented in the SPC during *cross-acceptance*: Treasury, Environmental Protection, Transportation, Community Affairs, and (alternatively)

Agriculture or Commerce and Economic Development. As a matter of fact, the copresence of the latter two has been almost constant, as the faculty of appointing two executive representatives besides the five cabinet members allowed the governor to anyway involve the 'excluded' department. Representation of the constituencies related to crucial areas of activity of these agencies was thus granted throughout the state planning process. Furthermore, an important aspect of continuity beyond the terms of appointment was due to the mechanism of nomination of 'official designees' at the meetings supplying regular cabinet members: the assistants of state departments' heads thus became the real pivots of the executive's involvement and – being normally redesignated even after a change of commissioners – the bearers of a 'memory' of the process. This was for instance the case of the representative of the rural community (Arthur Brown, secretary of the Department of Agriculture, who as official designee and later as cabinet member was in the commission from its establishment to 1995, throughout three legislatures), but also of other influent representatives, as Melvin 'Randy' Primas, mayor of Camden, who – along with the executive of Mercer County, Bill Mathesius – was the most active representative of the 'urban voice', and who entered the commission as a public member under governor Kean and, after becoming commissioner of Community Affairs, was reappointed as a cabinet member by governor Florio. A similar pattern of continuity developed among public members and local government representatives. Thus, new membership appointments often involved personalities previously involved in other ways in the commission' activity: a feature favored by the overlapping among two- and four-year appointments as well as by the frequent use of 'nondecision' by the governor during membership vacancies as a tactic for reacting to shifts in majority and for ensuring leadership and control to long-term members.

In summary, the identity of the SPC was that of a 'small-political' body, featuring a balanced representation of special interests and of related constituencies rather than a direct dependence on partisan politics.[32] This is of course the major reason for the importance of continuity in its membership, as well as of its continuity with the former Ad Hoc Committee that had been entrusted with working out a viable consensus on a state planning perspective as a response to Mt. Laurel.

Due to these factors, the constitution of a 'core-group' of members was to become a major feature of the activity of the SPC. Not surprisingly, various members of the SPC had been key actors in the process of building that consensus.[33] The core-group of the SPC was formed by four personalities whose appointment was renewed at each new term during the experience of *cross-acceptance*, as was the chairman who, first appointed by governor Kean, was confirmed by Florio until the new appointment by Todd Whitman, after approval of the SDRP. The core-group was itself the expression of a balanced mix of interests and biases:[34] the first chairman, James G. Gilbert, formerly a key member of the Ad Hoc Committee, was a banker; public member Fred Vereen Jr. (a member of a prominent dynasty of entrepreneurs from the mid-west of the state) was a housing consultant; Candace M. Ashmun, governor Kean's representative in the Pinelands Commission, was an environmental consultant; and Jay Cranmer was both a professional planner and (most importantly) a builder, appointed as such in 1987 as a crucial response to the uproar of the state's growth machine after the presentation of the first, unofficial draft of the state plan.[35]

Overlapping commitments – as, for instance, Cranmer's membership in the NJ Builders Association, or Ashmun's membership in the Association of NJ Environmental Commissions, of which she was former director – conferred to some of the commission members a privileged position in bridging institutional rationales and special interests. Due to the activism and the continuity of its members, but also to the reputation and to the influence held by its personalities within their respective constituencies, the core-group of the SPC was to become instrumental in ensuring throughout the *cross-acceptance* process a recognizable sense of mission and in building an effective working style and persuasive ability across differences in external commitments and substantive interests.

Notes

1   *Southern Burlington County NAACP v. Mount Laurel Township*, 965 N.J. 151, 336 A.2d 713, 720.22 (1975), known as (and henceforth) *Mt. Laurel I.*

2   The judicial branch has historically played a fundamental role in the evolution of planning practice and doctrine, however recursively balancing between extremes. As Hall (1977) has noticed, the courts' attitude towards zoning as a form of regulation of individual property rights varies from state to state as well as throughout time: in New Jersey, after a pioneering period between 1927 and 1930, following *Euclid v. Amber*, the Supreme Court remained hostile to zoning for almost two decades, until the reordering of the judiciary in 1948 and the assumption of an opposite attitude which, until Mt. Laurel, conferred on it the reputation of being 'pro-zoning'.

3   *Southern Burlington County NAACP v. Mount Laurel Township*, 92 N.J. 158, 456 A.2d 390 (1983), known as (and henceforth) *Mt. Laurel II.*

4   As in the case of other New Jersey Supreme Court decisions of the past, in several states, like California, Pennsylvania, New Hampshire, and New York, Mt. Laurel became a reference in jurisprudence on exclusionary zoning litigations (Anglin 1994).

5   The constitutionality of exclusionary zoning had been questioned in New Jersey in the 1960s and was reactualised by racially motivated urban riots at the end of the decade

(Buchsbaum 1993b). In *Vickers v. Township Committee of Gloucester Township* of 1962 Justice J.F. Hall, later to become the author of the majority opinion of Mt. Laurel I, had expressed his famous dissenting opinion against the upholding of the impugned zoning ordinance. The main antecedent, however, was the lawsuit presented by the Urban League of Greater New Brunswick against 23 of the 25 communities of Middlesex County. The lawsuit by the NAACP against Burlington County itself followed a first decision in the *Oakland at Madison Inc. v. Township of Madison* case of 1971, which had recognised the exclusionary character conferred to zoning ordinances by minimum lot and square footage and by maximum room requirements in detached housing zones (Martha Lamar Ass. 1988).

6   On Mt. Laurel, see: Buchsbaum (1977; 1985; 1991; 1993b), Hall (1977), Rose (1977), Payne (1983), Williams (1984), Rahenkamp and Rahenkamp (1986), Sureman (1986), McDougall (1987), Steinberg (1989), Hugues and Vandoren (1990); on its operational aspects, see: Burchell (*et al.* 1983; 1995), Listokin (1976), Martha Lamar Ass. (1988), Selig (1988); on its aftermath and on affordable housing in general, see: Rubin, Seneca and Stotsky (1990), Anglin (1990; 1994). The most intriguing planning-theoretical interpretation yet is that offered by Mandelbaum (2000).

7   On the issue raised by the epochal decision of *Robinson v. Cahill* in 1973 and renewed by *Abbott v. Burke* in 1985, in some sense a fundamental *pendant* to Mt. Laurel, see: Goertz (1988) and Weiss (1989).

8   On exclusionary zoning in the US, besides the classical studies by Babcock (1966; Babcock and Siemon 1985), see: Danielson (1976), White (1978), Lake (1981), Plotkin (1987), Logan and Molotch (1987).

9   On suburbanisation in the US, see: Jackson (1985), Lake (1981), Baldassare (1986), Fishman (1987); for its evolution in New Jersey, at different time thresholds, see: Bebout and Grele (1964), Sternlieb and Schwartz (1986), and Garreau (1991).

10   The average population density of New Jersey is the highest in the nation, 1,000/sq. mile. The 'Garden State' is the fifth smallest in the US in extension, but the seventh in total population, with just under 8 million inhabitants.

11   As an example, average population density in urban counties like Hudson is 12,000/sq. mile, in metropolitan-suburban counties like Essex 6,000, in rural counties like Cumberland, Gloucester, Warren and Sussex, however, only about 300.

12   The only relevant intervention, without however any significant practical consequences, was Governor Byrne's *Executive Order* of 1976 calling for a *Housing Allocation Plan* to be attached to the SDGP.

13   Of particular significance was the second decision in *Oakland at Madison Inc. v. Township of Madison* of 1997. In what appeared as a step backwards compared to Mt. Laurel I, the court took away from lower courts the burden of defining regional housing needs and of assigning fair share obligations to the municipalities, rather asking for an evaluation of their 'efforts in good faith' in providing opportunities for lower-cost (rather than low-income) housing. Other decisions in the meanwhile had restricted obligations to municipalities in growth areas, contributing to further complication in interpretations. This is an expression of the difficulties in operationally translating the 'ethical mandate' of Mt. Laurel, but also of a vacuum in political initiative. The decision, on the other hand, inaugurates the 'proactive' phase of judicial intervention, introducing the so-called 'builder's remedy'.

14   A crucial aspect of the decision, underlining its character of constitutional inquiry, was that it dealt unitarily with six exclusionary zoning lawsuits.

15   In particular, the decision "affirmed the right of government intervention to encourage low-income housing through government subsidy (if available) and tax abatements;

permitted density bonuses to private contractors [...]; limited mobile home construction as a mechanism to satisfy municipal fair obligation; precluded the use of Mt. Laurel decision as a base for encouraging middle-income housing (the court also limited the least-cost doctrine to extreme cases where low-income housing could not be built); and held that some exclusionary devices, such as large lot zoning, were legal if the municipality could satisfy its Mt. Laurel obligation" (Anglin 1994, p. 440).

16   Although anticipatory, the SDGP did not bear the strategic and proactive character of later state growth management approaches. The plan "listed six policy goals. Through these policy goals and map, the Guide Plan recommended where future development and conservation efforts should occur, and further determined where publicly-funded investments should be made. The map delineated 'growth', 'limited growth', 'rural' and 'conservation' areas throughout the state. The designated 'growth areas' consisted of nine highly urbanized areas and a transportation corridor, and eight rural centers in the less developed counties. For each of the four Planning Areas, the Guide Plan made policy recommendations along with suggested implementation strategies. State planners expected the Guide Plan to provide reasonable direction for a state investment strategy. Consistent with earlier state planning efforts, a major substantive thrust was to direct public investment and future private growth to areas within and contiguous to existing development. The Guide Plan was careful not to prohibit development in the limited growth areas, but rather urged that these lands be allowed to grow at their own moderate pace without the stimulus of public infrastructure expansion. Urban revitalization received emphasis, not only for distressed municipalities, but also for ageing urban areas that had significant risk factors for future distress" (Bierbaum and Nowicki 1991, pp. 13-4).

17   Edwards thus recalls the essence of this decision: "I cut out the planning division deliberately. The governor agreed to it, saying 'put something in its place' that would continue state planning but be more responsive to local concerns" (W.C. Edwards, interview 1990, quoted in: Luberoff 1993, p. 6).

18   Nine months after Mt. Laurel II, at a hearing of the Senate Legislative Oversight Committee on the case, commissioner Renna stated the absence of any initiative on side of the executive, and chairman G.R. Stockman was compelled to notice that the administration remained "strangely silent" (New Jersey State Legislature, 1983, p. 3).

19   The New Jersey Bar Association, for example, in a resolution of September 1983 called the executive to its responsibilities towards revising the SDGP (P.A. Buchsbaum, in: New Jersey State Legislature 1983).

20   E.J. Schneider, in: New Jersey State Legislature (1984).

21   The Ad Hoc Committee included over a dozen public and private interest groups, among them ingrained opposers of any form of supra-local planning, as the New Jersey Builders Association, and supporters of the *home rule* dogma, as the influent New Jersey League of Municipalities. Meetings were furthermore attended by representatives of planning practitioners and officials (New Jersey Federation of Planning Officials, County Planners Association), of regional entities (Regional Plan Association, Middlesex-Somerset-Mercer Regional Council) of consultants (New Jersey State Bar Association, Consulting Planners Group) and environmentalists (New Jersey Conservation Fund). An equal participation was ensured to influent state agencies, like the Departments of Transportation, Community Affairs and Environmental Protection. Among the outstanding personalities involved in the committee were chief counsel W. Cary Edwards, Gerald R. Stockman (Senator from Trenton district and chairman of the Mt. Laurel legislative oversight committee), Eugene J. Schneider (former director of the governmental commission on local

governments, of the planning department at Community Affairs, and at the time in force at the Office of Management and Budget of the Treasury), land-use consultant Peter A. Buchsbaum, and James Gilbert (former president of the New Jersey Federation of Planning Officials, and later to become chairman of the State Planning Commission).

22 Sources for the Committee's internal debates are: New Jersey State Legislature (1984), Gottlieb (1988), Bierbaum and Novicki (1991), Luberoff (1993), as well as interviews (C. Newcomb, 12.07. and 31.07.1995; T. Dallessio, 17.07.1995).

23 This position was typical, for example, of representatives of interests as different as J. Trafford and P. O'Keefe, respectively heads of the New Jersey League of Municipalities and of the New Jersey Builders Association.

24 F.G. Streckel III, in: New Jersey State Legislature (1984).

25 D. Moore, director, New Jersey Conservation Fund, quoted in: Luberoff (1993), p. 9.

26 E.J. Schneider, for example, in affirming the non-opportunity of compliance mandates underlines the constitutional as well as the constructive character of the process and its potential for the constitution of a political commitment: "There is enough power of persuasion in the way the plan is being made" (New Jersey State Legislature 1984).

27 Senate Bill No. 1505.

28 Senate Bill No. 1464.

29 L. 1985 c. 222, effective July 2, 1985: N.J.S.A. 52:27D-301 et seq.

30 A. Karcher, assembly speaker, interview 1990, quoted: in Luberoff (1993), p. 10.

31 L. 1985, c. 398, effective January 2, 1986: N.J.S.A. 52:18A-196 et seq.

32 It is not by chance in fact that the most serious tensions within the commission developed around the issue of relationships with the party system and particularly with the state legislature, determining differences among 'the politicians' and the 'good people of the world' (Kanige 1987) often expressed through contrasting communication strategies with elected officials and party representatives as well as through a sensitivity to changes in exogenous political conditions throughout shifts in executive leadership and in legislature majority.

33 On membership of the Ad Hoc Commission, see above, note 21.

34 With a tendency towards an unbalance to the right, however, as noted by C. Newcomb (interview 12.07.1995).

35 It is highly significant that organized political opposition to the state planning process arose mainly after the uproar determined by presentation of the *April '87 Draft* plan (see ch. 6). The main line of opposition came from the legislature. After a Republican initiative addressing the issue of an 'equal' legislative representation in the SPC, contrasted by Governor Kean according to the executive mandate of the commission, the main threat to the process was represented by the so-called *legislative oversight bill* initiated by chief Republican Senator R. Franck. The bill, which was readily approved due to bipartisan support in assembly, responds to an attempt to subordinate the adoption of state planning measures to parliamentary oversight and, thus, de facto, to legislative veto power. While certainly implicitly addressing a crucial issue, i.e. the importance of legislative support for any prospect of formal implementation of the plan, and while pointing to a lack of communication within the political system about the aims and features of the planning process, and thus to the importance of uncertainty in defining conditions for policy-making, the request for a legislative oversight was a measure clearly contrasting with the consensus-building mandate given to the SPC by the *Act*. Even if meant in operational rather than oppositional terms, the involvement of the legislature would have in fact impaired the process of mutual adjustment addressed by the negotiating approach of *cross-acceptance*

(Kanige 1987). The reality of political opposition to the state planning process was in fact much more prosaic than political arguments, and support groups like New Jersey Future and the Middlesex-Somerset-Mercer Regional Council readily denounced them as rhetoric devices which aimed in reality at blocking the new course of planning. The huge lobbying campaigns which backed legislative initiatives indeed offer a major explanation for the significant bipartisan support they received. The Political Action Committee of the New Jersey Builders' Association, for instance, appropriated during 1987 $211,765, almost double as much as during the 1983 election year; in sum, the committees representing builders and developers paid in 1987 $611,759 to single assembly members opposing the *Act*, like Senate President B. Russo and Senator C. Hardwick, who received respectively $130,360 and $64,000 (Kanige 1988). Two coalitions opposing the state plan, albeit from completely different cultural-political positions – the Foundation for the Preservation of the American Dream and the Coalition for Intelligent Growth – were also supported by builders and developers.

Significantly, after exerting a strong pressure on the SPC following the scandal of the April '87 Draft, even the League of Municipalities explicitly rejected the hypothesis of a legislative oversight, seen as an intrusion in the power to negotiate and to directly influence the process granted by the Act to municipalities in the state-local framework of relationships envisioned by *cross-acceptance*. It will however take several years until, after various failed attempts, the last initiative is finally dropped in early 1992, under the new Governor J. Florio, shortly before adoption of the plan.

6 Cross-Acceptance: Plans and Procedures

Introduction

At its beginning in January 1989, the self-representation of *cross-acceptance* offered by the 'institutional entrepreneurs' of the New Jersey state planning process appeared as an elegant formalization, pointing to the main innovative statutory features of the approach. At the same time, it expressed an openness towards learning and evolution which could hardly be defined or formalized. The process was thus conceived as a statutory procedural framework which entrusted the involved actors, through the operationalization of negotiation, with significant margins of internal revision.

Procedural aspects, related to the design of interactive processes, constitute therefore a major reason of interest in understanding the experiential dimension of *cross-acceptance*. They cannot be detached, however, from the evolution of the plan's representations clustered into the 'single text' to which interactive processes and negotiating procedures made constantly reference.

As has been noted, "[t]he plan development process in New Jersey has been difficult and filled with controversy, but it has also produced creative and innovative ideas and concepts that survived as draft followed draft" (DeGrove and Miness 1992, p. 42). Moreover, seldom the drafting of a plan so clearly turned out as a process which, through its provisional sedimentation in texts, accompanies the evolution of a field of collective practices, as it has in the course of *cross-acceptance* in New Jersey.

In this chapter, the *cross-acceptance* process is reviewed pointing to some of its crucial developments, highlighting the connection between the evolution of the plan's representations, throughout its different drafts, and the phases of the negotiating approach. Sticking – as is done in this chapter – to the *formal* phases of the process and to the formal plan's version, while responding to a necessary descriptive function, is nonetheless only one aspect of the interpretive strategy pursued in this case-study: it represents, so to speak, an analysis of the sequence of *static frames* which mark the evolution of the formal-statutory aspects of *cross-acceptance*.

This needs to be complemented by an analysis of the coevolution of *formal* and *informal* aspects of the process, which will be the subject of chapter 7.

Starting the Planning Agency's Activity: the First Versions of the Plan

The drafting of the plan was initially constrained by delays in completing the composition of the SPC and by the necessity to build up an internal working style in the OSP, which defined the core of its staff in 1987. It is therefore not surprising that the first phase of the agency's activity was prevailingly characterized by the role of consultants called in by the director to contribute to a new growth management approach amenable to sustaining the proof of consensus required by *cross-acceptance*.[1]

The choice of consultants, mostly acquainted with OSP's director, J. Epling, and all external to the political and professional establishment of the state, descended from some basic conditions. The New Jersey state planning initiative was able to profit from a background of experiences which was consolidating nationwide a new approach to growth management issues. Experiences in state planning in Oregon and Florida constituted in this sense an obvious but critical reference. In fact, the specificity of the task, albeit controversial in its operational interpretation, was soon clear to the planners. The New Jersey approach grounded its legitimation on building a political commitment to state planning in a much more explicit way than in any other case: a commitment, it must be noted, still questioned even at the highest levels of political leadership, where the new planning task and its consequences for intergovernmental relationships were still kept 'at arm's length'.

The first steps in the process therefore addressed a recognition of positions on the possible issues and representations of the plan as well as a comparison of available experiences. A statewide poll commissioned in 1986 and carried out by Gallup seemed to confirm an encouraging general level of consensus on issues of control and guidance of growth, as already gained through the activity of the Ad Hoc Commission, and the support of watchdog groups, which were progressively organizing into a pro-state-planning coalition.

The first document produced by the SPC in February 1987 (SPC 1987) was a synthesis of the consultants' work, intended primarily as a means of acquiring preliminary opinions from the state agencies; its achievement was to bring about a first framing of state planning goals according to the outcomes of the first rounds of intergovernmental consultations and public hearings conducted.

The first determinant input in the evolution of the planning process, however, was given by a semi-official document not formally approved by the SPC and produced in a short timeframe by the OSP in order to summarize analyses and suggestions contributed by the planning consultants. Despite its provisional character, the timing and modes of presentation of this document – which became (in)famous as the *April '87 Draft* – made it a turning point in the reactions and strategies of the actors towards state planning, influencing their attitudes throughout *cross-acceptance*.

For the first time, the principles of 'sound statewide planning' were translated in this document into specific strategies embedded in physical representations. At the heart of the approach is a system of *tiers* defining the boundaries of eight categories of development, ranging from consolidated urban areas to planned growth areas, down to conservation areas, backed by policies generally aiming at concentrating development on territorial nodes and axes structured around consolidated development patterns and infrastructure systems.

A conception of planning based on decisional and operational decentralization, calling for locally negotiated development choices, is thus embedded in a macroterritorially defined vision of development perspectives, explicitly connecting public funding strategies (through a concentration of infrastructure investments in designated growth areas) to a broad interpretive zoning of the state.

The approach proposed by the *April '87 Draft* was bound to raise a 'wicked' mixture of conflicts of interests, misunderstandings and sheer ideological opposition which changed the state planning climate. Of paramount importance was certainly the shock represented by the mapping of growth management strategies, interpreted (according to a misunderstanding which could later hardly be clarified) as a regulatory attempt in macrozoning and as the infraction to the *taboo* of *home rule*. In fact, this interpretation was a major cause of the general uproar generated in the state's polity, with radical consequences on the forms of opposition as well as on the structuring of the planning process.[2]

Hence it is interesting to comment on the reasons behind such a threatening move as the presentation of such a 'plan' in such a delicate phase in the consensus-building process.[3]

The reasons that motivated the planners' approach must first be recalled. These were reasons of a legal, a moral, a strategic and a political order. In legal terms, the opinion was diffused that, based on the interpretation of the *State Planning Act*, the SPC was empowered to a much larger extent than commonly thought.[4] In moral terms, this position was backed by the idea, typically held by consultants, of a sort of ethical mandate and obligation to adopt policies which could assure the effective pursuit of the act's goals. On the other hand, the feeling was rising of the need for a strategy to raise broader public attention to planning issues. And, finally, the feeling was developing, backed by the results of the Gallup poll, of a growing demand for environmental quality by citizens and organized groups, which led to the formation of cross-sectional coalitions in favor of state intervention and less attached to the *home rule* dogma.

The complexity of the planners' task must further be remembered. Their mandate was defined by an institutional compromise based on the demise of a regulatory approach to planning and on the call for an overall consensus. The question was therefore that of interpreting the margins of maneuver left by a formally 'teethless' planning mandate, the effectiveness of which was completely left to implementation means to be defined subsequently to the reaching of consensus. It may be questioned whether the line adopted by the planners with the *April '87 Draft* was "aggressive" (Luberoff 1993), but, in this sense, it certainly did not refrain from proposing a strong interpretation of development scenarios. Nonetheless, the legitimacy for this move was dubious, highlighting conditions of uncertainty and sensemaking on the new political course of planning which were common to all the actors involved.

Most observers, however, agree on the role played by J. Epling in this event. On the one hand, presenting the draft to the public responded to a strategy of the OSP in order to react to the pressure of planning deadlines: the draft was in fact little more than a collection of the consultants' contributions, forced into provisional coherence in the shape of a 'single planning document'. While this initiative responded to an aim of communication, it led to the SPC being substantially overrun, producing the sensation of an 'apocryphal' plan, not legitimated through complex mediations. In this situation of lack of accountability for the planners' work and of uncertainty in the structuring of consensus-building processes, Epling apparently played the card of a 'normative shock'[5] in order to gain attention and to clarify positions about the plan, according to a view of the perspectives of success of negotiated approaches relying on the constitution of coalitions around the principles and the procedures of the plan and around identifiable issues and interests.

Above all, the *April '87 Draft* lent a new visibility to state planning, influencing the whole of its process. Its most relevant result, intentional or not, was the climate of radical uncertainty it determined. By this, the polity of the state gained a sense of the planning process as a problem, as an enterprise defined by a multiactor concertative context. Roles and positions of the actors involved were consistently shaken by this: and the only possibly shared view became that of state planning as "the only game in town", a game that anyone was bound to play in order not to lose.[6]

After this event, the working style of the OSP also radically changed: the role of external consultants was redefined and largely exhausted, drafting the plan became more and more a task related to a fieldwork experience, comprising the promotion and facilitation of interactional settings throughout the state and nurtured by the feedback produced.

The Preliminary Plan: New Jersey's Growth Management Approach

The first version of the plan, presented in November 1988 after the completion of intergovernmental review, is as we have seen the result of a consolidated – albeit controversial – debate on the purpose of state planning aims.

The plan consists of three large-size volumes showing the attention given to the communicative function of the 'visions' proposed. However, its core bears a basically technical-disciplinary component.

The first volume, *A Legacy for the Next Generation* (SPC 1988a), introduces in discursive, non-technical terms the frame of the recovery of a sense of territorial belonging expressed by the title given to the plan: *Communities of Place*.

This expression introduces one of the most original features of the New Jersey experience, the explicit reference to a pragmatist tradition of the communitarian ideal.[7] This ideal is assumed as the framing concept for an approach – defined as *regional design* – which combines a transactional conception of regionalization processes[8] with the lesson of morpho-typological approaches to the guidance of development experimented with in other growth management initiatives. *A Legacy for the Next Generation* is devoted to presenting the components of this growth management approach:
- a structure for the statewide guidance of development, called the *tier system*;
- a set of principles for the active guidance of development patterns, called *regional design system*;
- a system of state strategies and policies for the pursuit of objectives common to different *tiers* through coordinated actions of all governmental levels involved;[9]
- a system of monitoring and evaluation for the assessment of the effective pursuit of planning goals;

- an intergovernmental process of a concertative and negotiated nature, called *cross-acceptance*.

A single cartographic representation, the *Preliminary Cross-Acceptance Map*, identifies the tiers and constitutes the basis for the conduct of comparisons and negotiations on planning issues during *cross-acceptance*.

The second volume (SPC 1988b) defines the substantive aims of the growth management approach, with the support of 16 thematic maps but still in the language typical of a 'policy plan' of a strategic, non-technical nature. Its core is represented by the linking of strategies and policies to the main tools of the planning approach, the *management structure* and the *regional design system*.

The Management Structure

The adoption of the *tiers* concept dates back to inputs by planning consultants and is justified on the basis of a comparison of four ideal-typical alternatives to growth management (see: OSP 1987), a status-quo and three scenarios responding to different planning paradigms:

- a system based on horizontal coordination of local plans and actions of the 567 municipalities of the state, with the exception of areas under a special regime (Pinelands, CAFRA, Hackensack Meadowlands);
- a 'facility driven' approach oriented towards the coordination of services and infrastructures principally through local public agencies, with limited objectives of allocational efficiency;
- a system based on 'critical areas' governed by agencies placed at each level of administration, limited to the protection of environmental and natural resources; and, finally,
- a system of *tiers*, defined as "gradations of levels of public service based on desirable intensities of use necessary to achieve both public service efficiency and environmental quality goals", "in which growth can be managed at multiple levels of government" (SPC 1988a, p. 19).

The tier system is thus a fundamental component of the approach:

In its purest form, a tier system distinguishes among land areas on the basis of their potential to support growth in beneficial ways. The term 'beneficial ways' means ways that assure 1. the most cost effective, efficient use of tax dollars in the provision of public services, and 2. the protection of the environmental and other quality of life features. High quality public services and facilities, particularly roads and sewers, tend to attract growth. The absence of such services in the face of growth leads to environmental degradation and a lowering of quality of life. Governments at all levels can use the tier system to agree on where growth should occur and then, in a coordinated fashion, strategically invest tax dollars to guide the location and intensity of growth. Other financial and regulatory tools can supplement these strategic decisions. This feature – the timing and sequencing of growth with public services – is the heart of the tier system of growth management. (SPC 1988a, p. 19)

The Regional Design System

The definition presented highlights the relationship between the *tier system* and the 'timing and sequencing' approach central to other referential experiences in growth management. However, the original feature of the New Jersey approach lies in its connection to the concept of *regional design*: it is the correlation of the two which defines the proactive and concerted idea of regionalization at the basis of its growth management strategy. Paradoxically, however, precisely the understanding of this relationship was to become a central bone of contention throughout the *cross-acceptance* process.

The *regional design system* addresses the multidimensional integration of issues, which is one of the core features of growth management programs. 'Regional design', in this sense, combines the idea of coordinated guidance of growth factors, aiming at an efficient allocation of resources by state agencies, with a dimension of local identity as a foundation for a 'sound development'. In this dimension, the legitimation resides also for a concerted approach to policy-making oriented to the constitution of domains of regional cooperation (OSP 1988c).

Regional design is thus intended as the framework for the development of New Jersey communities (SPC 1988a, p. 45). Its primary intention is to connect elements of the territorial development structure in a networking mode as a condition for the regional redistribution of growth potentials and for their direction towards patterns inspired by criteria of sustainability and of the protection of local identities.

The principle of a hierarchical ordering of centers is thus assumed as the design criterion in support of a broad repertory of disciplinary contributions, from urban design to conservation and the design of open spaces, through the coordination of governmental actors as well as through proactive and concerted forms of identification of the linkages between the aims of localities and the regional dimension of planning issues.

The *regional design system* is conceptually articulated into three components. The first is a typology of central places represented by five idealtypes of communities – *cities, towns, corridor centers, villages* and *hamlets* – which should constitute nodes and axes of regional networks of development patterns as the conceptual reference for physical planning. Communities are not asked to identify precise boundaries for each type of center, which are intended rather as criteria for subsequent assessments of the coherence and compatibility between development needs and state policies. This conception, which calls for an interpretation and for an innovative adjustment of local development frames according to physical patterns, is sensibly inspired by elements of a neo-traditional planning wave strongly diffused particularly in east-coast states in the 1980s.

The second component is represented by *development corridors*, linear patterns of development structured along regional transportation axes with plurimodal potentials, effectively anchored to urban agglomerations and secondary accessibility networks. Growth in corridor centers is seen as a means of rationalizing development according to criteria of functional (i.e. in terms of public investments in services and infrastructure) as well as physical integration (i.e. to objectives of sprawl and land-use containment).

The third component, intended as the 'negative' of development choices and as the counterpart to their networking pattern, is the non-urbanized territory surrounding central places. Attention is given here mainly to the safeguarding of open and rural space, relying on a tradition of experiences in agricultural land preservation.

The reason for the most serious controversies on the proposal is however represented by the third part of the plan, a 550-page volume devoted to guidelines and standards for local development planning (SPC 1989). Each of the plan's policies is translated in it into a series of indicators, in an attempt to 'objectivize' issues in terms of a technical-disciplinary language which apparently clashes with the general framework of the plan. Its criticism, as we shall see, will prove to be a major force in shaping the *cross-acceptance* process.

The First Phase of *Cross-Acceptance*: *Comparison*

The Formal Definition of Comparison: Rules and Guidelines

Comparison was the only phase of *cross-acceptance* to be clearly defined from the beginning of the process: it was in fact the only one definitely regulated by the first version of the *State Planning Rules*, ten months before *cross-acceptance* was started. In that, *comparison* operationalized some of the fundamental features of the idea of consensus-building pursued by the *State Planning Act*. The object of comparison is, in strict terms, the *consistency* or *compatibility* between *state* and *local* plans and regulations; county plans, themselves subject to comparison, are not treated as an intermediate level of planning, but as *local,* without hierarchical ordering.

The New Jersey model may thus be correctly defined as *state-local negotiated* (Gale 1992; Bollens 1992); there is however more to it than this. The setting introduced by *comparison*, as one of its most original features, focused directly on the relationship between state and local plans. An important mediating function in the process was nonetheless entrusted to county governments. Counties, which assumed the role of negotiating entities, were called to manage technically and operationally the relationships between state and local authorities throughout the entire *cross-acceptance* process, through the elaboration of working programs framing concertation with localities and finalized in the drafting of a county report to be assumed as the basis of negotiations (fig. 6.1).

The statutory representation of localities entrusted to the counties by the *State Planning Act*[10] was subordinated to the explicit formulation of a procedural model of interactions with municipalities and the general public, as well as to conformity to a model for addressing the comparison and for reporting its outcomes, called *Cross-Acceptance Manual* (OSP 1988a). Involvement in the process remained however voluntary at both county and community level. The counties' option and their right to access formal negotiations was made subject to a notice of participation or waiver to the SPC;[11] similarly, municipalities which formally dissented from county reports were entitled to present an autonomous report, to constitute a negotiating body in their own right, and to access autonomous negotiating sessions with the negotiating entity of the state.[12]

day 1	45	90	120	150	180
- notice of participation or waiver	- county informational meetings	- review *Preliminary Plan maps*	- draft *cross-acceptance rpt.* - publ. hearings	- final *cross-acceptance report* - notice of transmission to SPC	

Figure 6.1 Recommended negotiating work schedule in *comparison*
Source: adapted from: SPC (1988b, p. 83)

The mediating function attributed to the counties in *cross-acceptance*, far from being taken for granted in the framework of New Jersey's intergovernmental relationships, was a choice of extraordinary importance in the perspective of a renewal of their institutional role. It had its origins in two considerations of political realism: the need to rebuild their authority in light of the weak tradition and effectiveness of county comprehensive plans, but also the need to mediate confrontation with merely local issues and to effectively manage consensus-building efforts in a highly fragmented governance environment. In relying on the counties, however, only rarely could the state planing agencies rely on a given technical or political capital: the new formal role attributed to the counties, while respectful of local prerogatives, thus became an occasion for building this capital through renewed practices in intergovernmental relationships.

Risks and opportunities entailed in the form of self-commitment requested from the counties were defined by the margins for self-organization provided by the rules to local processes. Through this counties potentially become key actors in the mobilization and in the definition of the subsequent negotiation process. Moreover, the involvement of the counties realized an indispensable form of organizational decentralization in support of the formal activity of the SPC and of the much more informal engagement in consultancy and information by the OSP. Whereas, in absence of a real political will, this was bound to lead to a merely formal level of involvement, it became in general an important occasion for the shaping, renewal, and expression of a local political identity.

The Approach to Comparison: the Cross-Acceptance Manual

Comparison aimed at achieving a maximum of *consistency* and *compatibility*[13] among all administrative levels of territorial governance: its pursuit was therefore comprehensive and explicitly based on the complex of goals, objectives, strategies, policies, standards and territorial definitions proposed by the plan. Inevitably, however, the ability to understand and to respond to such an aim according to the rules of the process was to turn out to be highly differentiated and only partial. The *State Planning Rules*[14] in fact conceived the outcome of *comparison* as the identification of 'areas of agreement and disagreement', related to three components of the *Preliminary Plan*, i.e. *policies, standards* and *tier delineations*, which could be expressed in a protocol form. By articulating comparisons on each component around the concepts of consistency and compatibility, the outcomes were meant to polarize on single statements of the plan and to be accordingly framed into three categories, as to render the expression of differences flexible and thus amenable of negotiation:

- compatibility should be acknowledged in such cases in which, despite formal inconsistency, contents of local plans and regulations granted equal effectiveness in the pursuit of goals, objectives and strategies of state planning;

- compatibility should be acknowledged in such cases in which it could be achieved through modifications of corresponding provisions of the *Preliminary Plan*;

- compatibility was to be achieved through modifications of the corresponding local plans and regulations in all further cases.

The progressive attitude of this model cannot be ignored. It expressed in fact an approach far removed from a preemptive as well as from a merely consultative idea of intergovernmental relationships: it represents rather something like a procedural translation of the central meaning attributed to argumentation in defining planning contents. The precondition of comparison was the involvement in an *argumentative exercise* which, on the one hand, generated a negotiating dimension and, on the other hand, possibly entailed a mutual reframing of codes. However, its formal procedural expression also implied its limits: in fact, precisely the attempt of a reduction of this exercise into a protocol form represented by the *Cross-Acceptance Manual,* and its substantial rejection, contributed to the procedural 'explosion' of the *comparison* phase and to the largely autonomous development of the rationale of negotiations.

Comparison was in fact originally intended to cover a proportionally extended timeframe compared to the following phases, in the framework however of a process meant to cover about 12-15 months in all.[15] The price for the importance attributed to *comparison*, as was soon to become apparent, was therefore the taken-for-grantedness of a sequential and linear development of the subsequent phase of *negotiation*. The *Cross-Acceptance Manual* was precisely the tool developed to frame the results of *comparison* according to this expected schedule.

The core of the manual was represented by checklists which framed the process of comparing local plans with objectives, policies and strategies of the *Preliminary Plan*. While apparently necessary and obvious, the role of checklists proved to be highly problematic. They certainly above all responded to a need for reduction of operational complexity, required by the diffuse inadequacy of local and county planning bodies as well as by the objective of ensuring effective management of feedback expected by the state planning staff. The operationalization of the *State Planning Rules* offered by the manual was moreover quite obvious: the contents of county reports were supposed to be summarized by means of two checklists related to *goals* and *objectives* (*checklist A*) and to *strategies, policies* and *standards* (*checklist B*) and broken down respectively into the expression of agreement or disagreement and into defined typologies of compatibility or incompatibility. This procedure was to be repeated at the level of the county, of each municipality, as well as of a *summary checklist* which should have framed the position of local negotiating entities.

While *cross-acceptance reports* were not exhausted by the checklists, the centrality of the latter in *comparison* determined significant consequences and a broad array of reactions across the local authorities involved. Paradoxically, the richness of arguments involved in the piecemeal justification of each item foreseen by the checklists rendered them almost non-practicable. Thus, besides the considerable level of management complexity involved and their vulnerability to forms of rejection and opposition, the function of the checklists was basically overshadowed by the argumentative importance assumed by county reports. The progressive consolidation of a local policy style and of strategic linkages between counties and their municipalities, often favored by the state planning process itself, significantly contributed to expanding the argumentative apparatus of reports and to marginalizing the result of the checklisting exercise. Moreover, the problem of comparability between planning issues and planning scales, and their embodiment into local situations, frequently raised and often clearly motivated, itself implied broad margins of arbitrariness or latitude in interpretation which, rather than allow a reduction into the form of a state-wide protocol, highlighted the pervasive negotiable character of meanings. Far from resulting in useless outcomes, the comparison of checklists thus raised a highly differentiated variety of positions and interpretations which marginalized the idea of a direct transposition of the outcomes of *comparison* into issues of negotiation.

Outcomes and Problems of Comparison

If, as has been stated, *cross-acceptance* revealed at least a dozen different 'identifiable paradigms' of negotiation,[16] then their foundation lies in the counties' response to *comparison*.

Several structural constraints contributed to this differentiation. Among the most important of them was certainly the generally low preparation of county and local planning offices: this was particularly apparent in case of the former, called through *cross-acceptance* to an usually complex coordination task.

Moreover, the financial burdens involved were often inversely proportional to local fiscal capacities, as typically in the case of urban areas: funding of the process thus became a paramount issue since the first *cross-acceptance* meetings.[17]

But, above all, the actual conduct of *comparison* questioned the timing of the process: once the deadline set for the plan's adoption by the *Act* had been delayed, and the equivocal interpretation originated by the first *State Planning Rules,* according to which the six-month timeframe would cover the whole of *cross-acceptance,*[18] had been finally overcome, the period envisaged by the SPC for the conduct of *comparison* and the start of *negotiation* turned out to be clearly inadequate in view of the political as well as the technical complexity of the process. As a consequence of this complexity, the phase of *comparison* was actually extended to over fourteen months.[19] While determining a general shift in timing and a more complex procedural design of *negotiation,* this extension was also an important occasion for making sense of *cross-acceptance* and, to a certain extent, for reinventing its nature. As we shall see, the period following the formal start of *negotiation* and overlapping with the extension of *comparison* was devoted to reframing the process according to the experience already made.

The Second Phase of *Cross-Acceptance*: *Negotiation*

Designing Negotiation

Both the sense and conduct of *negotiation* were clarified in relationship with the progressive definition of the timing and phasing of *cross-acceptance.* The main features subject to evolution in its conception were to be its internal articulation and the broadening of the subjects involved.

The preparation phase of *negotiation* was formally started July 10, 1989, the day after the schedule for presenting county reports. *Negotiation* thus had an official albeit fictitious start. In fact, the time sequence foreseen by the *State Planning Act* and by the *State Planning Rules* was probably the most evolving aspect of the entire state planning process. The temporal dynamics of *cross-acceptance* became an important dimension of its internal evolution towards reflexivity and learning.

The extension of the *comparison* process at the local level was the formal reason for the provision of an instructional phase meant to cover the overlapping which was emerging between the extension of *comparison* and the start of *negotiation.* The intention of the SPC was clearly to grant maximum input to the revision of the *Preliminary Plan* while keeping control on timing of the process and preventing it from being arrested due to indefinite delays in delivery of some *cross-acceptance reports*: in this sense, *pre-negotiation* had the status of a procedure allowing for the beginning of "substantive discussions" (OSP 1989, p. 4) according to expected schedules without impeding the completion of *comparison* in late counties. Out of a critical evaluation of responses to the first phase of *cross-acceptance* the idea of a *living plan* was thus developing,[20] i.e. of a plan development process not only internally flexible, but extended towards an evolutive and recursive dimension.[21]

The timing of *cross-acceptance*, accordingly, ceased to be treated as a procedural input and became a mainly political issue. *Negotiation* itself was to be defined by an endogenous temporal extension similar to that of the previous phase and in large measure overlapping with it. In September 1989, shortly before *pre-negotiation*, the OSP proposed containing its timeframe to four months and that of *negotiation* to six months, in order to allow the presentation of outcomes for assessment in August 1990 and the start of *issue resolution* in November 1990. In December 1989, however, while *negotiation* had been completed in 17 counties, the end of the phase was more realistically expected in January 1991, the presentation of the *Interim Plan* in April, the drafting of the impact assessment in a timeframe of seven to nine months, and the adoption of the *Final Plan* in December; the introduction of a new phase of negotiations between the presentation of the *Interim Plan* and *issue resolution* determined a further time extension.[22]

The Formal Definition of Negotiating Procedures and Issues

The beginning of the *negotiation* phase represented a significant occasion for reflecting on the previous experience and for (re)defining the formal components of the negotiating approach. Its first steps went towards a distinction of types of negotiating issues and of their modes of treatment as well as towards the definition of a process-design allowing for collective deliberation on each of the issues according to an appropriate and viable sequence.

Negotiation was thus procedurally defined by two distinct processes:
- a *pre-negotiation* phase articulated into processes to be conducted at both a state and a multi-county level;
- a *negotiation* phase in the strict sense, articulated into negotiating sessions between the state committee and the negotiating entities of the counties or of dissenting municipalities.

The internal evolution of the process later led to the formal establishment of a third phase, *continuation of negotiations*, as an outcome of the overlapping between delayed negotiations and the plan's revision.

Pre-negotiation The main stage of *pre-negotiation* consisted in the identification and categorization of issues emerging from *comparison* to be dealt with in negotiations. For this purpose, the OSP defined a master list of issues organized according to a cross-section of two distinct kinds of categorization. The former was related to the problematic scope of the issues according to three main categories:

- *policy issues*, i.e. issues concerning principles and concepts of the plan;[23]
- *application issues*, i.e. issues implying differences in opinion on the appropriateness, applicability or meaning of specific plan provisions;[24]
- *fact issues*, i.e. issues amenable of resolution based on corrections and clarifications of facts and representations in the *Preliminary Plan*.[25]

The latter was related to the territorial scope of the issues and distinguished between them according to their state, regional or local relevance for the plan's provisions. The master list was based on formal documents, like the *cross-acceptance reports*, but also on requests and comments from public hearings and from the advisory committees: it thus represented an open repertory amenable of integration throughout the extension of the process.

The second stage of *pre-negotiation*, defined as 'exploratory' or 'consensus-building', was intended as preliminary to the start of the statutory negotiating sessions. In actual fact, it represented an important anticipation of negotiating rationales. In this stage, issues of state relevance were dealt with by the state negotiating committee in the framework of multicounty settings, through a definition of agreed alternatives and a selection of issues to be negotiated at the local level, in pursuit of a reduction of complexity of formal negotiating sessions.

The effective conduct of negotiations is thus pursued through a new 'hierarchization' of issues: its role to ensure that issues and concerns characterized by an impact potentially exceeding the specificity of single countries may be considered and treated as such, in a state or at least cross-county perspective and that, at the same time, issues and concerns may be resolved with the broadest base of consensus at the local level. While the former aspect stresses the importance of the activity of framing issues, the latter aim in particular calls for inclusive criteria for the definition of local negotiating committees in order to ensure access to local stakeholders and representativeness to local civic community. Both, as we shall see, play a relevant role in defining the constitution of informal processes parallel to the conduct of formal negotiating sessions.

Negotiation The start of the formal county-municipal negotiation process defined by the *State Planning Act* was therefore subordinated to the achievement of an appropriate level of consensus in the framework of statewide and multicounty processes in *pre-negotiation*. During this phase, the OSP provided the state negotiating entities with comments on the categories of each of the issues, assuming an important framing and mediating role; this was later to be continued through the diffusion of information and the informal monitoring and evaluation activity realized through the draft of county *negotiation updates*, by which also an ongoing update of discussions in the advisory committees was provided, and which frame the agenda for three *regional forums* conducted in order to collect reactions to the negotiation process.

The design of the formal negotiation process was again organised into stages, which are significant for an understanding of the shifting roles assumed in them by the actors:

- a first round of *interstaff meetings* in which the planning staff was frontstage: in each county (or dissenting municipality) the OSP area planning managers met local planning staff (director of planning or designated members) in order to agree on alternative solutions recommended for the issues raised by *cross-acceptance reports* and to identify unresolved issues to be dealt with by the negotiating entities. The aim was to identify any possible change in the *Preliminary Plan* which may not require detailed examination by the state negotiating committee: consensus thus reached was subject to ratification in the course of negotiating sessions;

- the negotiating committee of the SPC and the local negotiating entity met in order to negotiate previously unresolved issues;

- municipalities which had filed dissenting reports negotiated issues directly with the negotiating committee, county representatives attending;

- *agreements* and *disagreements* emerging from the negotiating sessions were published in periodical reports and filed in an *interim statement*.

Continuation of negotiations The complex procedural articulation and the circularity of the negotiation process at the state level as well as the definition of the nature of agreements became the reasons for an internal extension of *negotiation* and for a progressive overlapping of activities, in some sense reproducing what had already happened during *comparison*. The result was the need to formally extend negotiations beyond the presentation of the *Interim Plan* as well as of the *Impact Assessment*, i.e. of the documents which should have summarized their outcomes. In fact, *continuation of negotiations* was to become a quite specific phase in the process, distinguished from *issue resolution* by its insistence on the very issues deferred during formal negotiating sessions.

Dynamics and Formal Outcomes of Negotiation: Towards the Living Plan

Pre-negotiation was introduced by the constitution of the negotiating subcommittee of SPC, called *Plan Development Committee* (PDC). After the OSP-counties workshop of November 1989, the PDC started in early 1990 a series of informal meetings with counties, municipalities and citizens in order to present and discuss each single comparison report. During these meetings the formal composition of local counterparts, called *Local Negotiating Committees* (LNC), was defined and later ratified.[26] At the same time, the OSP promoted a series of training sessions in interest-based negotiation techniques particularly targeted to county planning staff.

The conduct of *negotiation* strictly followed the procedural sequence: in only one county (Essex) was a dissenting municipal negotiating entity established, while all other counties included representatives of dissenting municipalities in the composition of their LNC.[27]

As in the case of *comparison*, the results of *negotiation* were problematic, and their interpretation cannot be reduced to its formal outcomes. Their analysis nonetheless offers important clues to further interpretive steps.

The format of negotiation, as previously mentioned, was determined by the identification of *common issues* and by applying to them five kinds of possible resolutions: its structure thus allowed for a summary representation of the negotiating outcomes.[28]

During *continuation of negotiation*, deferred issues were addressed. In this phase a particular focus of activity was represented by mapping both negotiations and the deferred resolution of some issues resulting from, but also the important novelties introduced by, the *Interim Plan*.[29]

Issue resolution was again dominated by the mapping of negotiated issues: while addressing changes introduced by the revision of the *Interim Plan*, mapping in fact revealed the evolutive character and the interpretive content of representations, and thus their inherent negotiable connotation.

Two kinds of issues in particular emerged. The first concerned *mapping applications*, i.e. controversies related to misunderstandings of criteria introduced by the *Interim Plan* (and quite different from those of the *Preliminary Plan*) or to lack of documentation or timing, and called for technical clarifications in the framework of *issue resolution*. The second however, called *mapping concerns*, again related to an interpretation of the plan's principles, and sometimes raised new kinds of disagreements. The final *Statement of Agreements and Disagreements* (SPC 1992c) could only record the emergence of this problem,[30] which was to contribute to mapping one of the main fields of the ongoing evolution of *cross-acceptance*.

The Final Versions of the Plan

Between the *Preliminary Plan* and the subsequent version, the *Interim Plan*, the main innovations were introduced in the structure and contents of the state growth management approach. The revision of the plan actually occurred in a rather coevolutive way during *comparison* and *negotiation*, whereas the final version realized through *issue resolution* and the assessment procedure, as required by the planning statute, paradoxically represented a less innovative outcome, formalizing developments already corroborated in the course of the *negotiation* phase.

It is hence not only its contents that makes the *Interim Plan* (SPC 1991) more interesting than the *Final Plan* itself (SPC 1992a). Drafting the *Interim Plan* resulted in fact in significant innovations in the conception of plan development, rendering, as we shall see in more detail, a sequential interpretation of its relationships with negotiation rather inadequate. It is thus not an exaggeration to talk of it as an exercise in the collective reading and drafting of a plan.

The innovations introduced in the plan were certainly mainly a result of the difficult phase of *comparison*, which had been crucial in reframing conditions for consensus. As in the case of the design of the process, the thematic and procedural dilatation of concertation could not have left the drafting of the plan unaffected. The main features of its revision however relied in its coincidence with the conduct of negotiations and, moreover, in its becoming an integral part of them, contributing to their transformation into an informal concertative setting rather than formalized bargaining.

Parallel to the organization of the setting of negotiations, the SPC had selected the issues emerging from comparison which defined possible amendments of the plan. The resolutions adopted according to this selection constituted both a political response to the lessons from *comparison* and an integrative platform for *negotiation*. Distinguishing these dimensions in the actual development of the process is almost impossible and requires rather a relativization of interpretative frames biased by a presumptions of exchange and compromise mechanisms developing in a predefined structure of relations among involved actors.

The most critical issues emerging may be summarized in the following terms (OSP 1988b; 1990; 1991):
- controversies on definition criteria for problems and policies, questioning

their sense and their identification with specific situations: one of the most crucial aspects of the confrontation between 'generalized' and 'local' representations of territorial realities, confronting the designations of the plan with a broad array of reactions (from a perception of stigmatization by *tier 1* areas to the abstract definition of redevelopment policies; from difficulties in sharing definitions on suburban development patterns to a diffuse opposition to extended growth containment criteria in rural areas) and stressing constraints in the identification of 'regional' and local communities with generalized representations of development;

- controversies on the territorial designation of issues, questioning the analytical and instrumental apparatus centered on mapping activity and leading to the territorial definition of policies represented by *tier delineations*; at stake was their meaning and interpretation, which allowed for various local variants and called for articulated mutual learning processes in order to render the achievement of a local consensus possible;

- controversies on issues of implementation and compensation, addressing the whole problems, ranging from technical and political argumentation, focusing on the uncertain availability and definition of means for implementation and of their margins for negotiation;

- finally, and strictly related to this uncertainty, controversies on the legal means for implementation and compensation.

The connection between negotiating issues and concerns raised is apparent in some of the lines of revision of the plan proposed at the beginning of *negotiation*. Among them, at least the following must be recalled:

- a reconsideration of *municipal distress criteria* as indicators for *tier 1* designation, stemming from a demand for less stigmatizing and generalized categorizations and for more problem-oriented thematization and targeting;

- a reconsideration of criteria for *tier 4* and *tier 5* definition, often of a quite problematic distinction and tied to the controversial issues of normative constraints to development in rural areas, demanding more flexible articulations in relationship with the system of centers and with the possibility of identifying local low-density development potentials compatible with the safeguarding of non-urbanized areas;

- an urgent request for policies for rural areas capable of legitimizing norms of development containment through the provision of locally negotiable compensation tools tied to active state rural economic policies.

The main issues to emerge from *negotiation*, requiring policy interventions which went beyond the scope of bilateral agreements, further amplified this catalogue. At the end of negotiations, a diffuse demand had thus arisen for clarification and intervention on a broad array of issues,

such as:
- the instruments, procedures and funding criteria ruling intergovernmental relationships in implementation matters;
- the relationships between agreements reached through state-local negotiations and the action rationales of state agencies;
- the formulation of adequate criteria for joint action in distressed urban areas as an alternative to rigid *tier 1* definitions and designations;
- the excessively direct identification of development patterns in the definitions of tiers with the features of existing infrastructure;
- the problematic relationship between the aims of densification and concentration of development and existing structures;
- the need for a both operational and symbolic 'translation' of the normative apparatus of volume 3 in guidelines for local action, according to the role attributed to morphotypological means of urban design in the development of the centers system and in urban renewal;
- and, finally, sheer political issues, like 'equity' in the normative framework of development in rural areas, as well as a persistent reference to *home rule*.

Not all of these questions appeared immediately amenable of conciliation, and less so in the framework of a single planning document. Negotiated agreements tended therefore to focus no more on the issues in themselves, but on a commitment to action addressing the problems related to the issues as the only possible condition for a relative reduction of uncertainty and for a mid-term consensus. Through the process, often accompanied by diffidence and frustration, a shared perception developed that what was at stake in negotiations lay beyond the wording of the plan.

It is hence not surprising if, in the context of such a dispersion of issues, the main innovations introduced in the revised plan appear to be of an eminently symbolic nature.[31]

The *Interim Plan* presented in July 1991 is a much leaner single-volume document, without the analytical and normative apparatus which was the subject of the most contested third volume of the *Preliminary Plan*, and which had proved to be untractable outside of the rhetorical dimension of negotiations.

Most notably, and perhaps surprisingly, the plan stated the demise of the most controversial focus of negotiation, the *tier system*. The new version of *regional design*, renamed *Resource Planning and Management System*, explicitly underlined its adhesion to a principle of active *direction* rather than of *control* of development, and did so by substituting tiers with five *planning areas*,[32] which, through the *centers designation process*, are based on a diachronic, evolutive and negotiated perspective of definition of the links between strategies, policies, and areas of the state's territory. By stating the centrality of principles of strategic guidance and of concerted intergovernmental coordination as implementation tools, the revised approach expressed the measure of political compromise needed in order to gain consensus, but also addressed an idea of policy development and implementation which called for an ongoing concerted planning attitude.

Thus, although clearly revealing the persuasive instrumentality of its change of denomination,[33] the revision of the *tiers* concept introduced a new understanding of the domain of growth management policies, which, along with the coevolution of the negotiating exercise, constitutes the most original feature of New Jersey's *cross-acceptance* experience.

Notes

1   The main consultants in this phase were Freilich, Leitner, Carlisle & Shortlidge of Kansas City, Siemon, Larsen & Purdy of Chicago, Wallace, Roberts & Todd of Philadelphia, and Rogers, Golden & Halpern of Philadelphia. Among them, one may recognize the protagonists of important experiences in growth management (like state planning in Florida and the *growth boundaries* approach of Ramapo and San Diego in the case of Siemon and Freilich).

2   On reactions to the April '87 Draft and on the new wave of oppositional activism raised by it in legislature, see ch. 5, note 35.

3   The draft had been 'mysteriously' circulated before being presented during a conference at Rutgers in April 1987, raising demands for copies from external observers and causing evident embarrassment in the SPC, which was almost uninformed of its contents. These circumstances contributed to a climate of public scandal and to a revival of lobbying and counterinformation campaigns denouncing the 'secret plot', leading to legislative initiatives and even to vetoes from interest coalitions which had backed the *State Planning Act* like the League of Municipalities.

4   This was based on a phrase of the *State Planning Act* stating that "the Commission *shall* take all actions necessary and proper to carry out the provisions of this act" (emphasis added). This interpretation seemed to be backed by Mt. Laurel III, which had stated the regional orientation of the act.

5   Opinions diverge on Epling's intentionality in doing that. Whilst Bierbaum (interview 11.08.1995) talks of an explicit 'will to fight', it may be more appropriate to understand this choice according to the limited, strategically and situationally bound rationality of the actors. Significantly, Epling himself has recognised the dimension of personal experience involved in his way of conducting *cross-acceptance*.

6   "It's the only game in town so we played not necessarily to win but certainly not to lose" (J. Michaelson of Hudson County Planning Board, in: Luberoff 1993, p. 34).

7   The influence of the idea of *community of place* in framing the planning process should not be overemphasized but is nonetheless highly significant. Certainly related to the personal background of some influent OSP member, it has played an effective function in labeling the growth management approach, contributing to its communicative commitment, and has constituted a consistent argumentative grounding for its developments. The pragmatist ascendancy of the concept of

community of place has been underlined by H. Blanco (1994), herself an important actor of the first phase of the New Jersey state planning process.

8   Reference goes of course to Mumford (1938) and to revitalizations of his conception of *regional planning* proposed since the 1970s (e.g.: Friedmann and Weaver 1979).

9   It may be useful to remind of them here by quoting the original wording:
"a. Protect the natural resources and qualities of the State, including, but not limited to, agricultural development areas, fresh and saltwater wetlands, flood plains, stream corridors, aquifer recharge areas, steep slopes, areas of unique flora and fauna, and areas with scenic, historic, cultural and recreational values;
b. Promote development and redevelopment in a manner consistent with sound planning and where infrastructure can be provided at private expense or with reasonable expenditures of public funds. This should not be construed to give preferential treatment to new construction;
c. Consider input from State, county and municipal entities concerning their land use, environmental, capital and economic development plans, including to the extent practicable any State plans concerning natural resources or infrastructure elements;
d. Identify areas for growth, limited growth, agriculture, open space conservation and other appropriate designations that the commission may deem necessary;
e. Incorporate a reference guide of technical planning standards and guidelines used in the preparation of the plan; and
f. Coordinate planning activities and establish Statewide planning objectives in the following areas: land use, housing, economic development, transportation, natural resource conservation, agriculture and farmland retention, recreation, urban and suburban redevelopment, historic preservation, public facilities and services, and intergovernmental coordination" (*State Planning Act*, N.J.S.A. 52:18A-200).

10 N.J.S.A. 52:18A – 202.

11 N.J.A.C. 17:32 – 3.2. In case of waiver the SPC is empowered to nominate an alternative negotiating authority upon consultation of county governments (N.J.A.C. 17.32 – 3.4). This did not occur during the process, nor did the presentation of joint reports by counties involved in common regional planning bodies (N.J.A.C. 17:32 – 3.3).

12 N.J.A.C. 17:32 – 3.13.

13 According to the *State Planning Rules* (N.J.A.C. 17:32 – 1.4), *consistency* means that policies or standards of a local plan are substantially equal to those of the state plan; *compatibility* means that they are considered equally effective.

14 N.J.A.C. 117:32 – 3.11.

15 OSP (1988a), p. 13.

16 M. Bierbaum, interview quoted in: Gale (1992), p. 434.

17 Counties were granted by the state a *una-tantum* financial contribution to management costs of $ 20,000 each: this was clearly a quite modest, if not minimal contribution compared to real costs implied by an active conduct of the process.

18 See: 20 N.J.R. 673.

19 At the end of *comparison* and at the statutory starting date of *negotiation*, only one county (Mercer) had presented its *cross-acceptance report*. During the same year, nine counties presented reports, followed in the first half of 1990 by a further nine. Only at the beginning of 1991, i.e. 25 months after the start of the process had the whole of the 21 county reports needed in principle for starting *negotiation* been submitted. The latest counties in complying were Warren (September 1990) and Essex (February 1991); in March 1991 submission of dissenting municipal reports was completed (SPC 1992b).

20 J. Epling, introduction to a seminar of the *Peer Review Committee* in New Brunswick, December 6-7, 1990 (video).

21 This progressive change in attitude is well expressed by the following statement: "The purpose of the State planning process is to coordinate and integrate planning at all levels of government, with maximum participation of State agencies, counties, municipalities and the general public. The ultimate aim of the process is to arrive at State, county and local plans that are compatible, or consistent. The Act did not envision that total compatibility among these plans should be achieved during formulation of the first state plan. Compatibility would be achieved over time, as plans at all levels are periodically revised and updated. As part of this broader and lengthier process, each three-year cross-acceptance period is designed to result in a written statement specifying areas of agreement or disagreement and areas requiring modifications by parties to the cross-acceptance. These statements become the bases for revisions both during the negotiation phase and in later cycles of planning at all levels of government" (OSP 1989, p. 2).

22 The timing actually realized was March 1991 (excluding *continuation of negotiations*), July 1991, February 1992, and June 1992 respectively.

23 For example: the request for generalized redesignation of *tier 5* and *6* areas according to principles not related to tiers definitions, such as 'equity'.

24 For example: the need for infrastructure improvements in hamlets and villages, in contrast to Policy 2.3. of the *Preliminary Plan*, in order to implement the *regional design system*.

25 For example: redesignating an area from *tier 6* to *tier 4* following in-depth inquiries into the state of infrastructure (sewers and water supply).

26 R. 1990 d. 336, effective July 2, 1990; see: 22 N.J.R. 621 (c), 22 N.J.R. 2033 (a).

27 This was the record of negotiating sessions: after 18 *pre-negotiation* meetings between February and June 1990, 45 OSP-county planning staff meetings for framing the negotiating agenda were held between June 1990 and March 1991; 25 LNC-OSP working sessions were held between August 1990 and March 1991 for revising local agreements reached by staff and for approval by LNC; formal LNC-PDC negotiating sessions on disagreements followed between September 1990 and March 1991 and were concluded with negotiations on deferred issues between January and March 1992. On average, each county went through a minimum of three negotiating sessions.

28 The 64 common issues identified in *pre-negotiation* amounted throughout the 21 counties to a total of 548 resolutions, 54 of which were deferred, resulting in 487 'final' resolutions thus articulated: 123 *explicit agreements*; 186 *agreements in principle*; 124 *agreements on implementation* items; 54 *agreements to identify concerns*. Only seven issues resulted in explicit statements of disagreement.

29 Negotiation of deferred issues involving 18 counties led to six new disagreements, for a total amount of 13 against 535 agreements.

30 Besides non-compliance of two counties (Sussex and Hunterdon), seven groups of *mapping applications* and five groups of *mapping concerns* (each with two formal disagreements) emerge in this phase.

31 Albeit in different ways, this symbolic dimension has been experienced in equally strong terms by most of the actors; particularly planners with a normative conception of territorial representations offered by tiers saw it as a radical sign of the weakening of the plan's approach.

32 Redefined as: *PA1 metropolitan planning area; PA2 suburban planning area; PA 3 fringe area; PA4 rural planning areas; PA5 environmentally sensitive planning area.*

33 This interpretation is shared by many interviewees (T. Dallessio, 14.07.1995; D.

Maski, 19.07.1995; D. Brake, 10.08.1995), who agree in underlining the symbolic obstacle to consensus-building represented by the diffuse identification of territorial policies with a land-use planning model and by the 'phantom' of a statewide zoning.

7 Cross-Acceptance: The Coevolution of Processes

Introduction

The chapter introduced here plays a central role in the interpretation of *cross-acceptance*. In reconstructing the procedures of the state planning process, a process-oriented and evolutive perspective has already been stressed in the previous chapter. However, their evolutive dimension has been followed only with reference to the *formal* aspects of the process, thus addressing only one dimension of their internal dynamics and isolating it from other dimensions of the process. As for the plans, they have similarly been treated as situated outcomes of a complex and evolving deliberative process, however analyzing their contingent definitions as *static frames*: observing, so to speak, the *static properties* of the representations and images embodied in each of their stages.

A further dimension must now be addressed in which the evolutive interpretation of the processes turns towards a perspective centered on their *generative* aspects. A decisive function is played in this by an interpretation of the coevolution of *formal* and *informal* aspects of the processes, through a joint analysis of modes of interaction, representations and actors involved. Here, the reciprocal relationships are stressed between the internal evolution of the formal statutes of the processes and the amplification of their informal aspects. This distinction allows us to understand the role of the forms of interaction enacted with regard to the evolution of the arguments displayed in the planning process.

A crucial stage of this interpretive exercise is clearly represented by a close analysis of the negotiation process. Formal negotiation reveals itself as a quite specific action situation: however, its belonging to a circular domain of relationships is also highlighted, which calls for an understanding within a broader system of action frames and rationales. The key to this is the nexus between the formalized and structured conduct of negotiation and the expansion of a field of interactive practices of a mainly argumentative and informal nature. Their mutual interplay bears a significant influence on the plan itself and on the features of the institutionalization of an interactive planning approach.

135

The Evolution of the Statutory Dimension of *Cross-Acceptance*

Formal and Informal Domains of Processes

Let us first interpret the evolution of the *statutory definition* of the *cross-acceptance* process in its intertwining with the actual evolution of action situations. This will be done with the help of two representations of the process as defined at two different temporal thresholds. The former (fig. 7.1) offers a self-representation of the process according to the provisions of the *State Planning Act* and *Rules*, while the latter (fig. 7.2) is a representation of how it actually developed. In both figures, formal-statutory aspects are distinguished from non-statutory, informal aspects of the process.

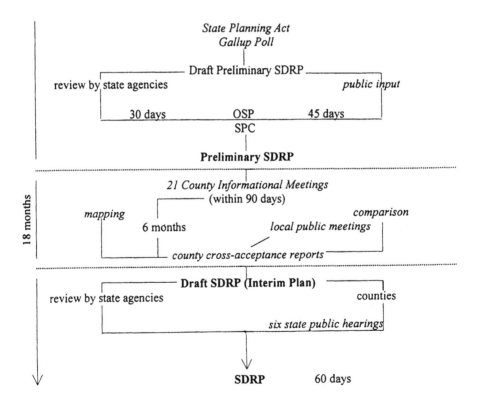

Figure 7.1 The *cross-acceptance* process according to the *State Planning Act* and to the first draft of the *State Planning Rules*
Source: adapted from a scheme by OSP, August 1987

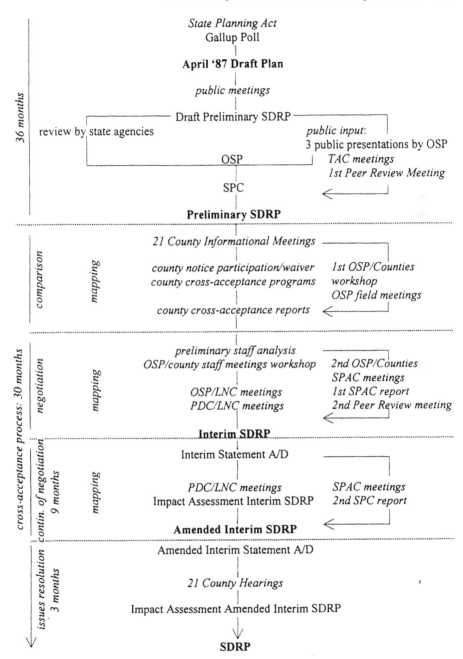

Figure 7.2 *Cross-acceptance* **as actually conducted as of 1992**

The former representation develops at a time in which the operational nature of the negotiation approach of *cross-acceptance* is still to be defined: in some sense, it could be stated that the formal consultation framework envisioned in the scheme, rather than of a defined design rationality, is the expression of the imaginative repertory of possible processes available to the 'institutional entrepreneurs' at this stage.

Some significant design features may nonetheless be noticed. What is represented here is a tree-like process converging into a central decision-making sequence of a linear kind, developing through *comparison* and *negotiation,* and contained in a relatively short timeframe. This linearity corresponds to the idea of a procedural consequentiality between the comparison of policies and plans and the issues and forms of negotiation, as postulated in particular by the application of the *Cross-Acceptance Manual*.

The actual development of *comparison*, however, produced an 'explosion' of this sequence in both time and statutes, which may be observed in the following scheme. Not only an expanded diachronic sequence of statutory processes emerged, but also a progressive generation of interactive activities, not or only loosely formalized, in coincidence with significant turns in the process. These activities tend to constitute a series of group processes parallel to the statutory decision-making process, intersecting it in a circular manner and contrasting its consequentiality.

Besides the evolution of the statutory procedural sequence, a parallel, basically informal process thus developed based on interactive settings of political and technical-disciplinary argumentation, constantly retroacting on formal decision-making. Thus, a third dimension was introduced into the process and became its crucial evolutive motor: the circular dimension represented by the development of decisional feedback loops, determining a radical shift away from a consequential conception of the negotiated state planning approach.

The Function of Rules

The *State Planning Act* entrusted the SPC with the normative and procedural definition of the *cross-acceptance* process through the adoption of rules "including procedures to facilitate the solicitation and receipt of comments in the preparation of the preliminary and final plan and to ensure a process for comparison of the plan with county and municipal master plans, and procedures for coordinating the information collection, storage and retrieval activities of the various State agencies".[1] This quotation expresses the overt but fruitful contradiction of a piece of legislation which, while envisioning a methodical approach to consensus-building, left its modalities largely undefined. The *State Planning Rules* became in fact the real nucleus of the process design, the tool through which the idea of consensus pursued by the *Act* was procedurally and substantially defined, but also the primary subject of feedback from its evolutive trajectory.

The rules, based on the centrality of intergovernmental coordination issues stated by the *Act*, defined *cross-acceptance* as "the process of comparing local, county and regional plans with the Preliminary State Development and Redevelopment Plan and the dialogue which occurs among participants during and after this process to achieve compatibility or consistency between local, county, regional and State plans": a process hence "structured so as to establish vertically integrated and compatible local, county, regional and State plans".[2] Through this argumentative process and its evolution, however, the rules themselves became a stake, and their availability for a negotiated redefinition became a condition for an integrative outcome of negotiations. The conception of rules adopted in the course of *cross-acceptance* therefore bears a crucial meaning for the consensus-building exercise and is worth analyzing in its connection to the main stages in the evolution of the process.

The Evolutive Character of Rules

The first version of the *State Planning Rules* became effective March 1988, i.e. over two years after the *Act* established the new state planning process.[3] This delay, certainly related to organizational and operational factors connected with the constitution of the new planning bodies, requires further observations. On the one hand, the definition of the most innovative and qualifying aspect of the process, the negotiating approach, remained for a long time overshadowed by uncertainty. *Cross-acceptance* was defined as a public negotiation on the contents of an intermediate plan, but its negotiating modalities remained open to speculation. This uncertainty certainly played a role in a general underestimation of the potential of *cross-acceptance* in the initial phases of the process, in particular at the time of the uproar against the *April '87 Draft*. On the other hand, the first rules adopted at the formal beginning of *cross-acceptance* already bore the results of an informal experience of group processes and discussions around the meaning and scope of the planning process. The formalized rules thus in some sense represented a document already corroborated by a practice of interactions with involved actors.

In the 1988 version the *Rules* comprised three of the eight subchapters, which, through three revisions, would have constituted the version adopted in 1992 at the end of the process. Their main function was to translate the provisions of the *Act* concerning *cross-acceptance* into a clear timing and phasing: this however soon appeared possible only in a non-definite, rather evolutionary sense. The rules of 1988 in fact still identified the whole negotiating process with a single phase between the draft of the *Preliminary Plan* and the adoption of the *Final Plan*, based on a comparative approach between plans and on the subsequent definition of negotiating issues. The *Cross-Acceptance Manual* issued to support this vision, in some sense, is a testimony to the relative underestimation of the scope of a negotiating approach to growth management. In fact, as we have seen, the progressive development of negotiating attitudes during *comparison* caused an extension of both the timing and the internal articulation of the process. It was therefore only the first integration of 1990, introducing the fourth subchapter on negotiating procedures, that delineated the structuration into phases which appears retrospectively as a key feature of the state planning process.[4]

In doing so, the 1990 *Rules*[5] respond to a crucial concern raised from public debates on means and objectives of the process, i.e. the inadequacy of timing. While a formal schedule was set for the beginning of the 'first' phase of *cross-acceptance*, identified with the presentation of the *Preliminary Plan*, a predefinition of its timing was formally dispensed with: the conclusion of *comparison* was in fact identified with its acknowledgement by the SPC, endowed in this decision with an autonomy similar to that of the acknowledgement of the level of consensus required for its mandate to be completed. However, the meaning of this is primarily that of a change in attitude: from now on, containment of timing for the effective completion of the state's mandate would be pursued in different ways.

In 1990 also the most important amendment to the *State Planning Act* became effective, i.e. the request for a two-tier impact assessment of the plan, as a condition of its adoption and as an ongoing activity backing implementation. The importance of this new provision determined the final articulation of the process into three phases, introducing an *Interim Plan* functional to the formalization of the outcomes *cross-acceptance* and to the conduct of the assessment. The second integration to the *Rules* of September 1991, while responding to this input, faced the emergence of further claims.[6] The overlapping of negotiating levels which the actors of the *living plan* (as it was to be termed) were experiencing during *cross-acceptance* forced a prolongation of *negotiation* in order to deal with *deferred issues*: its deadline was therefore also delayed until after the end of the *additional negotiations* following presentation of the *Interim Plan*.

Another overlapping of phases was thus acknowledged and formalized: in this case, it reframed the meaning of the public conclusion of the process, to be legitimated by the multidimensional assessment procedure. The new subchapter 5 thus formalized a third and final phase, named *issue resolution*, conceived to allow for revision and public comment of the apparatus of documents leading to the adoption of the SDRP.

The rules did not conclude their evolution with it, however: a few days later, in June 1992, at the formal conclusion of *cross-acceptance*, the concepts and procedural tools introduced by the new subchapters 6, 7 and 8[7] projected *cross-acceptance* towards the dimension of plan implementation. Negotiated outcomes themselves, the rules thus point towards a prolongation of negotiating practices and their institutionalization in a continuing dimension as an ongoing process.

The Evolutive Dimension of Formal Processes: the Dynamics of Negotiating Sessions

Redefining the Issues for Negotiated Agreements

A major outcome of interactions during *comparison*, as previously seen, has been the need for a conceptual redefinition of the issues of negotiation.

The identification of issues introduced a significant novelty in the process. Their definition no longer overlapped the structure of the plan and its discursive articulation for two main reasons. First, it questioned the methodical and somehow artificial distinction between goals and objectives, strategies and policies, and rather redefined issues as both a definitional-representational and as an operational problem. Secondly, and accordingly, it questioned the thematization itself proposed by the plan. In fact, not only did issues connected to the identification of *tiers* appear to be often cross-sectional, weakening the possibility of identifying clear boundaries related to unitary representations of development (and thus also conferring a more then technical meaning to mapping): this identification itself became questionable. The reasons for this are complex and related to a combination of frames which have dominated public debates. It is, however, significant that this phase of *cross-acceptance* was entered in a climate of uncertainty which bore the seeds for an internal reframing. The relationship between the substance of public debates and the structure of the plan, as postulated by the rationale of *comparison*, had in fact become quite labile, whereas a determinant, almost paralyzing role threatened to be played by confrontations on the wording of the plan.

Reframing the issues, on the one hand, according to problematic categories and, on the other hand, to procedural and methodic categories which differentiated themselves from the letter and structure of the plan appeared at first glance as a definitional regression, and for the first time downplayed the importance of the planning document itself: in fact, most advocates of planning interpreted the final revision of the plan as such a regression. Three aspects must however be considered in order to understand the meaning of this reframing. First, a reflective dimension was explicitly acknowledged within the process. Second, reframing the issue of negotiation implied reframing the meaning of negotiated agreements themselves. Thirdly, as we shall see, the reframing of *cross-acceptance* issues occurred in an interactive context which in some sense inaugurated a collective procedure of revision of the plan.

Redefining the Nature of Negotiated Agreements

Let us first consider the profound reformulation of the *meaning* of agreements. The most relevant task facing the SPC in the phase preceding formal negotiating sessions, besides the evaluation of issues emerging from *comparison*, is the redefinition of the conceptual nature of agreements and disagreements in order to render controversies tractable. This task went beyond the selection and redefinition of issues operated by the OSP staff, however strategically important that was. It was in fact a task which questioned the very nature of negotiating practices and impinged on the possible meaning attributable to agreements as well as disagreement.

The main concerns of the members of the negotiating committee of the SPC emerged as a result of pre-negotiation consultations with county officials and stressed the focus of negotiation. Two problem areas emerged from these consultations, rendered sensitive by some features of the plan, and possibly distortive of the meaning of negotiations: both, however legitimate, tended to shift the focus of negotiations away from the provisions of the plan towards issues external to it: "this drift away from actual substance threatened any productive discussion of the Plan itself"[8] (SPC, 1992b, p. 3). The first was a tendency to move away from specific concerns towards principles, 'philosophical concerns' such as *home rule* and the safeguard of property rights. The second was represented by frequent reference to implementation issues.

The committee addressed them by deliberating on the adoption of a hierarchy of types of resolution for the issues raised during *comparison*:
- *explicit agreements*, meaning that divergences on specific issues had been resolved with mutual satisfaction in terms of wording or mapping;

- *agreements in principle*, signaling the achievement of a formal consensus to be translated into specific terms in the framework of the *Interim Draft*: these kinds of agreements resulted from the the difficulty of achieving an exact formulation of amendments in the course of the negotiating sessions;

- *agreements on implementation issues*, implying the assurance that, albeit external to the mandate an authority of the SPC, the issue would have been specifically addressed in an *implementation report*, an advisory document addressed to the parties involved in implementing the plan to be attached to the *Interim* and later to the *Final Plan*;

- *agreements to identify concerns*, representing a form of identification of concerns on general planning principles, which allowed for their assumption as *matters of public record* in the statement of agreements and disagreements required by the *State Planning Act*;

- *agreements to defer*, finally representing a tool for exploring, filing and submitting to further elaborations in the framework of supralocal settings the issues deemed amenable of alternative policy solutions.

Disagreement, moreover, was itself framed as a form of agreement: it was an *agreement to disagree*. Whilst slightly humorous, this definition stated the will not to exempt from formal recording any of the confrontations raised in a broad sense by the wording of the plan: at the same time, it signaled the necessarily radical semantic reductionism of any record of conflict which was not reducible to it.

The basic meaning of this outcome, however, has to be seen in the internal differentiation and argumentative articulation conferred to the definition of the *nature* of negotiated agreements, and in the reference this implied to the domain of interactive processes other than those defined by formal negotiating sessions.

Redefining the Nature of the Negotiation Process

From an implicit assumption of consequentiality between protocols of comparison and the identification of negotiating issues, as embodied by the checklists of the *Cross-Acceptance Manual*, the definition of negotiating procedures was thus moving towards a sophisticated form of issues differentiation, related both to the substantive contents and to the procedural kind of consensus which might have been expressed on them.

The semantic articulation of negotiated agreements should not however be separated from the situation which defined them, i.e. from the segment of the negotiating context in which they were situated. In order to interpret their meaning in the evolution of *cross-acceptance*, the internal dynamics of the formal design of the negotiating process must be seen in light of its relationship with the dynamics of the other components, both *formal* and *informal*, of the *cross-acceptance* process.

By establishing a relationship between the procedural rules of *negotiation* and the redefinition of agreements, a schematic interpretation of its articulation may be introduced (figs. 7.3 and 7.4). Along the vertical axis is displayed the diachronic sequence of negotiation phases, summarized in the two stages of *pre-negotiation* and in the three sessions which constituted the formal phase of *negotiation* in a proper sense. Along the horizontal axis, on the other hand, the definitional field is displayed from which the issues amenable of a preliminary agreement and, by elimination, the issues requiring formal negotiation are selected; among the latter, as we have seen, the definition of the issues is further split according to the nature of the agreements.

In the former phase (fig. 7.3), the issues oscillate along the horizontal dimension between two extremes of complexity, related to the nature of the situation and to the actors involved. *Pre-negotiation* is therefore characterized by a peculiar combination between the definition of the issues and the aim of relating their treatment to broader arenas.

phases of negotiation and their actors:	*field of negotiating issues*		
1. **pre-negotiation**	*1*	*2*	*3*
preliminary issue analysis: OSP-counties	single counties: *fact issues* *(application issues)*	more/all counties: *(fact issues)* *application issues* *policy issues*	more/all counties: - *(application issues)* *policy issues*
exploratory "consensus-building" phase: PDC-counties	- -	- -	*cross-county issues* *statewide issues*

negotiation

Figure 7.3 The articulation of issues and negotiated agreements, 1

The pursuit of a control and of a non-contingent legitimation of the negotiation process leads to the search for a cross-sectional and conjoint consensus throughout negotiating contexts. This implies a practice of deconstruction and reframing of the issues raised in *comparison*, which occur both through their merging into thematic clusters and through their joint discussion in multicounty informal negotiation settings. Elementary issues, or *fact issues,* deemed amenable of local resolution through supplements of inquiry were hence isolated, while more complex *policy* and *application issues* were addressed informally in multilateral negotiating arenas. This phase, which introduces formal negotiation, and which is significantly defined as 'consensus-building', tended therefore to extract issues from a bilateral negotiation setting, linking them to a perspective of internal reframing of the plan legitimated by a supralocal, multilateral level of agreements.

In the latter phase (fig. 7.4), on the contrary, formal negotiations as defined by the *State Planning Act* were reduced to bilateral settings and to the exclusive deliberative prerogatives of the negotiating entities representing counties and municipalities before the negotiating committees of the state. However, even in *negotiation,* the procedural treatment of the issues followed a similar semantic pattern, aiming at distinguishing areas of clarification or amendment amenable of a preliminary agreement which might have exempted them from a dispute on principles.

phases of negotiation and their actors:	*field of negotiating issues*					
2. negotiation	*1*	*2*	*3*	*4*	*5*	*(6)*
OSP-county staff	*(agree)*					
OSP-LNC	*(agree)*					
PDC-LNC	*agree*	*agree in principle*	*agree on implem.*	*agree to id. conc.*	*agree to defer*	*disagree*
		expl. *agreement:* formal acknowl. →	explicit disagreement: formal acknowledgment informal treatment			→ disagreement: formal acknowl.
			embedded negotiations in an iterative dimension: 3-year *cross-acceptance*			

SDRP

Figure 7.4 The articulation of issues and negotiated agreements, 2

The first two stages of *negotiation* thus explicitly aimed at a semantic reframing of conflict, leaving non-reducible conflicts to the third formal stage of the negotiating sessions, the confrontation between LNC and PDC.

Pre-negotiation established a strict linkage between the identification of generalizable areas of negotiation, connoted by the insistence on principles and on implementation issues, and the need for supralocal and cross-sectional modes of public deliberation, thus introducing a sensible informal extension of negotiations and opening up new forms of structured interaction. The formal phase of *negotiation*, however, was also affected: what was assumed as the subject for bilateral confrontations between negotiating entities was the result of a selection, itself articulated into a distinction of simple issues, the residual nucleus of which was given by the irreducible matter of a confrontation between the comprehensive principles of the plan and the preferences of local communities.

The expressions of agreement were therefore the outcome of a semantic reduction of conflict, operated horizontally (i.e. through selection at a supralocal level) and vertically (i.e. through a progressive refinement at the local level). The expressions of disagreement, which turn out to be few in number, represent the extreme of this articulation, that of an absolute irreducibility. Between the extremes, agreements are dispersed in a field of discursive formulations, of the type 'agreement for...', which state their irreducibility to a binary pattern of decision as that defined by formal negotiating sessions. In this way they revealed their non-self-sufficient and non-decisive character. *Agreements* and *agreements in principle*, which exhausted the issues in the situation defined by negotiating tables, were to be paralleled by forms of agreement which, in contrast, implied further, alternative contexts of argumentation.

The joint consequences of this redefinition of the strategic function of negotiation have been the *organizational opening* and the *recursive turn* of the procedural sequence. The achievement of local consensus on a part of the issues was possible only by allowing further discussion on issues in other situations which directly impinged on the task of the plan's revision.

The horizontal axis of the scheme represents what might be called the dimension of the *discursive reduction* of negotiating issues. The evolutive design of *negotiation* emerging from the progressive practical and discursive consciousness of the process may be interpreted as a strategy for extracting the search for consensus from an entropycal tendency towards a minimalist and proceduralist assumption of the negotiating task, which would have entailed on the formal level the risk of an 'infinite regression' of the process and on the substantive level of a non-integrative reduction of confrontations, at the same time safeguarding the autonomous faculty of expression and deliberation on the state plan's provisions granted by the *Act* to each single negotiating entity.

The horizontal dimension of a tendential *discursive reduction* of the negotiating issues was therefore backed by a vertical dimension, which might be called that of a tendential *formalization* and *operational closure* of the negotiating contexts. Both, however, contributed to highlighting the relativization of a formal conception of negotiation. The final stage of the negotiating exercise achieved a definite agreement on a particular typology of negotiating issues, seemingly defined by the very statutory need for achieving consensus. But far from being exhausted, the process of consensus-building was rather amplified by this reduction: it became situated in interactional settings defined by the development of informal negotiating attitudes, embedded in the argumentative interplay of the actor's roles and preferences in the further group processes displayed in the course of *cross-acceptance*.

The Internal Dynamics of the Negotiation Process: the Living Plan

Let us now address the relationship between the internal articulation of negotiations and the other dimensions of the process. What has been observed as an evolution of *negotiation* in fact occurred at different stages in different contexts throughout the state: what from the perspective of each single negotiating setting might appear as a *sequential* process thus turned into a *coevolutive* process, with significant influences on the actor's reflective behavior.

It is not arbitrary to state that the metaphor of the *living plan*, coined by the director of the OSP, J. Epling, expresses the frame-for-action of the *institutional entrepreneurs*: it represented in fact the reflective answer to a peculiar condition of political malaise, determined by the state of evolution reached by *cross-acceptance* at the beginning of the most delicate stage for its overall effectiveness and legitimation, the *negotiation* phase.

What were the consequences of the evolution taken by the process on its management and on the actor's attitudes? Let us recall its main features.

The primary one was the enormous growth of information requirements. Concern about their complexity and on its implications for the effectiveness and legitimation of the negotiation process may be found at all levels of responsibility and involvement. The in-progress definition of the negotiating approach had always been a primary matter of uncertainty and controversies during public concertation, raising procedural claims as well as attempts at delegitimation. Opposing interpretations of policy-making were at stake, which involved radically different images of the regulatory and/or cooperative role of planning. At the beginning of *negotiation*, the issue raises mainly procedural questions, which however bear significant strategic consequences. The dominant question concerns the possibility of managing a process involving 21 counties, a much higher number of dissenting municipalities, and, as interested observers, the whole of the 567 communities of the state in the pursuit of a decentralized consensus on issues of a comprehensive state plan, and all this in conditions of sensitive shifts in timing and in political and management culture in most local settings. How could it be possible to conduct negotiations in conditions of equality and to achieve legitimation under the constraint of defining issues based on *in fieri* documentation and of gaining local consensus at different stages of timing? How to ensure homogeneity criteria, how to enhance mutual learning in the course of a process which tends to assume circular features, like a 'chasing-your-tail kind of process'?[9] The answer offered by the managers of the process is articulated on different levels, including a rhetorical one: the metaphorical image of a *living plan*, seemingly trivial, summarized this tension between the urgency of the political-institutional mandate, of reaching a consensual *state*, and the progressive evidence of the recursive and coevolutive implications of the consensus-building process. Building consensus came to be identified with the achievement of a difficult balance between contrasting requirements.

The first move towards such a balance was the separation of the issues: a strategy which, in order to avoid the risk of a deadlock, tended to restrict the field of reference for negotiated agreements to the specificity of the plan's provisions,[10] at the same time acknowledging their limitations and opening to further terrains of argumentation. By this, the negotiating committee could focus on treatable issues and keep to a strict timing, however relativizing the outcomes thus obtained by reference to a broader framework of a public obligation towards non-resolved issues.

The second move directly addressed the implications of the diachronic dynamics of the process. At stake was avoiding the constraints of a sequential conduct of negotiations, which would have set the timing for the completion of the planning mandate out of control. For this purpose, the OSP staff forced the timing for technical revision and addressed the redrafting of the plan in the course of negotiation itself by adopting their partial results, by assessing their direct effects, their influence on the plan's framework, their interpretive variants, and by realizing an interface with the PDC through an ongoing exchange of information on the effectiveness and feasibility of the outcomes emerging from negotiations. Technical consultancy thus again became an active and pervasive component of the process, enacted through continuous interactions with the SPC and its negotiating body, realizing a revision process of the *Preliminary Plan* parallel to the conduct of negotiations.

In light of these observations, a further interpretive step becomes possible. The negotiation process may be represented (fig. 7.5) along two dimensions: along the linear axis of the *diachronic sequence* of *cross-acceptance* phases, leading to the plan's adoption, and along the *recursive loop* introduced by the articulation of negotiated agreements as well as by the shifts in timing occurring during the simultaneous conduct of negotiations in the 21 counties of the state.

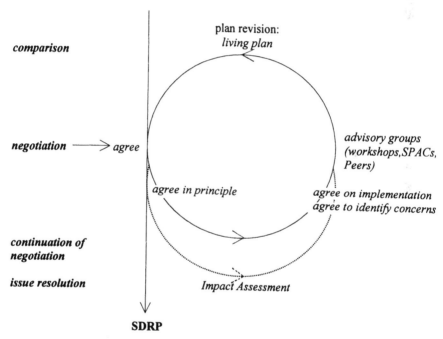

Figure 7.5 The *living plan*

The scheme indicates that, in the framework of the process, only a minimal proportion of the issues emerging from *comparison* were directly amenable of resolution in the formally defined phase of *negotiation*. Only *agreements* and *agreements in principle* were achieved along the linear sequence of the *cross-acceptance* process. Issues which implied discussions on the principles of state planning entered a complex plan revision process, which bore several different connotations: a comparative function, pursuing homogeneity in the response to local inputs; a technical-informational function, linked to the consultancy role of the OSP; a strictly strategic function, responding to the need for the political system to display interactive feedback channels on the positions concerning policy orientations expressed by the plan. Moreover, timing delays in the conduct of the process throughout the state became a crucial factor in iteratively shifting such a process along the diachronic sequence: feedback loops, in fact, were reproduced in the phases which followed *negotiation*, i.e. in *continuation of negotiation* and even in *issue resolution*.

The two concluding phases of the *cross-acceptance* experience were therefore dominated by a circular pattern of relationships between formulation-reformulation of the plan, formal negotiation, technical consultancy, political argumentation, and public involvement. This should, of course, neither imply homogeneity of the actors' behaviors nor neglect of the 'paradoxes of public involvement' which emerged from *cross-acceptance*. This however makes it the more important to analyze the specific responses to the opportunities provided by this turn in the process.

Before introducing these aspects, the internal and external conditions for the emergence of such a turn towards a reflective development of the process must be pointed out. The former refer to the modes of relationships within the universe of actors involved: the correlate of the consolidation of such a pattern of relationships was the constitution of a system of reception and treatment of communication fluxes which determined their self-regulatory potential. In this sense, action situations and modes of involvement of organized interests displayed through the group processes initiated in the course of *cross-acceptance* bore a great importance: strategies of inclusion and cooptation into advisory bodies, the cross-sectional representation of state agencies and their respective constituencies, the openness of local negotiating entities to social interests were their most evident aspects. The latter conditions refer to the nature of relationships among the institutional planning bodies and the specific functions enacted by the planning staff: the reciprocal fluidity of consultancy, mediation and negotiation tasks and the informal articulation of roles which overlapped the formal design of the processes are decisive for understanding the non-formalizable reasons for 'success' of the *cross-acceptance* experience.

The Actors of Formal Negotiations: the Negotiating Committees

Contrary to what could be expected, the PDC represented neither a monolithic body nor a unitary negotiating style. Its organization was rather functional to the emergence of different 'negotiating paradigms' as well as of the need to provide a unitary image of negotiations. The PDC (fig. 7.6) was formed by a certain number of SPC members, not necessarily including its chairman; the single state negotiating committees entrusted with holding sessions with the counties' negotiating entities were therefore a filiation of the PDC. The minimum number of members of PDC required at each negotiating table was three. The committees were further attended by OSP staff with an assistance function, led by the director or by an assistant director and by an area manager or associate.

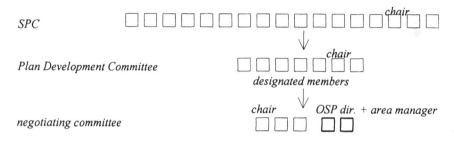

Figure 7.6 Composition of PDC and of state negotiating committees

Thus, at each county negotiating table a different composition of the state representatives was in charge; at the same time, however, throughout 21 counties, several overlaps occurred. In fact, beyond formal schemes, the negotiating committees were attended by almost any member of the SPC and OSP who could usefully be involved in light of specific competencies or contextual needs. While designated membership was stable inside each single committee, members could be substituted according to dominant issues and personal abilities. Thus, the local counterpart was granted stability, but also competence in the negotiating authority it faced. Only in the last phase of negotiations between the PDC and the local negotiating entities was involvement of staff and administrative actors restricted primarily to elected officials.

Similarly, the 1990 reform of the *State Planning Rules* extended the LNC's composition beyond the provisions of the *State Planning Act*, which implicitly identified their membership with county freeholder boards, and thus responded to a demand for broader representation which had emerged during the *comparison* phase of *cross-acceptance*. LNCs were designated by freeholder boards and included at least three of their members as well as two members of the county planning board and a member of the planning staff, but were in principle unlimitedly open to the involvement of representatives of civil society.[11] Negotiation tables were thus formally attended by designated members of the SPC, of the OSP, of county freeholder boards and planning boards and staff, as well as by planning board members and staff of municipalities which had presented dissenting reports; informally, however, attendance was open to any representative whom the counties would have liked to involve; it was hence common to have municipal officials and interest representatives at the table. through their openness, the composition of LNCs has therefore also contributed to reducing to a minimum the recourse to autonomous negotiating sessions by dissenting municipalities.

The SPC since *pre-negotiation* followed a policy of public announcement of each PDC meeting, without distinctions regarding the nature of the meetings: each of them was thus formally open to the public.[12] No professional facilitator or mediator attended negotiating sessions: their involvement was limited to the more informal advisory phases of the process as well as to *pre-negotiation*. There were practical reasons for this, of course, mainly associated with a lack of resources for external consultancy. However, this also expresses one of the peculiar features of the environment created: at the negotiating tables, elected officials prevailed, and political connections were therefore the more sensitive. Whereas the role of facilitators in advisory committees was to keep specific and generalizable planning-related issues at the forefront of argumentations and to guarantee their prevalence over a confrontation of pregiven positions, negotiating sessions focused rather on the pursuit of a formal, locally based consensus to the system of intergovernmental relationships proposed by the state planning process.

The Actors of Informal Negotiations: the SPACs

The *State Planning Advisory Committees* (SPAC) were constituted in 1988 according to the Act's provision by resolution of the SPC[13] in order "to contribute to an effective formulation of the SDRP through a structured multi-disciplinary dialogue"; they in fact became active in 1990, during the crucial phase of *pre-negotiation*, in which the negotiating approach of *cross-acceptance* was defined.

The new *State Planning Rules* of July 1990 acknowledged their role, while the first meetings of the new committees had already begun in January, in parallel with the start of *pre-negotiation*: again, the rules did not institute, but rather institutionalized an already established practice. Its institutionalization is nonetheless a noteworthy witness of the key role assumed by SPACs during *cross-acceptance*.

The SPACs were an evolution of the experience in the involvement of technical consultants previously conducted by the OSP. In the course of the policy development phase leading to the *Preliminary Plan*, the staff involved several external experts beyond the four consultant groups which drafted the studies which lay down the basis for the *April '87 Draft*. The nature of relationships with them were rather of an internal nature: they basically provided technical consultancy at the service of the planners' staff, which was transmitted in a mediated form to the SPC through the plan's partial drafts and contributed to a clarification of issues implied by the comprehensive approach of the first standard-oriented version of the plan, and to a learning process internal to the staff. However, members of the SPC were not excluded, and mainly during the debate arising after presentation of the *April '87 Draft*, technical meetings became important for assessing commitments to the process and for a confrontation between institutional representatives and the general public.[14] Expert groups, still termed *Technical Advisory Committees*, gathered *ad hoc* according to the staff's needs for knowledge resources and were representative of the dispersion of issues involved, growing to a number of twelve and requiring specific coordination tasks in the OSP. Demand for external communication soon consolidated the role of a *Peer Review Committee*, which assumed a prominent role in framing public debate and in involving key personalities who became stable actors of the planning process. During *cross-acceptance*, however, managing such an array of advisory bodies became too complex. In restructuring them, moreover, the OSP acknowledged a change in their role: following *comparison*, the issues raised required a form of synthesis to which only a more integrated 'bias group' of experts could contribute. Although in formal continuity, as they continue to be understood as consultative bodies of the OSP, the advisory committees thus assumed a more explicit role of supporting the SPC in public argumentation on planning issues and in the building of an 'overall consensus' on the planning process. The issues addressed were therefore reframed as *policy issues* and the committees, renamed as SPACs, were reduced to five, including the peer.[15]

The committees always followed a distinct path from the formal-statutory definition of *cross-acceptance*: even if they substantially contributed to it, they did this, so to speak, from the outside of the negotiation process. The explicit aim pursued in their composition was to offer the broadest possible representation of positions relevant for debating the plan, and in particular to include voices which could actually or prospectively produce an active opposition to the plan. In defining membership in this way, an intention of cooptation was followed, albeit not explicitly expressed: emphasis was rather put on the relevance of information brought by participants for plan development and for the fulfillment of the SPC's mandate.

Both the sensitivity and the persuasive meaning of the composition of membership required all the relational expertise of the SPC members, and extended over six months, during which project managers and commissioners exchanged suggestions on special interests and groups as well as on personalities to be involved. In the meantime, previous experience suggested that the number of participants to each committee should be kept manageable, which was identified at about twelve members: in actual fact, membership of some SPACs turned out to be broader according to the convenience not to exclude relevant voices. A contribution to conciliating the aims of operational leanness and representational amplitude was made through ample recourse to invitations and auditions of external personalities, with the effect of establishing a second-level form of cooptation no less important than the former: this practice in fact frequently involved the same coopted members, conferring to them an active mediating role which consolidated their reputation both internally and relative to their constituencies, and determining a sort of self-empowerment of the committees due to the strengthening of the members' own stake in the effective participation to the process. In this sense, involving interest representatives and prominent personalities constituted a success in itself for the staff, and in fact most of those involved were finally to become at least informal supporters of the planning process.[16]

Composition of the SPACs turned out to be extremely balanced[17] by design: they were conceived as a microcosm of political debate in order to grant adequate treatment to each of the issues of public controversy. Only the *Peer Review Committee* emphasized an academic and disciplinary focus of discussions.[18] Conferring to each SPAC a maximum of balance did not therefore so much mean emphasizing a bias in favor of the plan and to downplay internal conflict (which was on the contrary often quite consistent), but, on the contrary, to guarantee equal representation to each different position involved. In constituting a balanced argumentative setting, attention was expressed towards the informal connotations of consensus building. For this reason, the aim became paramount of avoiding the dominance of a position on the whole of the assembly, which would have caused a loss of its credibility as a discussion forum, but also of avoiding forms of personal isolation, as the constitution of 'lone wolves' would have forced members to stick to confrontational positions and accordingly to retreat into the rhetoric of their own constituency. Meetings were hence the occasion for the building of a collective commitment towards planning as a process, built through mutual strategic adjustment of positions on the principles of state planning.

The environment of meetings was, accordingly, the result of an attitude and partly of a design aimed at favoring effective group work. Many anecdotes might be adduced on this aspect. But what was the meaning of the climate in which debates were embedded? To understand it, their development in parallel to that of formal negotiating sessions must be kept in mind. An important feature of this climate was the open character of exchange, an explicit effort to grant an equal opportunity for expression to group leaders in a neutral setting: the prevailing atmosphere was therefore that of a seminar, where nothing alluded to a possible negotiating function, which, if expressed, would have contributed to lock-ins in the expression of positions.

Whereas SPAC meetings nurtured such an atmosphere, their working environment nonetheless is reminiscent of that of a negotiation setting. The opinion among participants is in fact diffuse and holds that, behind this facade, SPACs in fact played a consistent negotiating function. But what was its specific nature then compared to that of formal negotiations?

The functions attributed by the *institutional entrepreneurs* to the SPACs were diverse. First and most explicit was their informational function: committees provided the plan development process with inputs coming from diversified public and private actors, in parallel to the formal conduct of *cross-acceptance* but in much more ample and flexible ways. The meaning attributed to this function implied therefore an acknowledgement of the shortcomings of formal negotiations in dealing with informational feedback, constrained by their procedural and representational limitations as well as by their relatively strict adherence to the wording of a contested planning document: during the crucial phase of plan development, the richness of issues and arguments emerging was thus likely to be devalued in the framework formal negotiating contexts.

Moreover, while *negotiation* dealt with the building of a *formal consensus to state planning*, the most profound dimension of the argumentative setting represented by SPACs concerned the constitution of a *commitment towards its implementation*. The committees' mission was therefore seen by the *institutional entrepreneurs* as something very different than a procedural routine, and rather, in a proper sense, as a contribution to the development of a plan which could be effectively implemented. In this sense, they represented to a certain extent a step beyond the mandate of policy development defined by legislation and a statement of the necessarily contextual and constructive nature of the conditions for effective policy implementation. As such, it also progressively moved away from a formal-consultative understanding of *cross-acceptance*, serving as an introduction to a specific coevolutive dimension of the institutionalization of interactional practices.

The issues of implementation as addressed by the activity of SPACs referred in the first place to a technical-operational dimension: advisory meetings allowed contacts with an array of professionals and political agents who were experiencing 'in real time' the challenges of implementation of the plan's issues throughout the state as well as in other contexts. A complex clustering of multidisciplinary contributions was at stake in the involvement of experts from municipal planning boards, developers, environmentalists, housing advocates, land-use consultants: again a microcosm of the general public debate and, at the same time, a significant cross-section of the state's growth coalitions.

Contributions of this kind were particularly influential on the evolution of the *regional design system* introduced by the plan, one of its most heavily contested strategic aspects.[19] The function of acquiring information thus also took on the role of a process of selection and development of the relevant knowledge base. But not less important than the use made of information inside the SPACs was the role their activity played in disseminating them on the outside. Identifying and selecting in the most accurate way the stakes and their representatives meant ensuring a capacity of diffusion of information relevant for plan implementation, dealing with distorted information at its source, and continuing an 'educational' action already started with the beginning of *cross-acceptance*, however with much stronger potential effects of cooptation. Realizing a cognitive as well as strategic feedback from committee members to their constituency, moreover, was as important as gaining their direct involvement. Achieving support from committee members thus became a key goal for implementation. The kind of consensus pursued here, however, may be distinguished from the kind pursued through formal negotiations for its peculiar integrative and coevolutive nature, based on the role played by cognitive and learning processes in the communicative-strategic environment created. SPACs did not only represent the occasion for a conceptual reframing of planning issues emerging from the negotiation process: they also directly embedded it in a circular communicative flow. Its circularity was enabled by diffuse relationships of filiation between the actors, as well as by the presence of multiple connections between their roles both at the institutional and extra-institutional level. Participants in SPACs were often in strict relationships of an either political or professional kind with the SPC or were informed or directly involved in meetings of the SPC or other committees; as for the

OSP, the involvement of staff members was cross-sectional to deliberative and argumentative settings, and often entailed overlapping roles in consultancy and organization throughout situations, ranging from field-work to the redrafting of the plan.

As previously stated, the integrative working style of the SPACs did not entail the exclusion of conflict from its domain. Issues on the contrary were usually presented to highlight tensions between extreme contrasting positions, in order to make them a matter of discussion.[20] The climate could in some cases prove to be as conflict-laden so to require the intervention of professional facilitators.[21] In fact, tensions inside the SPACs could become extreme even in personal terms: representatives of a regulatory vision of the plan as expressed by the *April '87 Draft* exclusively interested in the safeguard of natural areas and in urban revitalization stood besides advocates of the standard-oriented approach expressed by the third volume of the *Preliminary Plan* strongly skeptical of any version which loosened its normative strength, and both faced opposers who participated only to provide testimony of their own and their constituency's refusal of the very idea of planning, and of course a whole range of intermediate positions. Tensions could therefore emerge even inside the field of planning advocates, and planning staff were often caught in the crossfire of contrasting disciplinary cultures. Concepts like *regional design* and *centers* directly involved an evaluation of the state's development in post-war years, as well as taking side on notions as controversial as 'community', 'growth', 'suburban sprawl', and on their sociological, ethical and esthetic connotations: were planning inputs valuable alternatives to be opposed to continuity with given patterns of community development? At stake in the debate were of course different material interests, but also different and often irreconcilable images of development, which involved a radical confrontation with a cultural and disciplinary heritage and, even for planners, i.e. for 'experts', a possibly more radical confrontation with the arguable and negotiable nature of the representations of the territory.

Notes

1   N.J.S.A. 52:18A-203.
2   N.J.A.C. 17:32-1.4 and 1.2.
3   R. 1988 d. 121, effective March 21, 1988; see: 19 N.J.R. 1971 (b), 20 N.J.R. 673 (a).
4   The first version of chapter 3 was in fact entitled "Procedures for conducting *cross-acceptance*", later changed into "for conducting the *comparison* phase of *cross-acceptance*".
5   R. 1990 d. 336, effective July 2, 1990; see: 22 N.J.R. 621 (c), 22 N.J.R. 2033 (a).

6   R. 1991 d. 457, effective Sept. 3, 1991; see: 23 N.J.R. 1778 (b), 23 N.J.R. 2654 (a).

7   These are respectively the *letters of clarification*, the *consistency review*, and the possibility of amending the *Resource Planning and Management Map*: R. 1992 d. 253, effective June 15, 1992; see: 24 N.J.R. 1241 (a), 24 N.J.R. 2287 (a).

8   SPC (1992b), p. 3.

9   J. Epling, introduction to a seminar of the Peer Review Committee in New Brunswick, December 6-7, 1990 (video).

10   Epling (intr. quoted) pragmatically summarizes *policy issues* for the audience of the Peer Review Committee as 'plan issues', distinguishing them from 'concerns' and 'implementation issues'. OSP staff, according to Epling, used to ask: what is there in the plan that exactly raises your concern? If the answers concerned policies or strategies, the issues were addressed and eventually resolved as such; but if it did not concern specific provisions of the plan or of the planning process, they would acknowledge and record it as a concern and move on to further issues on the agenda.

11   Dissenting municipalities had the right to ask for autonomous negotiating sessions sending their own representation, backed by a member of the county's committee.

12   According to New Jersey's *sunshine law* (*Open Public Meetings Act*, N.J.S.A. 10:4-6 et seq.), a *committee* is distinguished from a *deliberative commission meeting* by the possibility of being held in private form in case of absence of the members' majority.

13   Respectively, N.J.S.A. 52:18a-204 and Resolution No. 88-014.

14   According to Kanige (1987), however, the subcommittees held closed meetings during a period following the uproar caused by the presentation of the *April '87 Draft*.

15   These are the denominations which express their thematic focus: *Business and Labor SPAC*; *Housing SPAC*; *Natural Resources SPAC*; *Regional Design System* (later *Resource Planning and Management*) *SPAC*; *Peer Review SPAC*.

16   It is therefore understandable that any shortcoming of the SPAC's activity, as of group processes in general, is seen by staff members in the difficulty of dealing with some key issues due to the impossibility of achieving involvement of some related personalities (T. Dallessio, interview 17.07.1995; B. Kull, interview 13.07.1995).

17   T. Dallessio, interview 17.07.1995.

18   The chairmanship of the *Peer Review SPAC* was assumed from the beginning by J. DeGrove (Florida Atlantic University). Eight of their 16 members were academics or representatives of research institutions: besides the *chairman* , these were B. Chinitz and R. Einsweiler (Lincoln Institute of Land Policy), J. Hughes and M. Lapping (Planning School, Rutgers University), R. Liberty (University of Pittsburgh, representing One Thousand Friends of Oregon), I. Reed (Woodrow Wilson School, Princeton University), and R. Roper (Council on New Jersey Affairs, Woodrow Wilson School, Princeton University).

19   The outcomes of SPACs had a substantial influence on the revision of the plan. The *Regional Design System* SPAC in particular was key in developing the *resource planning and management* structure and the concepts of *centers* and *planning areas*: the expertises in the committee helped in understanding and defining the operational meaning of the defining parameters and the design criteria for centers. It was, at the same time, the stage of one of the most relevant evolutions of the plan, i.e. the demise of the *tier system* in favor of *planning areas* and of the progressive prevalence of a weak steering of growth through the principles of *regional design*.

20   C. Newcomb, interview 12.07.1995.

21   This role in the *Regional Design System* SPAC as well as in the *workshops* of *pre-negotiation* was assumed by J. McGuire of the Center for Dispute Resolution of the Office of the Public Advocate, assisted by D. Brake of MSM Regional Council.

8 Cross-Acceptance: An Interpretive Summary

Most of the previous pages have been devoted to discussing the evolutive articulation taken by interaction processes in the course of the *cross-acceptance* experience. The analysis has focused in particular on the relationship between the emergence of a confrontation between conflicting frames and the correlation between their nature in relation to the text of the plan (i.e. their *rhetorical* and *argumentative* nature) and in relation to the procedures regulating interaction (i.e. their *strategic* nature). In this way, the form taken by this relationship has been indicated as a coevolution between action situations, images and representations of problems and actions, and identities of actors.

At the conclusion of this discussion, which mainly concentrated on the coevolutive dynamic of the processes, it is necessary to summarize in a more systematic manner the analysis of the organizational and structural nature of the interactions which constituted both a condition of development and a specific manifestation of this evolution.

The relational setting among the actors which developed during the *negotiation* phase cannot be understood in a unitary way, as shown by the progressive constitution of two distinct, parallel levels of interactional practices, a formal and an informal one. Moreover, it must be interpreted in light of the non-linear nature of its evolution: in fact, it is the outcome of a 'stratified' system of mutually interrelated games, of a multidimensional nature, implying the enactment through time of multiple roles and their embeddedness into the evolutive definition of the actors' identities.

To anticipate some elements of this interpretive framework, let us introduce the discussion of such a relational setting by loose reference to a representation offered of it, albeit in a quite different context, by Ostrom (1990) (fig. 8.1).[1] The analysis of interconnections between the nature of roles taken by the actors and the networks of relations on which they rely will allow for some generalization.

This representation is only apparently amenable to be interpreted in a synchronic sense, since the bilateral nature of relations and their extension in the form of a multilateral network, which grounds its character as a self-governed and self-monitored system, are given only in a diachronic perspective, intended as an outcome of evolution, as well as a future-oriented projection of expectations on relations and roles. Its temporal domain, therefore, is that of an iterative perspective of interactions. At the same time, as has been implicitly indicated by analyzing the roles of individual actors in the process, its networking nature is defined by the copresence of multiple roles: the multilateral character of relations is connected to the (not necessarily intentional or conscious) enactment of a multidimensional role profile by each of the actors. The temporal dimension for the constitution of this form of relations is, of course, cyclical, i.e. grounded on processes of mutual strategic recognition and learning and on their feedback on the nature of relations: it is grounded in the process of mutual learning, which constitutes one of the dimensions of this experience in structured interaction.

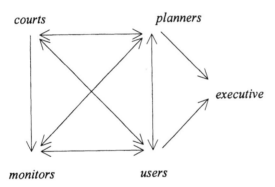

Figure 8.1 Patterns of monitoring and accountability among actors of the *cross-acceptance* process
Source: adapted from: Ostrom (1990, p. 74)

In order to reconstruct the microconstitutional foundations of this framework of relationships, the scheme presented must however be supported by an interpretation (which, in the case of some of the actors, takes the form of a summary of previously described features) of the nature of the roles assumed by the actors in the different situations enacted in the course of the interaction processes.

The following representations must be understood according to an implicit but fundamental reference to previously discussed aspects: the internal differentiation of the negotiation process (of which the *formal* negotiating sessions constitute only a specific stage), the non sequential, but intersected, 'stratified', and tendentially circular dimension taken by the processes, and the inclusionary, pluralistic, overlapping nature of roles and identities of the institutional and non-institutional actors involved.

Let us address the latter aspect first through an analysis of the internal composition of the 'institutional entrepreneurs'. This is particularly true – as we have seen in ch. 5 – in the case of the SPC, within which the peculiar constitution of multiple commitments determined by the features of its composition and by the overlapping of roles affecting most of its members has been underlined. The same may be recognized in the LNCs and in their progressive opening to non-institutional forms of representation, as well as in the nature of composition and participation to the SPACs, which developed them into an arena for exchange of a broad array of positions.

Similar albeit more specific and complex observations are possible with regard to to the apparently most homogeneous institutional actor, the OSP. In describing its composition and activities, it has been underlined that the presence of strong internal leadership was combined with a great programmatic flexibility of roles. While this excludes the aptness of a hierarchical interpretation of its internal organization of activities, a careful examination of the roles assumed by its individual members in the course of their diverse activities shows how the definition itself of working areas, in itself non-hierarchical, was characterized by a continuous intersection of thematic levels and operational competencies. A representation of the internal role-sets of the agency according to the – by itself not linear – definition of its functional competencies and working areas, as shown in fig. 8.2, offers thus a quite reductionist image of the nature of its activities.

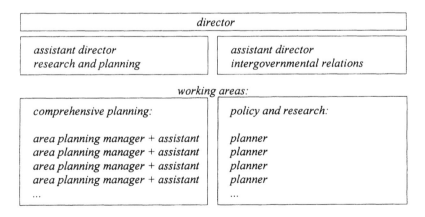

Figure 8.2 Scheme of functional-organizational composition of OSP

The basically non-hierarchical character of relationships and the overlapping of roles within and between the two planning agencies, as they emerge from their 'thick descriptions', is better illustrated by a joint functional and organizational analysis of their activities (fig. 8.3).

1. design and management of processes, drafting of the plan
functional roles:
- *coordination of decision-making*
  (SPC, OSP)
- *supervision and technical-*
  disciplinary input (OSP)

organizational roles ('departments'):
- *area-based competencies (OSP)*
- *interdisciplinary and interagency*
  working groups (OSP, SPC through
  SPACs)

2. consultancy
functional roles:
- *information and 'education'*
  at county and municipal level
  (OSP)
- *information and education*
  at the level of the general public
  (SPC, OSP)

organizational roles ('departments'):
- *area-based competencies (OSP)*
- *policy-areas (OSP, SPC through*
  SPACs)
- *advisory meetings (SPC, OSP*
  through SPACs)

3. consensus-building and negotiation
functional roles:
- *framing of issues (OSP, SPC)*
- *conduct of negotiations (SPC, OSP*
  through PDC)

organizational roles ('departments'):
- *supervision (OSP)*
- *monitoring (OSP)*

Figure 8.3 Main areas of activity and according functional and organizational roles of the state planning agencies (SPC, OSP)

None of the activities schematized in the figure, particularly if one thinks of the complex set of informal mandates coming from the SPC, is amenable to an exclusive attribution to a determined set of actors. If the main functional and organizational areas of activity of the OSP, in the framework of the division of labor between the planning agencies, are framed according to a unitary definition (here assumed under the label of 'departments'), it may be observed that their specificity was defined by the relative autonomy of roles assumed within them by the actors, which are not reducible to decisional and organizational relationships (both internal and external) of a hierarchical type; for this very reason, in connection to the diversification of activities carried out by the same actors in other contexts, a system of multiple, overlapping roles was thus constituted.

The meaning of the expression by one of the planners, which many of her colleagues might have shared, that in fact they had been 'finding their way as they did it',[2] may thus be referred to a less subjective dimension: the experiential dimension of the planning process carried on by the actors is itself an expression of the coevolution of multiple roles and identities.

These observations, however, acquire meaning particularly if the analysis of the nature and composition of internal roles of the collective actors of the *cross-acceptance* process is related to the evolution of their reciprocal forms of interaction, thus leading to an examination of the external relationships established by their internally defined roles.

Let us examine these aspects first with reference to the external relations which involved the planning agencies in the framework of the progressive broadening of the field of interactions between the formal and the informal domain of negotiation. This may be illustrated starting from the observation of the networks of actors which are constituted in the most formal (and at the same time most decentralized) moment of negotiation, i.e. the final sessions between PDC and LNCs in the *negotiation* phase. For this purpose, the composition of the state negotiating committees previously presented in fig. 7.6, is reframed here in a bilateral schematization, further articulated in its multilocalized dimension (fig. 8.4).

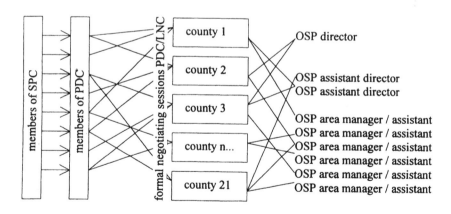

Figure 8.4 Composition of the state negotiating committees: schematic representation of relationships between the actors of negotiations

The formal decisional task assigned by the *State Planning Act* to decentralized negotiating practices between representatives of the state and of localities developed in a dense network of internal relations, conceivable both as intersections of roles and as overlappings of partial outcomes, which in the course of their temporal deployment allowed for the development of an internal feedback loop and of a dynamic of learning.

The formal sessions of negotiation, however, as has been shown, were only one of the relevant nodes of 'decisions', of the definition of shared representations and of agreements on the action perspectives which were the aim of the planning process. A fundamental factor for feedback and assessment was in fact represented by the external relationships displayed in the field of informal practices which were strictly correlated to the reaching of formal negotiated agreements. Let us therefore schematically represent the forms of structured interaction developed in the course of *cross-acceptance* by framing them – in analogy to fig. 8.1 – into four spheres of activity (themselves internally characterized by specific interaction situations), which may be defined as the conduct of *discussion forums*, of *formal negotiating practices*, of forms of expert and technical *consultancy*, and of explicit practices of *evaluation and self-monitoring*. If the composition of the actors involved in these activities is examined, the result appears as both a simple and a complex framework (fig. 8.5). On the one hand, the diffuse presence of different actors throughout different action situations confers to the latter a highly inclusionary character. On the other hand, the multiplicity of roles which the situations 'assign' to analogous and sometimes apparently identical compositions of actors highlights the function assumed by this copresence of roles in building 'bridges' and thus in constituting the overall network of relationships.

Each of the action situations composing the framework of these activities thus maintains its own irreducible specificity, but this specificity is conjointly defined by the nature of relationships, of a prevailingly informal type, which it establishes with other action situations through the network of multiple roles enacted by the actors in their involvement in different action situations.

The outcome of this complex relational system, which, in the words of Innes (1991, p. 2), may be called a "mixed system of shared power and joint deliberation", is a form of consensus which cannot be understood as 'abstract' since it does not allow for a distinction between its *procedural* and *substantive* components: it is an *agreement for action* which emerges from a virtually equipotential, circular form of relationships between action situations and self-assessment practices.[3]

This kind of consensus, however, is a complex outcome: it entailed the construction of forms of relations which could render its emergence possible; and this construction was realized in a conflictual environment.

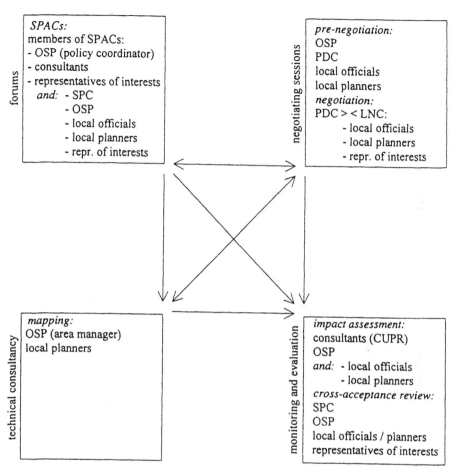

Figure 8.5 Pattern of relationships among actors and roles in formal and informal areas of activity in the course of *cross-acceptance*

As Ostrom (1990) reminds us, in discussing the scheme previously introduced (fig 8.1), the framework of relations which the analyst is confronted with is characterized by an extremely high component of conflict. It is, however, as such a *staged* conflict: it is made explicit, recognized, and treated as such. It is, moreover, a conflict whose expression is embedded into institutional conditions which could minimize the threat of an overdetermination of its expression (in the form, for example, of a hypostatization and of an irreducible confrontation of positions). This has been made possible by empowering the process itself to pursue the very definition of the issues at stake (thus transforming the issues at stake themself into an outcome of interaction), by determining an environment which blurs the actors' roles and strategies by coping with the uncertainty and ambiguity of the issues at stake (thus affecting the very constitution of their roles and strategies), and by putting the dimensions of identification, sensemaking, strategic self-monitoring, and mutual learning of the actors in the forefront. Interactional practices hence could develop into an ongoing dimension of *embedded negotiations*, oriented towards an *evolutive* dimension of consensus, the always precarious realization and institutionalization of which was decisively influenced by conditions framed by the sharing of rules, the flexibility of roles, and the availability and improvement of networking relationships.

Notes

1   In our scheme, the 'courts' does not only refer to the background of Mt. Laurel but also to the quasi-judicial functions assumed by COAH in the framework of the 'administrative' solution provided to the Mt. Laurel issue by the *Fair Housing Act.*

2   "We were finding our way as we did it": Teri Schick, interview 18.07.1995.

3   The only exception in this sense is represented by the formal evaluation and monitoring activity, which evidently was a subsequent and conclusive moment in the process; however, in a perspective of periodical iteration, as provided for by the idea of *cross-acceptance*, the circular dimension of relationships between action situations becomes real.

9 Conclusion: The Fortune of Cross-Acceptance

Introduction

Since the beginning of a process seen by most actors as "very, very scary, and in some sense very dangerous"[1] or, at least, as highly unlikely to succeed, surprise was widely shared about the actual participation in the formal concertative procedures: in fact, no substitution to the counties' role in negotiations was required, nor did most municipalities refuse their mediation. *Cross-acceptance* had resulted altogether in an unquestionable procedural success.

However, at its conclusion, and even before the formal adoption of the plan, feelings among the actors involved were, perhaps understandably, quite contradictory. A generalized weariness prevailed towards a process which has been characterized to a large extent by a dominance of formal procedures which had not been completely compensated by the flexibility and openness of its informal aspects, and which, above all, had required unprecedented forms of engagement. Its burdens had been sensible for local authorities and, in light of the state's weak financial support, in particular for the counties. Responses to the process had been, accordingly, highly differentiated.

The reasons for this feeling may certainly be found in the exceptional features of the process carried out. For over five years, state planners, local officials, pressure groups, professionals, and the general public had fostered a process summoning up the most diverse areas of interests and coalitions of the state to deal with overt conflicts and complex mediations, recording an impressive record of field meetings, from the very first preliminary hearings to the formal negotiation sessions.[2] Two further and more specific reasons must however be added. It is, first of all, undeniable that in its development and in the perception which resulted from it, especially at the local level, *negotiation* had became a cause of frustration. A feeling was diffused among participants to negotiations that the formal effort in defining the agenda and the tables had not been backed by real exchange and substantial agreement. On the other hand, among state planning advocates and watchdog groups, the sensation prevailed that the

167

plan had been consistently weakened in terms of its political as well as technical meaning,[3] and a rather negative evaluation was given of the balance between the transaction costs of the process and the results achieved.[4] Both these positions seemed, moreover, justified by a largely uncompleted agenda of reforms deemed necessary for the plan's effectiveness, which had often been referred to as formal commitments to implementation during *cross-acceptance*. Above all, the commitment of the executive to state planning was overshadowed by uncertainty.

At the same time, a diffuse conviction had nonetheless developed that the political game has been enormously enriched, even transfigured compared to the political customs of the state, by the constitution of opportunities, channels and networks of relations unprecedented for most of the actors involved, and that a general learning process had contributed to reframing mutual relationships among positions and strategies involved. This is first of all true at the local intergovernmental level, where an apparent result was the foundation of new forms of relationships between previously almost unrelated governmental bodies, separated by lack of communication and mutual ignorance of their interests and action rationales. At the level of the state's executive agencies, new 'embedded' forms of negotiation had introduced opportunities for innovating decision-making processes where formal mandates had or would probably have failed. And a similar perception was shared beyond resistances and oppositions by many members of the general public, for whom *cross-acceptance* has simply changed parameters for the definition of interests and strategies of action and for their access in the political sphere.

Cross-acceptance, in this sense, undoubtedly meant a radical change in the political scenario of the state. It was, however, a change which had been largely generated in an endogenous way, through the dynamics developed by the process itself. In order to understand the peculiarity of the role played by planning processes in fostering change and in enhancing innovation in the forms of institutional action, however, it is first necessary to look to the features of responses given by the political system.[5] A brief account will then follow on the peculiar development in planning practices which has in contrast emerged as an endogenous outcome of the *cross-acceptance* experience.[6]

The Uncompleted Agenda of State Planning

At the beginning of 1994, about one and a half years after signing the adoption of the plan, and shortly before handing over to the newly elected

Republican C. Todd Whitman, Governor J. Florio issued *Executive Order No. 114* (State of New Jersey Executive Department 1994). With this typical end-of-mandate initiative, the executive for the first time explicitly required state agencies to adopt the policies of the SDRP as a reference for sectoral policy-making, to conform to the resource allocation criteria for *planning areas*, and to undergo an annual revision process of their plans and programs under coordination of the OSP in view of the presentation of the *Capital Improvement Program*. Moreover, it required agencies to formally report every six months on the measures adopted for implementing the plan.

Florio's initiative finally answered a crucial question regarding the actual support state planning by the executive; it did so, however, only in formal terms. While the new governor never denied its contents, rather publicly stating her support for the state planning process in her first legislature address, an *executive order* is nonetheless like 'law on the books'. Its enforcement is, as such, dependent on the will of the chief of the executive.

At the same time, innovations requiring formal legislative intervention were delayed. This was, for example, the case in respect of elaborations over more than one decade for a reform of the fiscal system, which had highlighted the perverse relationships between the local tax system and the constitution of segregational as well as functionally unbalanced patterns of development;[7] this was, moreover, the case in respect of tools providing legal legitimation and financial compensation for growth management policies, like transfers of development rights.[8] An active consensus by the political system to the new planning course was still forthcoming at the threshold of legislative reforms.

An emblematic example of resistance to formal changes in the balance of governmental powers and jurisdictions, in contrast to a change in the actors' attitudes developed in the course of the planning process, is provided by attempts in the reform of supralocal planning competencies.[9] The case of the *County-Municipal Partnership Bill* revealed in fact both the integrative potential represented by the change in strategies induced by *cross-acceptance* and the prevailing obstructionist power of traditional patterns of decision-making and political representation. The new activism of the counties had been the occasion for a reform movement aiming at their formal empowerment in matters of supralocal coordination and programming. This was in particular the rationale of a bill proposal by the County Planners Association; moving in the same direction, however, had been the Department of Transportation, the most powerful state agency and one of the most committed to a perspective of regionalization of policy-making, with a previous proposal which had been frozen in legislature in 1986. The renewal of initiatives in this direction had forced even the League of Municipalities to a defensive reaction, which resulted in fact in quite an advanced elaboration: the League's policy advisors acknowledged a considerable shift in formal prerogatives in favor of the counties as a condition for a perspective of concerted intergovernmental coordination on a regional scale, however centering this perspective mainly on the achievement of bi- or multilateral intermunicipal agreements. As such, the proposal expressed a peculiar – albeit moderate – form of appropriation of the policy-making style of *cross-acceptance* and of the plan's regionalist perspective by the municipalities' representatives. Nonetheless, it also appeared too daring for the lobby of municipal interests in the legislature, and it was also frozen.

It was thus a typical feature of this phase of implementation having been marked by a progressive practical acquisition of principles, enhanced by the new meaning conferred by the planning process to given practices, rather than by formal innovations in terms of a fulfillment of the agenda of reforms consensually defined through *cross-acceptance*.

This aspect may be further exemplified with reference to the redefinition of budgeting procedures and of their role for state planning, a crucial factor of uncertainty during the *cross-acceptance*.

State budgeting rules in New Jersey provide for the consultative role of an executive commission in the process of political evaluation of capital improvements. Following the start of the state planning process, legislation was amended introducing an explicit reference to the provisions of the SDRP and to the activity of the planning agency.

The current interpretation of these provisions was thus that of a requirement for consistency between state budget and the state plan.[10] However, the procedures through which the SDRP might influence budgeting are anyway subject to considerable mediations: in fact, the Commission on Capital Planning and Budgeting may consider strategies and policies of the plan only for recommendations to the executive before submission of the *Capital Improvement Plan*, which is itself a merely advisory document, to the legislative. Thus, the possibility of the plan's influence on budgeting is quite indirect, depending on the commission's interpretive ability, on the governor's will to consider its advice, and on the dynamics of legislative decision-making.[11]

In this sense, interpretive positions which stress the dependence of New Jersey's formally 'powerless' consensus-building approach on the actual power of achieving consistency residing in budgeting decisions must be highly relativized.

A much more effective influence could, on the contrary, be exerted on the budgeting activity of executive agencies, through the ongoing consultative interactions with the planning agency and in particular through the procedures adopted by the Department of the Treasury since the start of the state planning process for revising the budgets of state agencies, with the involvement of OSP staff. Assessing each single agency's budget on the basis of SDRP policies is the way through which the difficult matter of consistency could be pursued between the rationale of a suborganization's policy-making and the plan's objectives: the OSP was to revise capital improvement requests for the commission, which may adopt its advisory opinion. The process thus envisioned is, again, strictly consultative, but this emphasizes even more the importance assumed by the quality of informal relationships between planners, executive commissions, and state agencies.

The Reflexive Dimension of the Institutional Experience

Although statutory reforms represent a most important dimension of institutionalization, the main outcomes of the consensus-building process conducted in New Jersey are to be found elsewhere. In fact, the dimension in which the most significant forms of institutionalization of interaction forms introduced by *cross-acceptance* may be found concerns the level of awareness and appropriation of the processes themselves. The acknowledgement of the integrative character of *cross-acceptance* and of its potential for collective learning in fact found its rooting in various forms of self-monitoring and self-evaluation by the actors.

The first and most institutionalized forms of this was represented by the plan's *Impact Assessment*. The demand for this had been originated in the framework of a counterreform initiated by representatives of the state's *growth machine* in the legislature after the 'scandal' of the *April '87 Draft*. The aim was to bind the plan's provisions to a peculiarly defined focus on the state's economy, which, however, clearly contrasted with the comprehensive aims of the plan. The issue at stake (implying, it must be remembered, a binding power on the plan's adoption) had been correspondingly reframed during legislative review according to the argument, raised by advocates of the state planning process, that a multidimensional set of planning objectives required an accordingly multidimensional plan assessment.

The outcome was an attempt at a disciplinary response to the complexity of this multidimensional aim, which basically resulted in an endorsement of the plan's provisions (CUPR 1992).

For our interpretive purposes, some of its aspects are most notable. First of all, the timing of its presentation makes it clear that, while responding to a key statutory function, the assessment could not really exert its implicit but crucial legitimizing function during *cross-acceptance*.[12] Nonetheless, dealing with the aims of the plan's assessment played a significant learning and sensemaking function precisely as the assessment exercise was deeply involved in the very making of the process; thus, it represented a means for consolidating a reflexive attitude towards the process and for the institutionalization of self-monitoring as a part of its strategic approach.

Second, a component which readily assumed a peculiar meaning in the framework of the assessment was the evaluation of qualitative improvements in intergovernmental relationships. Its actual result was a survey conducted in the course of the process which, while contributing to its legitimation, represented a still quite unsophisticated treatment of the issue: at the same time, it responded to one of the needs most strongly felt by the actors following the tiresome experience undergone in the previous three years.

The appropriation and conduct of self-monitoring and self-assessment practices was significant also in the framing of the activity of OSP following the adoption of the plan.[13]

A decisive role was played in this by the perspective of a revision of the procedures of conduct of *cross-acceptance*, for which the assessment exercise represented an important reflexive foundation. At the end of *cross-acceptance*, two main issues to emerge during the process framed considerations about possible improvements: the high transaction costs involved in the pursuit of a negotiated consensus, and the 'paradoxes of participation' which had been expressed by a sort of physiological decline in public interest and in the diffusion of information, despite the proliferation of informal group processes, in the course of the process towards formal negotiations. In light of the change in conditions and of the considerable social capital constituted through the process, the focus of a perspective of renewal of *cross-acceptance*[14] was mainly placed on a consistent interpretation of the ongoing connotations of concertative and negotiating practices. Besides aspects of transparency and effectiveness of the means of information, the most critical elements were identified in the flexibility of the process rather than in regulatory outcomes and in a consistent but not constraining reduction of its timeframe; above all, however, the acknowledgement emerged of the need for a further 'opening' of the negotiation process through the attribution of a broader meaning to the informal resolution of issues raised during formal negotiating sessions, the latter being themselves relativized by the embeddedness of significant negotiating attitudes in diffuse aspects of planning practice and, moreover, by the regularized iteration of the consensus-building process.[15]

The Renewal of Planning Practices

The Renewal of Regionalization Practices

Three interpretive perspectives on the innovation of *regionalization* practices at the state level may be referred to the outcomes of the *cross-acceptance* experience. The first, somewhat more traditional but strategically decisive, regards the revitalization of comprehensive planning attitudes by local authorities. What matters in this regard is, however, not so much the enforcement of tools for regulating local planning available to the counties, i.e. the (re)establishment of a 'normal' rational-comprehensive planning attitude, but rather the enactment of innovative practices. It is notable that various counties did not only enhance in the implementation phase the mediating and facilitating function gained through *cross-acceptance*, strengthening their role as enablers by consolidating relationships between local authorities and technical bodies, but even extended the concertative and consensus-building approach to their own style of planning, introducing a new understanding of comprehensive attitudes and a more effective anticipatory capability.

The second perspective relates to a shift away from traditional rationales of functional or area-based regionalization, coincident with the field of competencies of sectoral agencies, particularly in environmental protection and public services, towards forms of regional action based on agreements, cooperation, and the shared formulation of supralocal development orientations. Significantly, this area of innovation does not necessarily match the counties' jurisdiction, and requires rather the adoption of new institutional frameworks: even in this case, non-statutory and informal modes of agreement prevailed on statutory perspectives of reform related to the formal implementation of the plan. In certain significant cases, the nature of initiatives kept a certain level of functionality to defined problems and highlighted the role of forms of concertation with single executive agencies, but nonetheless implied a profound innovation in intergovernmental relations, often based on conventions and on the mediation of cooperative and associative forms of a voluntary character.[16]

The third interpretive perspective concerns the rooting of the interactive experience of *cross-acceptance* in the practices of governmental actors in terms of a progressive extension of its rationale to local concertative practices of regionalization. Reference should be made in particular to the diffusion of a local consensual style of policy-making, which owed its effectiveness and legitimation to the networks of relations and interdependencies constituted through *cross-acceptance*.[17] Similarly to the phenomenon called by neo-institutionalist scholars 'contagion of legitimacy', according to which learning processes may alter the perception of institutional relationships among the actors and, accordingly, their strategic attitudes, the diffusion of concertative attitudes as a source of legitimation of decision-making progressively developed a diffuse need for the adoption of a new style of territorial governance and policy even at the level of localities.

The Renewal of Disciplinary Practices

The most direct forms of disciplinary innovation introduced by *cross-acceptance* were of course related to the growth management approach and to the implementation of the concept of *regional design*. A most important aspect is the profound influence exerted by *cross-acceptance* on the elaboration of its tools. In fact, it may be stated that one of the most significant outcomes of the process was the introduction of a high component of negotiation in the very definition of the tools for the plan's implementation. This outcome may be understood as a concession to the need of achieving a viable consensus, but also as an acknowledgment of the inherently interactional dimension of problem-setting, which stresses the ineffectiveness of *a priori* categorizations and the importance of concerted rationales for local action. The consequence was the development of a much more differentiated conception of 'growth' and of its management, defined by a shift from the aim of control towards the aim of active direction of development tendencies and patterns. A no less important consequence was moreover the fact that the iterative conception of planning introduced by the mandate for revision of *cross-acceptance*, which rendered concertation and negotiation almost ongoing practices, was backed by the embedding of consensus-building attitudes into intergovernmental practices strictly tied to the plan's implementation and to its components of territorial representation and identification, constituting an important condition for the diffusion of mutual monitoring practices.

The fundamental feature of key elements of the plan's strategy may thus be summarized as their voluntary and negotiated character and as their foundation in the mutual acknowledgement of planning practices by the actors, expressed by the (certainly not least symbolic) role attributed to procedures for the 'designation' of centers of the *regional design system*[18] and for the 'certification' of 'consistency' of local plans.[19] The principle (albeit not clearly formalized) which inspired criteria for the choice and concession of certifications and designations are connected to the incentive represented by a potential attribution of priority in public funding of projects and development initiatives to certified and designated areas and jurisdictions;[20] most relevant, however, was the mid-term constitution of regularized forms of interaction between state and local planning bodies and their extension to related agencies, based on exchange of information on the nature of local problems, on the rationale of local initiatives, and on the orientation of related policies. In general, the centrality attributed to these tools for the plan's implemention contributed to conferring a strategic role to joint intergovernmental planning practices conducted under the supervision and technical support of the OSP.[21]

Despite evident contradictions, through these practices an attitude has been emerging towards a flexible, 'weak' form of institutionalization of negotiating approaches learned through *cross-acceptance*, connected to the diffusion of interactive forms of regionalization, contributing to change in the scenario and in the meaning of regional planning practices.

Summing up, one of the most decisive outcomes of the consensus-building experience appears to be identifiable in the ongoing adoption of diffuse interactive practices. Through the development of consultative and organizational functions in process management, the sharing of planning tasks, and the regularization of conventional forms of coordination among local actors and executive agencies, they have contributed to defining around the state planning agency a weakly institutionalized but potentially highly rooted environment marked by redundancy of information and by diffuse mutual monitoring and evaluation activities. A perhaps indirect proof of the potential represented by this towards the constitution of a 'thick' memory of territorial processes is the tendential inversion of relationships between the consultative activities of the planning agency and the policy-making of sectoral state agencies and local authorities themselves: the latter rely more and more on the former not only for its competence in process management, but also as a mediator of inherently interactional and relational forms of knowledge, connecting the actual lessons of diffuse practices of interaction to the forms and strategies of future local action.

The Renewal of Intergovernmental Practices

With all the due caveats, the most notable aspect of innovation in intergovernmental practices, besides those concerning regional and vertical relationships, was the constitution of a cooperative attitude at the level of horizontal relationships between the executive's agencies. As we have noted, the issue of the agencies' role in implementing negotiated outcomes has been endowed with much controversy throughout its conduct of *cross-acceptance*: although the agencies had been involved since the beginning in a parallel consultation process, with significant results in terms of mutual communication, the nature of relationships between their policy-making activity and the plan's provisions, seen as a fundamental lever for implementation, remained uncertain until the end of *cross-acceptance*, and was even then given a rather weak clarification, determining much uncertainty and ambiguity among the actors during debates and negotiations. This therefore makes an analysis of tendencies towards a change in horizontal intergovernmental practices introduced after *cross-acceptance* all the more interesting.

The main innovation may be identified in the constitution of embedded forms of negotiation among agencies. Even in this case, readiness and capability to cooperation have of course been unequally distributed, as well as differently motivated, and similarly unequal were therefore their outcomes.[22] It is nonetheless possible to register a new style of relationships, grounded on the constitution of regular channels of communication and on normalized forms of mutual consultancy, often formalized through conventions,[23] as well as in new procedures for the conduct of formal tasks which incorporate elements of informal negotiation among agencies.

A relevant – if not particularly generalizable – aspect of this renewal in policy styles, tied to specific initiatives in reform, was represented by a change in regulatory attitudes of state agencies. The most significant experiences in this direction were centered on measures for the alignment of sectoral normative frameworks to the consequences of the adoption of the SDRP, as for example in the case of the normalization of relationships between the plan's provisions and the regulatory activity of COAH, rooted in the mandate of Mt. Laurel II, and of the revision of coastal protection policies, which oddly had been excluded from the framework of competencies of the state planning process. The result of regulatory realignment went in most cases far beyond mere adaptation, however: relationships with a policy-making approach as peculiar as that experimented with through the new state planning process, with its non-mandatory attitude, its negotiated form of legitimation, and its attribution of centrality to exercises in the representation of territorial development scenarios, induced a sensible change in their operational rationales and, in some cases, such as that of CAFRA, even in the organizational form of planning practices.

Notes

1   H. Yeldell, interview 18.08.1995.

2   It has been estimated that at least 50,000 persons have been involved throughout *cross-acceptance* in the formal interactions provided by the process.

3   A vaguely disillusioned mood emerged among watchdog groups like New Jersey Future following the *Draft Preliminary Plan* of January 1988, and culminated with the outcomes of negotiations reflected in the *Interim Plan*. The official position prevailing among the environmentalist-regionalist coalition was that of an unconditioned support for the process and for the plan, seen as a remarkable progress in state policy-making (D. Brake and C. van Horn, interviews 10.08.1995 and 14.08.1995); disciplinary dissatisfaction, however, was clearly expressed by the results of an expert consultation promoted by New Jersey Future and MSM Regional Council in 1991, which significantly have never been published. Several contributions (Beauregard 1991; Liberty 1991; Seskin 1991), sticking to a traditional sequential conception of the policy development / policy implementation relationship, show a peculiar underestimation of the originality and of the integrative features of the *cross-acceptance* experience. A significant exception in this sense is represented by DeGrove (1991).

4   For an example, see: M. Danielson, in: Council on New Jersey Affairs (1989a).

5   The best reconstruction of these aspects is offered by Luberoff and Altshuler (1998), basically an update of a previous study by Luberoff (1993). As we have previously noted, however, the prevailing focus on the dimension of 'interest politics' in their analysis also constitutes a major shortcoming in their understanding of the interactive and coevolutive dimension of the *cross-acceptance* experience.

6   Our inquiry into the evolutive outcomes of *cross-acceptance*, limited to the nature of the processes, has been backed by a series of case studies, which unfortunately cannot be presented here extensively. Comparisons have been conducted in particular between the operation of four state agencies (Departments of the Treasury, Transportation, Environmental Protection, and Community Affairs) and between the conduct and results of the *cross-acceptance* process in three counties (Hudson, Hunterdon, and Mercer), chosen as a cross-section of differences in territorial features, political-administrative culture, and planning tradition given throughout the state. An analysis has furthermore been conducted of the evolution of activities of the agencies responsible for affordable housing (COAH) and for the *New Jersey Coastal Zone Management Program* (CZMP) of 1978 ruled – among others – by the *Coastal Area Facility Review Act* (CAFRA) of 1973 (N.J.S.A. 13:19-4).

7   On fiscal reform, see: SLERP (1988); Lederman (1989).

8   As of 1995, New Jersey did not yet have enabling legislation for adopting transfers and purchases of development rights. Although the state has been at the forefront of research (e.g. through the activity of B. Chavoosian as a professor at Rutgers and as

the long-term director of the Division of State and Regional Planning), transfers of development rights were practiced only experimentally through pilot-projects in Burlington County (1992). Experiences in state-funded set-aside policies in rural areas have been in contrast practiced since the 1970s, particularly in Hunterdon County.

9   D. Brake, interview 10.08.1995. On the *County-Municipal Partnership Bill*, see: Buchsbaum (1989a; 1989b).

10   C. Newcomb, interview 21.07.1995.

11   The governor herself has considerable power on budgeting through the prerogative of line-item veto, but her ability in directing it is nonetheless highly constrained and, notably, rarely directed by priorities related to territorial issues. Conflicts on budgeting between the executive and the legislative have focused rather mostly on the amount of financial means for the activity of the planning agencies, in this way indirectly questioning the meaning and very role of state planning.

12   The assessment was realized in a very short timeframe and presented in February 1992, briefly before the plan's adoption; the assessment should itself have been updated according to law on the basis of the *Amended Interim Plan*. The final assessment was in fact presented in April; in June 1992 the *Final Plan* was adopted.

13   Important stages in making sense of the *cross-acceptance* experience were the workshop conducted in 1992 (OSP 1992) and the internal proposals for revision of the process on which discussions started in 1995 (OSP 1994).

14   This concerned mainly the interplay of formal and informal activities regulated by the *State Planning Rules*, *cross-acceptance* still being defined according to the *Act*.

15   Two further important assessments must be recalled here, namely those realized by R. Anglin (1991) at CUPR and by M. Cidon (1993) during an OSP internship. While not promoted by the agency, their common focus on a dimension of learning, oriented towards the achievement of a greater effectiveness of the process, makes them significant contributions to self-evaluation. One interesting feature of the latter is particularly an approach oriented to both an *extrinsic* and *intrinsic* perspective of analysis of the effectiveness of interactive processes, as well as the highlighting of their inherent limits and of their 'wicked' and unintended outcomes.

16   An example is the connection between negotiated forms of planning and the shift towards area-based approaches based on the notion of non-point source pollution in environmental policies through the *Clean Air Act* and the *Clean Water Act* and their implementation at the state level. The meaning of development patterns for building indicators and for implementing non-point source environmental policies confers a beneficial role to their coincidence with 'negotiated' patterns, enhancing innovation towards joint regulatory practices (M. Bierbaum, interview 18.08.1995).

17   The most notable case was represented by the change in regulatory approach in the state's coastal protection program, which constituted a good empirical example of what has been called by L. Zucker (1977) a 'contagion of legitimacy'.

18   While constituting a *lapsus*, having been changed into *resource planning and management*, the original denomination bears an unmistakably peculiar flavor, related to its implicit reference to L. Mumford's idea of regionalism.

19   It unfortunately impossible to expand here on the features of these tools: it is however important to remember that regulation for *centers designation* and for *certification* procedures, as well as *urban complexes* and *strategic revitalization plans*, have all been introduced after and as a result of the *cross-acceptance* experience (OSP 1993a; 1993b; 1993c).

20   This aspect was developed particularly in the case of *urban complexes* and *strategic revitalization plans*, where the needs of urban redevelopment stressed the importance

of financial support of local initiatives (OSP 1993a; 1993b).

21 Particularly important in this sense is the growing role attributed over regulations to guidelines supporting measures in *regional design* according to the plan's view. The bilateral relationship between state consultancy and local inputs is particularly evident in this field (OSP 1993c; 1994; Buchsbaum 1993a).

22 Involvement was in fact statutorily defined in the case of COAH, but much more dubious in that of the department hosting it, Community Affairs, a traditional bastion of the builders' interests; it was strong in the case of Transportation and, at least in part, in that of Environmental Protection; it was nearly absent in most others cases (C. Newcomb and B. Kull, interviews 31.07.1995 and 23.08.1995).

23 The array of intergovernmental agreements reached after the plan's adoption are probably the most tangible results in this perspective of support for state planning.

Appendix 1: List of Interviewees

William Bauer, *Research Planner, NJ Office of State Planning,* Trenton, July 7, 1995

Martin A. Bierbaum, *Administrator, NJ Department of Environmental Protection, Office of Land and Water Planning, former Assistant Director, NJ Office of State Planning,* Trenton, August 11 and 18, 1995

Kenneth R. Blane, *Executive Director, Hudson County Improvement Authority, former Director, Hudson County Department of Planning and Economic Improvement,* Jersey City, August 21, 1995

Dianne R. Brake, *President, Middlesex-Somerset-Mercer Regional Council,* Princeton, August 10, 1995

Thomas G. Dallessio, *Area Planning Manager, NJ Office of State Planning,* Trenton, July 14 and August 7, 1995

David J. Hojsak, *Area Planning Manager, NJ Office of State Planning,* Trenton, July 17 and August 7, 1995

John W. Kellog, *Director, Hunterdon County Planning Board,* Trenton, August 7, 1995

Robert A. Kull, *Assistant Director, NJ Office of State Planning,* Trenton, July 13 and August 23, 1995

Donna Lewis, *Assistant Director, Mercer County Planning Department,* Trenton, August 15, 1995

David K. Maski, *Area Planning Manager, NJ Office of State Planning,* Trenton, July 19, 1995

Charles P. Newcomb, *Assistant Director, NJ Office of State Planning,* Trenton, July 12 and 31, 1995

William S. Purdie, *Area Planning Manager, NJ Office of State Planning,* Trenton, July 14, 1995

James Reilly, *Senior Research Planner, NJ Office of State Planning,* Trenton, July 14, 1995

Charles A. Richman, *Assistant Commissioner, NJ Department of Community Affairs,* Trenton, August 18, 1995

Teri Schick, *Special Assistant, NJ Office of State Planning,* Trenton, July 18, 1995

Herbert Simmens, *Director, NJ Office of State Planning,* Trenton, July 11, 1995

Caroline J. Swartz, *Principal Planner, Hunterdon County Planning Board,* Trenton, August 7, 1995

Christy Van Horn, *former Executive Director, NJ Future, and former Cabinet Member, NJ State Planning Commission,* New Brunswick, August 14, 1995

Steve C. Whitney, *Manager, NJ Deptartment of Environmental Protection, Division of Coastal and Land Planning,* Trenton, August 18, 1995

Helen Yeldell, *NJ State League of Municipalities,* Trenton, August 18, 1995

Appendix 2: List of Abbreviations

CAFRA	*Coastal Area Facility Review Act*, 1973 (N.J.S.A. 13:19-4)
COAH	Council on Affordable Housing, New Jersey State Department of Community Affairs
CUPR	Center for Urban Policy Research, Rutgers University
CZMP	New Jersey Coastal Zone Management Program
LNC	Local Negotiating Committee
MSM	Middlesex-Somerset-Mercer Regional Council
NAACP	National Association for the Advancement of Colored People
N.J.A.C.	*New Jersey Administrative Codes*
N.J.R.	*New Jersey Register*
N.J.S.A.	*New Jersey Statutes Annotated*
OSP	Office of State Planning, New Jersey State Department of the Treasury (now: Community Affairs)
PDC	Plan Development Committee
SDGP	New Jersey State Development Guide Plan
SDRP	New Jersey State Development and Redevelopment Plan
SLERP	New Jersey State and Local Expenditure and Revenue Policy Commission
SPAC	New Jersey State Planning Advisory Committees
SPC	New Jersey State Planning Commission

PART III
RETHINKING THE DIMENSION OF INSTITUTIONALIZATION IN PLANNING PRACTICE

10 Planning, Social Interaction and Institutionalization

Introduction: Questions from a Case Study

The coevolutive dimension of interactive processes displayed by experiences in consensus-building leads to a re-framing of the nexus between the *institutional design* of planning processes and the constitution of innovative forms of *institutional action*. Let us introduce a discussion of these aspects through a summary interpretation of the case-study and by introducing a further level of theoretical generalization with reference to recent debates in the field of institutional analysis.

The process in institutional reform previously analyzed highlights the *social-constructive* and *experimental* implications for a policy of innovation of planning practices. The originality of this experience may be seen, in the first place, in the *style* of response given to a crisis in traditional political routines and to the need, *de facto*, for a reform of policy-making. The path to reform is entered through a process which, though formally/statutorily initiated and structured, is strongly shaped by the potentials allowed to its internal flexibility and coevolutive dynamics.

We may thus talk, in the light of this kind of experience, of a political experience which is not directed towards the 'primacy of outcomes' (March and Olsen 1984; 1989), and which rather stresses reflective attention towards the coevolutive nature of the institutionalization processes it enacts.

The 'process-orientation' which characterizes a long tradition in the analysis of planning practice, understanding consensus, collective action, and effectiveness as possible outcomes of interactive processes which possibly transform the relationships between actors involved, and the impact of policies on the system of social interactions as an eminently process-like and emergent construct (Crosta 1990), is confronted here with an experience which assumes these aspects, at least in part intentionally, as experimental challenges, as a central problem of policy-making, from the very framework design of its processes. The institutional dimension of patterned interaction processes is therefore itself assumed as a construct, as one of the major outcomes of the processes themselves, according to a circular and recursive dynamics comparable to that of a multiple-loop of institutional learning. In this all but linear sense, the institutional design of planning processes may be interpreted as a vehicle for the self-reforming of institutions, impacting on the potential margins of transformation given inside institutional settings in a both experimental and realist manner.

The experience analyzed is therefore an exemplification of the fact that change may be more effectively induced in a political-institutional framework through prevailingly indirect, non-authoritative, informal ways, entailing significant components of decentralized initiative and participation 'from below', rather than through 'constitutional', hierarchical-conformative mandates of reform; it is, however, also an example of the strict interdependence of innovation and change with the determinants of the institutional settings (rules, routines, structures of roles and relationships, networks of communication). On the one hand, conditions for effectiveness and innovation are thus grounded in social practices developing through actual interactions, i.e. through on-going practices of communication, argumentation, negotiation and consensus-building, and on situated forms of self-organization and initiative, rather than on the mandatory forms of co-ordination. On the other hand, their development and fortune entails an on-going confrontation with aspects of the institutionalization of social practices. Conditions for policy effectiveness and innovation through argumentation, consensus-building and self-organization stress therefore their coevolutive relationship with aspects of the institutionalization of practices.

Planning, as an institutional practice, is set in a field of tensions where aims for effectiveness and potentials for innovation, on the one side, face resistance to change embedded in institutional settings and, on the other side, rely on the diffuse and non-institutionalized emergence of innovative social practices. Planning, however, understood as a body of procedures and representations belonging to a broader field of social practices pertaining to public policy-making, becomes itself a factor of institutionalization.

This consideration leads us to assume the forms of *institutionalization*, understood as a necessary process in the evolution of social practices and endowed with both constraining and enabling potentials, as the critical focus (in both an interpretive and normative perspective) for the capability of planning, as both an institutional practice *per se* and as a set of practices embedded in an institutional field, in opening to the possible dimensions of effectiveness and social innovation.

Its possible enabling capacity resides in the way it acts towards developing a fertile, generative connection between the expressions of social pluralism and creativity and the aims of institutionalization: between – in other words – the *intelligence of democracy* and the *intelligence of institutions.*[1]

The thesis of this work is that the progressive character of planning may reside in its reflective capability in addressing processes of institutionalization, in their constraining and enabling duality, in terms which combine conditions for *institution-building* and the provision of directions for *institutional design* along a collectively shared but ever renewed perspective of change.

Between Institutional Design and Institution-Building: Towards a Definition of Institutionalization Issues in Planning

Patterned Interaction between Forms of Action, of Representation, and of Institutionalization

Before addressing contributions to rethinking issues of institutionalization as a crucial aspect of planning practice, in the following the most relevant questions raised by the case study will be recalled, introducing the need for new interpretive perspectives.

1. First, a distinction between allegedly distinct 'phases' of *policy formulation* and *policy implementation* appears problematic and, above all, heuristically less fertile. Solutions to a policy problem may in fact anticipate or even create the definition of the problem itself: implementation (i.e. conditions for its effectiveness and for its very definition) is therefore indissolubly intertwined with the activity of defining and identifying problems, their sphere of pertinence, and the means of legitimation for their 'solutions'. In the first instance, this highlights the intimate linkage between the representation of policy problems and solutions and the identification, selection and engagement of actors relevant for their implementation. Processes in the 'identification' of actors bear a double meaning: that of their *individuation*, i.e. recognition and characterization with regard to the definition of a policy, and that of their own processes of *reidentification* within the sphere of a defined policy.

2. Reconsidering the link between policy formulation and policy implementation by overcoming an assumption in terms of a 'logic of consequentiality' bears a further specific implication. Implementation may be seen as a process which is pervaded, i.e. both defined and conditioned, by forms of *appropriation* of the policy by its subjects: an appropriation

(not necessarily of a conscious kind) which has its roots in the stage of anticipation of policy solutions which is performed in the framework of representations and agreements produced in the interaction and public argumentation, and which may develop into the constitution of a sense of obligation. The coevolution of representations, policy scenarios, and of processes of the actors' identity and preference formation, which develops through practices of communication and mutual strategic adjustment, in this sense bears a potential in the constitution of informal commitments to a policy-making environment which may endogenously change the terms of reference for policy implementation. This implies considering the centrality of aspects of action-learning in the framework of the very construction of policies, revising – if not reversing – the sequential nexus between policy formulation and policy action.

3. Consequences of such a perspective appear to be relevant for an understanding of the nature and role of argumentative and negotiating practices in the design of structured interaction processes. If 'negotiating policy issues' bears a crucial meaning for the *collective framing* of policy problems, i.e. for defining conditions for problem-setting as well as for addressing problem-solving alternatives, implying a dimension of mutual learning and identification, then the heuristic significance of a conceptual distinction between argumentation and negotiation, between 'arguing and bargaining' (Elster 1991) may be relativized. In an anticipatory, integrative understanding, which rejects an assumption *ex post* as a remedial activity, based on a paradigm of exchange, and rather stresses its inherent attachment to processes of policy development, negotiation may be understood as a specific dimension of a multidimensional and contextual process of public argumentation, tending to the constitution of a shared horizon of knowledge, of a *common sense*, and thus underlining the dynamics of the intertemporal evolution of preferences and of their representation through public discourse. Negotiating practices, rather than being identifiable as a specific, discrete 'functional' segment of the policy-making process, thus become understandable as a dimension stretched along a continuum between formal and informal determinants of structured interaction, questioning the problematic implications of any attempt at their institutionalization.

4. An orientation to the effectiveness of policy-making defined according to these assumptions implies two joint consequences in terms of the definition of substantive aims of policy-making, which is accomplished through the collective framing of policy problems and solutions, and in terms of the setting of roles relevant for the policy-making process, which cannot be subject to predefinition. The definition of both aspects is rather

linked by the dynamics of interactions and mutual evolution displayed within concrete action situations. In terms of our understanding of policy processes, assuming this linkage is crucial. Policy-making and planning, as the *enactment* of practices aimed at a shared perspective of action, far from representing a precondition for action become the framework for the constitution of (possibly new) action situations, of new settings for public argumentation and social initiative.

5. As far as the contingency and mutability of problem-setting and of the identification of the issues at stake is assumed as a central matter for policy effectiveness, however, the role of representation in the formation of preferences and strategies of behavior is itself at stake in patterned processes of interaction. Even more than in the formal accessibility of democratic forums and in the intersubjective transparency of public argumentation, the challenge for institutional interactive approaches lies in the bias embedded in policy representations. While questioning assumptions of a 'logic of consequentiality' in understanding the role of the 'institutional entrepreneurs' in framing policy-making, this also invalidates the image of a dominant, disjointedly incremental connotation of policy processes, whereas the only means of institutional input are identified in the provision of rules and contexts for exchange games. Rather, a major institutional dimension of policy processes may be seen precisely in the potential for the mobilization of policy representations.

A triad of theoretical issues concerning the experimentation of forms of patterned interaction in the framework of institutional policy-making initiatives may thus be provisionally advanced:
- on a phenomenological level, addressing the empirical dimension of policy issues (e.g. the definition of relevant territorial problems) is a task undetachable from that of addressing the form of their representation and their contribution to the constitution of a policy discourse (Hajer 1995);
- on the level of policy-making processes, understanding a policy and its prospective outcomes is impossible without considering the ideas, theories, criteria, and contingent strategies according to which a policy is analyzed and evaluated (Majone 1989);
- on the level of institutional determinants, understanding the nature of action and of interaction processes displayed in policy processes and their potential for effectiveness and innovation requires an inquiry into the enabling as well as constraining dimensions of institutional settings and of their patterns of institutionalization (Giddens 1984).

Addressing these questions, however, makes it necessary to make explicit the meaning which has been progressively attached throughout our reflection to the notion of *effectiveness*.

Effectiveness, Innovation and Discovery

The significance attributed to the symbolic-cognitive as well as to the behavioral and strategic dimension of social interaction in the framework of a 'cognitive model' of policy analysis entails a radical revision of the subjectivism and of the rationalistic assumptions inherent in an approach to the design of policy-making dominated by an output-orientation or, in March and Olsen's (1984) terms, by a 'primacy of outcomes'. Overcoming the limits of such a position requires going beyond model-like assumptions, and rather addressing experimentally the blurring of a dichotomy between notions of *substantive* and *procedural rationality*. The condition for such an experimental commitment to action, however, is an understanding of policy processes as part of a dialectics between a *logic of invention* (or, to keep to our terminology, of *innovation*) and a *logic of discovery*.

While an orientation towards policy effectiveness based on strategies of structured interaction poses the question of the openness of its frame of reference to the rise of innovation, the socially constructed nature of this frame of reference contrasts the idea that innovation, understood as a policy objective or even a as program for the innovation of policy-making, might be exhaustively defined *ex ante*, and that its main challenge is that of consistent implementation. This is all the more true in cases in which a shared vision of the nature of the innovation needed and of its aims is impossible, and the definition of objectives is the outcome of the mutual adjustment of emergent representations defined from the relative position of involved actors. Only in a radical, non-reductionist assumption of interaction processes, in fact, may a perspective be justified which does not understand a loss of coordination towards defined objectives in favor of the emergence of innovation as a 'problem', and the process of consensus-building as a cost or as a threat of loss of intentional control on action.

A perspective which assumes the dimension of structured interaction as a condition for the effectiveness of policy action rather tends to see its own practices as set in a tension between *effectiveness, innovation* and *discovery*. While the aim of effectiveness consistently calls for framing the 'possibility of innovation' (Häussermann and Siebel 1994), implying a focus of policies on forms and conditions for innovative action (Melucci 1987; Donolo and Fichera 1988), it also implies questioning the possible outcomes of policies, and assuming hermeneutically the ability to face and to grasp the possibility of innovation as an asset (Dryzek 1982), turning the emergence of unintended consequences of action into a resource and a condition for experience (Boudon 1977; 1984).

In this perspective, even the search for consensus cannot be referred to a single-dimensional conception of power: its features as an interactional construct are not reducible to a logic of political consequentiality, and rather stress the form and conditions for the sharing of policy representations. Reference to consensus in policy processes hence turns to practices of collective sensemaking and to their influence on practices of mutual strategic adjustment of the actors. An analogy may be drawn to multidimensional conceptions which move away from an assumption of consensus and power as an expression of constraining or exclusionary conditions in the representation of relevant policy objects towards a constructivist conception based on the coevolution of relevance criteria of policy representations.

Consensus and legitimation appear in this sense as two mutually constitutive spheres of meaning: both are inseparable from the dimension of symbolic-cognitive mediation enacted through the representations developed in the very argumentative settings on which consensus and legitimation are based (Douglas 1986; Benhabib 1990).

The Constitutive Relationship between Representation and Interaction

A privileged subject of inquiry on the role of interaction in planning processes may be identified in the intersection between the representations which frame the development of a policy and the social processes involved.

Assuming a linkage between representations and forms of interaction implies a series of hypotheses on the role played by aspects of institutionalization in framing their relationship: e.g. the existence of a mutual implication and influence between representations and interactive-communicative settings; the influence of 'expert' forms of representation in framing the construction of a shared cognitive background and in mediating with *ordinary* and *interactive* forms of knowledge (Lindblom and Cohen 1979); the role of this mediation in defining the level of effectiveness and the nature of legitimation pursued and possibly achieved.

Above all, however, assuming such a linkage implies a particular understanding of representation. Acknowledging the relevance of representations which are produced, combined, modified, and mutually translated in the course of the 'social construction' of policy processes means acknowledging their significance as constitutive elements of these processes. This is a necessary premise for an inquiry into interaction processes, which not only do not assume a given structure of policy actors and of their preferences, strategies and mutual relationships, but also do not reduce their outcomes to mere accidents of exchange and bargaining. As an expression of a refusal of both rationalist and relativist assumptions, a focus on the constitutive dimension of representations should rely on the interdependence of the notions of *knowledge, subject* and *situation*. Representations, as sediments rather than as results of a cognitive intentionality and, as such, privileged objects for an observation of the field of relationships between common-sense knowledge and expert knowledge, are at the same time products necessary to the production of those who produce them (Giddens 1984): they are bearers as such of a 'special argumentative virtue' which makes them crucial in understanding policy processes.

Between the constitution of an action situation to which interactions among social actors concur and the representational activities which are displayed in it, a form of circular relationship of a hermeneutic nature is thus established: representations, in their nature of 'situated outcomes', appear to be indissolubly tied to a conjunctural situation of action, as the result and a condition of interaction. Representations are in this sense always related to the situations in which they are produced: they are shaped by the mutual dependence among the subjects involved, their form of knowledge, and their situational conditions of expression. As structuring elements as well as issues at stake in policy-making and planning processes, representations are defined by a duality of procedural and substantive features which makes up for their *constitutive* dimension.

Institutionalization at Issue: the Constraining and Enabling Duality of Practices of Patterned Interaction

The interpretive itinerary previously outlined implies abandoning an exogenist perspective on policy-making (March and Olsen 1984; 1989), realistic or naturalistic conceptions of needs, as well as methodologically-individualist conceptions of interests, in favor of an assumption of preferences and volitions as social constructs, defined by the belonging of actors to institutional settings endowed with a constitutive dimension.

If we assume planning as belonging to a domain of policy as sensemaking, its horizon of effectiveness and legitimation has to be redefined beyond a contrasting distinction between *procedural* and *substantive rationality*, by consciously assuming the conditions of inter-subjective enactment of knowledge and the symbolic-cognitive nexus between effectiveness and legitimation established through interaction.

If thus, on the one hand, a normative intentionality is pursued, the focus of attention should shift from the definition of a unitary process of policy development to the concrete social mechanisms which influence forms of action and of symbolic-cognitive mediation. If, on the other hand, it is assumed that patterned forms of interaction in planning cannot constitute a legitimation in itself, but rather an occasion for the development of processes which display the symbolic-cognitive dimension of representations around which the object of policies is constructed, making it the centerpiece in a collective game, the relationship should be empirically inquired into between the specific conditions of interaction and their feedback and coevolutive outcomes in order to understand their actual degree of openness and integrative potential: patterned settings of interaction should be seen in this sense not only as a constraining framework but as an enabling condition, developing on a continuum of formal and informal determinants.

Effectiveness, innovation and discovery are set along a continuum defined by the problem of the ongoing construction and reconstruction of a background of common knowledge and of a shared horizon of sense. The institutional perspective pertaining to an inquiry of social interaction in planning is thus a conception of institutionalization processes stressing their constitutive nature. The dimension of institutionalization cannot accordingly be represented as a distinctive property of a formalized system of planning practices. In a perspective of policy processes as a way of making sense of reality, planning in fact assumes the features of a kind of 'generative action' (Lanzara 1993), of a course of action of non-foreseeable outcomes, open to innovation and discovery, which through its representational practices 'institutes' sense and generates knowledge resources. Its subjective intentional and purposive components are not erased, but absorbed in a process of learning and reidentification which transcends and codetermines them, not however within a single sphere of knowledge, but within a broader field of forms of knowledge, practice and institutions pertaining to policy processes and to which its processes of institutionalization are related.

As a component of a – by definition – not self-contained field, the institutional dimension of planning is thus best understood in terms of an issue of institutionalization, i.e. of the patterning and consolidating of forms of relationship between its 'own' practices and the practices displayed in 'its' field. Through these relationships, institutionalization processes develop an enabling and constraining dialectics which assumes a constitutive meaning for its appropriation and use of knowledge.

Institutionalization phenomena, in a perspective open to the aims of effectiveness as well as to those of innovation and discovery, should thus be addressed as an *evolutive problem*: not as a status, but as a process connoted by a continuous and undetachable form of codetermination between structure and action. The theoretical problem connected to this is hence that of inquiring not into a static and reified dimension, but into their forms and dynamics, i.e. into the way forms of institutionalization may allow planning processes to remain open and receptive to generative relationships with the field of practices to which they belong.

The challenge for an institutional perspective on planning is to think of its processes in a normative perspective which would not presuppose a strong identity or subjectivity, and which would rather build on experiences which bear in themselves the prospects for an ongoing process of reidentification.

Note

1   These expressions are inspired, respectively, by the titles of two influential books by Lindblom (1965) and Donolo (1997).

11 New-Institutionalist Perspectives

Introduction

This chapter is devoted to an essential outline of the interpretive strategy of new-institutionalism. In the course of our arguments, references to developments in the field of institutional analysis has been frequent, but rather incidental. While necessary at this point, a clarification of such references is anything but an easy task, however, if it is true that "[t]here are, in fact, many new-institutionalisms – in economics, organization theory, political science and public choice, history, and sociology – united by little but a common skepticism toward atomistic accounts of social processes and a common conviction that institutional arrangements and social processes matter" (DiMaggio and Powell 1991, p. 3).

Most observers of the field of research commonly referred to as *new-institutionalism* indeed face the difficult task of unraveling a complex web of ascendancies and differences in order to achieve a clear definition.[1] It is thus difficult to deny good reasons for a call for rigor, and for the claim that "[w]hen someone announces that he or she is conducting an institutional analysis, the next question should be, Using which version?" (Scott 1987, p. 501). At the same time, there is widespread recognition of the fact that, if there is a *new*-institutionalism, its feature is mainly the formulation of 'new answers' to 'old questions' about how individual action and social choices are shaped, mediated, and oriented by – and possibly constitutive of – structures of an institutional order.

There are hence good reasons that may justify the adoption of a more eclectic position and of an 'observer's perspective' on new-institutionalism in the framework of our arguments.

The first reason is the acknowledgement of a common field of essential theoretical problems and of the fruitfulness of hybridizations and cross-fertilizations among contributions to new-institutionalist analysis, which should definitely highlight the futility of attributions of the meaning of a 'supertheory' of social phenomena to any of its versions. The second and related reason – only seemingly paradoxical in light of often radically different theoretical options – is the need to avoid that single theoretical contributions become themselves 'institutionalized', i.e. codified and reified, as distinct and incommunicable 'theoretical strategies', impairing the richness and diversity of the field (Jepperson 1991). The third reason refers, finally, to the primary focus of our work. While certainly critically and selectively biased, the following considerations do not aim at an exhaustive overview of theoretical positions in the field of new-institutionalist analysis. Their aim is rather to identify contributions to an understanding of institutionalization phenomena and of their social-constructive dimension which are deemed critical in a perspective of experimentation of interactive planning approaches.

New Institutionalist Approaches: Towards a Definition

A Typology of Approaches to Institutionalization Processes

A useful introduction to our endeavor is offered by a classification of contributions to the analysis of institutionalization processes, in which institutionalist approaches are distinguished according to alternative interpretations to two fundamental theoretical dimensions (fig. 11.1).

The first dimension considered stresses the relative dynamics of institutionalization, alternatively defined as the reproduction of *properties* of discrete institutional forms (rules, procedures, organizations, and structures) or rather as a process of institutionalization of a continuous, path-dependent kind; the second stresses its relative degree of either social *objectivation* or *subjective* foundation.

The distinction introduced in the latter dimension is representative of differences in analytical attitudes rather than of dichotomic aspects of institutionalization processes themselves: both dimensions are in fact always copresent to some extent in the development of institutional phenomena.

The intimate relationship between aspects of a subjective order and components of objectivation, which is a property of institutional phenomena, is indeed the foundation for their very availability for inter-subjective experience:

> Institutions are composed of both material practices, resources, and rules which objectively direct and condition the action of individuals, and of symbolic constructs, cognitive maps, and principles of conduct giving sense and meaning to the action of individuals from a subjective point of view. [...] The fact that institutions may be seen both from a *subjective* and from an

objective point of view is perfectly coherent with the nature itself of institutions. In fact, at the center of the institutionalization process stands the creation of an *intersubjective* dimension of action. Institutions direct individual action, and in that they bear a subjective dimension, but they direct it as they are perceived as 'external' and 'objective' by the individual. (Lanzalaco 1995, p. 46)

The former dimension, on the other hand, concerns the dialectics between *generation* and *stabilization*, between *persistence* and *change*, which characterizes institutional phenomena at the various degrees of their occurrence. The interpretation of generative aspects of institutions and of the issue of change indeed represents a matter of significant differences in approach and in heuristic potential among contributions to the field of institutional analysis.

	institutionalization *as a property*	*as a process*
objective *approach*	institutions as *constraints to action*: emphasis on the *formal-juridical* component institutional economics public choice new institutional economics	institutions as *valid and persistent models* *of behavior*: emphasis on the *structural* component classical sociological thinking exchange theories positive political economy
subjective *approach*	institutions as *regulatory principles of action*: emphasis on the *prescriptive-normative* component classical sociological thinking new institutional economics	institutions as *constitutive elements of* *social reality*: emphasis on the *cognitive* component symbolic interactionism phenomenological sociology ethnomethodology

Figure 11.1 A typology of approaches to institutionalization processes
Source: Lanzalaco (1995, p. 47)

These distinctions, with all the caveats necessary in adopting them, mainly help in identifying the specificity of contributions coming from different strands of institutional studies, as well as the conceptual orientation they represent and the theoretical consequences they bear. Our interest here is in approaches placed in the lower-right section, characterized by prevailing attention to the *cognitive dimension* of institutional phenomena, understood as *collective*, *dynamic* and *coevolutive constructs*, as outcomes of *experiential processes*, defined by their belonging to a determined social reality and, at the same time, constitutive of their own reality.

Before addressing a conception of institutions as constitutive elements of social reality, some analytical-methodological implications must be underlined of the position taken in this classification by the approaches to new-institutionalist analysis we are primarily going to refer to.

The first implication is their focus on a *process-based* notion of institutional phenomena, i.e. on aspects of *institutionalization* rather than on static-comparative definitions of institutions. The main contrasting reference for new-institutionalist approaches, in this sense, is what has been defined as the 'metaphysical pathos' of classical institutionalism, i.e. the combination it implies between a restricted, formalistic, prevailingly static conception of institutions, and a tendency to determinism in the interpretation of institutional forms, which represents the reason for its inclination towards 'methodological constitutionalism' and for its heuristic limits. Recent contributions from sociological thought and from cognitive disciplines, on the contrary, have stressed a consideration of process components of institutional phenomena which make it possible to highlight the variety of their constitutive dimensions and to endow them with an autonomous analytical dimension. While the complexity of implications raised by this has certainly contributed to a differentiation of the field of institutional studies which is hardly to be reduced to a unitary perspective, and which is possibly characterized by low levels of internal coherence (Scott 1987), it has also evidenced significant heuristic potentials.

The second implication of an emphasis on the process-like dimension of institutional phenomena is the attribution of a crucial meaning to the nexus between *institutions* and *action*, which bears significant consequences. First, attention to the dimension of action implies making the *microanalytical foundations* of institutional analysis explicit. Second, the relevance assumed in the interpretation of institutional phenomena by their microanalytical foundations tends to highlight their cross-sectional features, contributing to a significant relativization of the autonomy of sectoral or scalar forms of specialization of inquiry. Third, as a consequence, traditional forms of 'weak' determinism in social analysis are revised and relativized as bearing an analytical rather than an explanatory function. This applies for instance to relationships between institutional analysis and organizational studies, tending to overcome a distinction between approaches assuming the environment in which organizations operate as an 'institution' (thus emphasizing its constraining factors) and approaches seeing organizations themselves as 'institutions' and as factors of institutionalization (thus concentrating on the genesis and reproduction of institutions through organizational channels).

New-institutionalism thus moves in the direction of rejoining the interpretation of *macroinstitutional* components (structures and organizations) and of *microinstitutional* components (procedures, rules, values) of institutionalization processes, particularly by focussing on the diachronic dimension of their mutual determination, and hence on the necessarily *nested* and *multidimensional* character of levels of analysis.

This specific attitude of new-institutionalism may be exemplified by switching from a reference to institutionalist schools of thought to the observation of analytical-methodological positions in organizational studies. Jepperson (1991) has proposed an interpretive scheme of idealtypical approaches to organizational analysis in the social sciences which distinguishes their position with reference to two main dimensions of inquiry: the degree of representation of social units as 'social constructs', and the levels of sociological analysis adopted (fig. 11.2).

degree to which units are socially constructed:	*featured levels of analysis:*	
	low order (individualist)	*high order (structuralist)*
high construction (phenomenological)	'organizational culture'; symbolic interactionism	institutionalism
low construction (realist)	actor and/or functional reduction attempts; neoclassical economics; behavioral psychology; most neoinstitutional economics; some network theory	social ecology; resource dependence; some network theory

Figure 11.2 A typology of lines of theory in organizational analysis
Source: Jepperson (1991, p. 154)

In the definitions proposed by Jepperson, the character of social construct denotes that the investigated social objects are thought of as complex social products, reflecting rules and interactions at a high degree of contextual specificity, in contrast to a position which conceives them as pregiven, in terms of a 'realism' of a quasi-naturalistic kind (i.e. as autonomous primordial elements of social structure) or of a structuralist kind (i.e. as elements defined by the position occupied in a given structure of social roles or functions); at the level of analysis, the prevalence of explanatory schemes at a macrolevel and of a multidimensional kind is opposed to the causal schemes of a microsociological and monodimensional kind typical of methodological individualism.

According to this scheme, new-institutionalism appears therefore to be characterized by an emphasis on the *conditional* rather than deterministic and causal character of institutions, i.e. on their constitutive and endogenously determined features, which implies both an accent on their character of social construct and a focus on their effects of a superior, structural or macrosociological order.

Consequently, in Jepperson's definition, (new-)institutionalism tends to assume both a *phenomenological* and a *structuralist* connotation, in the sense pointed out by Giddens in the framework of his theory of structuration:

> To say that structure is a 'virtual order' of transformative relations means that social systems, as reproduced social practices, do not have 'structures' but rather exhibit 'structural properties' and that structure exists, as time-space presence, only in its instantiations in such practices and as memory traces orienting the conduct of knowledgeable human agents. (Giddens 1984, p. 17)

In contrast with the idea of organizational structures prevailing in a 'realist' approach, tending to identify their hierarchical features as a pattern of subunits, 'constructivist' and phenomenological approaches do not understand social organizations as units decomposable into fundamental units, but rather as *nested systems*. Similarly, the tradition of methodological individualism, idealtypically situated at the lower degree of analytical approaches, refers in its explanations prevalently to microlevel elements of social organization, to its 'primigenial units', attributing a form of causal primacy to phenomena of a microsociological order on phenomena of a macrosociological order, and thus tending to focus on single levels of observation and of theoretical explanation. Approaches defined by a 'structuralist' line of analysis, on the contrary, "allow for independent and unmediated effects of multiple orders of organizations, and often, though not necessarily, see higher orders as having greater causal potency than lower orders" (Jepperson 1991, p. 154).

In a phenomenological perspective, this aspect opens to a fundamental consideration of cultural determinations of an endogenous order in organizational phenomena: organizations, no more 'black boxes', are the expression of an organizational culture; i.e. of processes which are constitutive of identity, roles, and attitudes (conceivable as 'stories', as situationally and contextually determined paths), which both theoretically and methodologically exclude the possibility of a distinction between endogenous and exogenous explanations of social phenomena, and which rather tend to a diversification of interacting levels of analysis.[2]

The role of basic elements of the analysis of social structure is no more assumed in this approach by primary units, but rather by behaviorally determined aspects of social organization, like *rules*, *frames*, *scripts* of action, i.e. by preanalytical connotations of behavior, the structuring property of which is indissolubly tied to their symbolic-cognitive and to their interactive character as well as to their substantive content.

By emphasizing the constructive and constitutive dimension of exchange and of interconnection between diverse levels of social order, the new-institutionalist approach thus explicitly distances itself from approaches based on prevailingly aggregational or ecological-demographic explanations, i.e. from interpretations of collective phenomena as outcomes of forms of aggregation and distribution of microlevel phenomena, but also from positions adopting single analytical dimensions or purist forms of analytical self-limitation.[3]

Institutionalization as a Property and as a Process

In introducing a typology of new-institutionalist approaches, the specificity of contributions from social psychology and organizational analysis has been indicated in a prevailing interest in the symbolic-cognitive dimension of factors of institutionalization in the framework of a dynamic and process-based conception of institutionalization phenomena.

Properties and processes, in this sense, are not intended as dichotomic variables of institutionalization. The crucial aspect of new-institutionalist positions understanding institutions as 'constitutive elements of social reality' resides in the role attributed to symbolic-cognitive processes in the definition itself of the notion of institutionalization.

According to this perspective – as developed by some of its main interpreters (Zucker 1977, 1987; Meyer and Rowan 1977) on the basis of contributions from the sociology of knowledge, and particularly from ethnomethodology and from the analysis of the cognitive dimension of institutionalization and modernization conducted by Berger and Luckmann (1966; Berger, Berger and Kellner 1973) – institutionalization as a property or a status of a social entity may be understood along an evolutive continuum, and institutions may therefore be detached from an identification with formal structures or with specific elements of a social environment. Institutionalism, in this sense, is intended as a sociological inquiry into the determinations of a superior order of socially constructed realities: if institutions may be defined as social patterns or orders endowed with a certain status or property, institutionalization may thus be defined as the peculiar process of reaching this status (Jepperson 1991), and a meaningful institutional analysis must therefore intersect the definition of the social sphere to which an institutional phenomenon pertains with its relational degree of definition along a contextually and temporally dependent path of institutionalization.

In this sense, the cognitive theoretical focus of institutional analysis privileges a process-based notion of institutionalization rather than a static definition of institutions. The assumptions of a conception of institutions centered on processes may thus be further defined in the following terms:

- institutions are seen as social patterns repetitive in time, as stabilized patterns of repetitive sequences of activities, which owe their stabilization and survival to relatively self-activated and self-sustained processes; their reproduction is therefore not primarily dependent on processes of social 'mobilization', i.e. on intentional forms of collective action, but rather on routinary reproductive procedures (Jepperson 1991, p. 145);

- institutions tend therefore in the course of the process of institutionalization to assume a character of *taken-for-grantedness*, understood as a form of reification and externalization (Berger and Luckmann 1966; Zucker 1977), distinct as such from comprehension, and autonomous from an individual and collective dimension of consciousness (i.e. as 'discursive consciousness': Giddens 1984);

- institutionalization however is at the same time to be understood as a *relative* property, dependent on the analytical context and on the relative position of the issue analyzed within a *field*[4] or within a specific domain of institutionalization processes;

- furthermore, institutions and processes of institutionalization, in their cognitive foundations, may be understood only in terms other than as exogenous constraints, and rather represent a dual form of copresence of factors of empowerment and control; in Jepperson's words, "institutions operate primarily by affecting persons' prospective bets about the collective environment and collective activity" (Jepperson 1991, p. 147).

Some analytical consequences of this conception should be mentioned here. First, intended as a tendential process of stabilization of cognitive frames of social action, the notion of institutionalization does not coincide *per se* with defined processes of reproduction, nor does it stand in contradiction with the notion of change; it signals rather the need to inquire into processes of change at the intersections between dynamics of institutionalization relative to the diverse spheres of knowledge and action which compose the situated framework of a specific social reality. Second, it does not exclude from its interpretative horizon the role of action, but implies its redefinition in opposition to a conception of action centered on the subject and on an institutionally decontextualized profile of its identity. Third, the interpretative function of the notion of institutionalization leads to the adoption of further important strategies of inquiry, which constitute perhaps the main aspects of innovation of new-institutionalism compared to the tradition of classical institutionalism:[5] particularly, a rejection of the identification of the notion of institutionalization, on the one hand, with a specific analytical stance (as the adoption of a single, e.g. 'macrosociological' level of inquiry) and, on the other hand, with single issues (as specific domains of social activity) or with their assumption to primary explanatory criteria (as in determinist versions of the interpretation of contextual and environmental conditions).

Elements of a New-Institutionalist Strategy of Inquiry

Innovations introduced by new-institutionalism compared to its classical sociological tradition bear major implications for its strategy of inquiry and for a definition of its objects. Building on DiMaggio and Powell (1991) and on their account of the field of organizational theory – which constitutes a major strand of development of new-institutionalist thinking – their foremost features may be summarized as follows:
- a change in conception of organizational phenomena at the center of inquiry into processes of institutionalization: it is forms, rules, and structural components of organizations which are assumed as being subject to institutionalization, rather than specific organizations as such. These are for their part no more conceived as coherent, functionally integrated systems, but as non-homogeneous aggregates of 'loosely coupled' standardized elements, subject prevailingly to influences of a non-instrumental, and rather of a cultural, symbolic, relational kind. The *locus* of institutionalization is therefore no more seen in the organization *in se*, but in the *interorganizational field*[6] or in the *social sphere* in which it is contained. New-institutionalist research thus favors a prevailingly

supralocal conception of the organizational environment, emphasizing its interorganizational determinants and their embeddedness within complex social *fields*. The nature of *embeddedness* of organizational phenomena is not identified with reified entities and with traditional or quasi-natural conditions of the sharing of forms of knowledge, but rather with complex relational fields, sectors, or domains of society, in which conditions of plurality of forms and spheres of knowledge constitute the primary connotation facing institutionalization processes; similarly, its nature is not understandable by reference to mechanisms of cooptation led by interests or by intentional forms of action, but is rather of a constitutive and reflexive order;

- a general deemphasizing of the role of interests and of inter- and intra-organizational conflicts as causal factors for the origin of constraints to organizational rationality – which is, on the contrary, central in the approach to organizations by classical institutionalism – in favor of alternative explanatory criteria. Attention shifts from a centralized conception of conflicts, seen as being directed to the core of the 'organizational mission', to a decentralized conception, which sees them as diffused throughout its relational field. Similarly, attention shifts from an attribution of disfunctionality to the presence of vested interests to a consideration of the imperatives of legitimation inherent to organizations.[7] Reasons for the 'irrationality' of organizational phenomena are therefore no more identified in the 'twilight zones' of informal interactions deriving from particularistic strategies and conflicts of interests – appraised on the contrary in their function for the identification, reproduction and persistence of organizations[8] – but rather in the formal structure itself of organizations, of which the symbolic and ceremonial role is analyzed;[9]

- a shift from a prevailingly psychological conception of the cognitive bases of action, relying on notions like the 'infusion' of values into organizations and the identification and internalization of commitment towards values, to a conception emphasizing the everyday, process-like connotation of cognitive activity, oriented by *typifications*, i.e. rules, scripts, and taken-for-granted procedures, which constitute the preanalytical foundations of practical forms of action and of behavioral attitudes, of 'dispositions' (Bourdieu 1994). A shift is therefore introduced from cognitive patterns based on theories of socialization towards learning models of an attributional and paradigmatic type; the understanding of commitments to rules changes from that of a form of 'moral' obligation to norms and values to that of a practical conformation to conventions and routinary practices and to legitimation procedures;

- a prevailing emphasis on aspects of reproduction, stability, and resistance

to change over an emphasis on the endemic transformation of organizations: these are no more, or not only, seen as 'recalcitrant instruments', defined by a 'mission' intended as an institutionally defined structure of ends, but as systems of practices constitutive of forms of shared knowledge and action, tendentially non-reflexive, taken-for-granted, and of the preferences and strategies themselves of the individual and collective actors involved. Within this conception, emphasizing the symbolic-cognitive dimension of organizational behavior, innovation and change are rather seen as an interpretative as well as normative problem;[10]
- finally, a different conception of politics and power, defined by the fundamental ambiguity of knowledge and of the scopes of conjoint action, where conflicts and interests, as well as the identities of actors themselves, are defined in relationship to the objective of reducing uncertainty in the definition of the sphere of policy, rather than in relationship to predefined, distinct, and alternative assumptions of its ends.[11]

Ascendancies and New Perspectives of Institutional Analysis

In the following, a critical and selective review is presented of new-institutionalist research. Far from being a complete account of the strands of new-institutionalism, our review aims at exploring lines of research (in particular new-institutionalist contributions from the sociology of knowledge and from organizational and political studies) assumed as key references for taking position on issues of institutionalization.[12]

In schematic terms, the theoretical sources for a renewal of institutionalism are identifiable with two broad areas of research.

The first of them, *new institutional economics*, develops essentially in autonomy from sociological thinking and – despite bearing reference to it in its name – moves from quite different theoretical premises than the tradition of 'institutional economics', relying on a neoclassical paradigm of rationality and of models of economic behavior.

Nonetheless, the renewal of an institutional focus in economic theory, particularly through the influence of contributions to the *new economics of organizations*, has represented a crucial factor in a revision of neoclassical orthodoxy, adding "a healthy dose of realism to the standard assumptions of microeconomic theory" (DiMaggio and Powell 1991, p. 3) and establishing important channels of communication with approaches directed towards a revision of the explanatory tradition of methodological individualism and towards the formulation of an 'enlarged' notion of the actor's rationality.

Neglect of an explicit discussion of the new economics of organizations is of course a major shortcoming of this brief review. Without ignoring the importance of its contribution, as well as of the heuristic potential of attempts in a conciliation between sociological and politological traditions of institutional analysis with lessons from rational choice theory (e.g. Ostrom 1991), this nonetheless expresses a basic agreement with major critical appraisals of development in institutional analysis within the field of economic theory (Sen 1977; Granovetter 1985; Douglas 1986; for a criticism of a functionalist conception of institutional persistence, see: Meyer and Zucker 1989). Among the foremost of them is certainly the criticism formulated by Granovetter (1985), in the framework of a discussion directed towards establishing a bridge between hypo- and hypersocialized conceptions of action, at the origin of an increasing dichotomy between macro- and microsociological research traditions. Granovetter points in particular to the 'pervasive functionalism' he uncovers in the position of authors who, like Williamson (1985) and Schotter (1981), interpret the embeddedness of organizational phenomena of economic activity in terms of a rational disposition towards contractual equilibrium and who, while acknowledging the role of institutional factors like trust in governing the economy, seem unable to account for the preanalytical and precontractual conditions of contracting, and thus apparently imply a conception of institutions that equates them to 'functional substitutes' of trust rather than to structural conditions for its constitution.

Precisely in pursuing a line of research which aims at accounting for factors of an ideological and traditional order in politics, North (1981; 1990a) has fostered a revision of positive theories of political and economic institutions, moving from a criticism of the inability of rational choice theory in explaining institutional persistence and change; his position aims in particular at bridging the gap represented by the intrinsic inability of methodologically-individualist theories in dealing with the actual constitution of preferences. In this direction moves North's application of transaction cost theory to the understanding of political and economic institutions in an historical-evolutive perspective (North 1990b). Again, however, North's incremental interpretation of institutional change is dependent on the assumption of a rational paradigm of action: the explicative adequacy of such a paradigm remains thus questionable in both theoretical and empirical terms, along with the issues of paradigmatic change and of its possibility of endogenous insurgence. As has been noted with regard to this tendency to a paradigmatic reduction and of its explanatory shortcomings, which associates North's theoretical approach with a traditional tendency to the exogenization of explicative variables, "precisely in measure as it leads to conclude that even political institutions – and the more so all other types of institutions – must ground their legitimation (i.e. the motivations which account for conformity to institutional rules) and cogency (i.e. the mechanisms which render conformity to rules enforceable) on the presence of other institutions, economic theory highlights its own limits or, at best, the tautological structure which supports its analytical construct" (Lanzalaco 1995, p. 51). Again, one of the reasons for this may be seen in the residual rather than constitutive role attributed to factors related to the symbolic-cognitive mediation of preferences and choices.

The second broad area of research of new-institutionalism, usually referred to as 'sociological institutionalism' (Powell and DiMaggio 1991; Hall and Taylor 1996),[13] is defined by much more differentiated disciplinary matrices, developing on the basis of a direct confrontation with the outcomes of the 'behaviorist revolution' in the social sciences and of the resulting reduction of the interpretation of social phenomena to aggregations of individual behaviors. In carrying on this confrontation, sociological institutionalism builds on two important schools of classical institutionalist tradition, both constitutive of new approaches to the interpretation of relationships between political systems, social structure and economic behavior, as well as of radical alternatives to methodological-individualist paradigms: the *political economy of institutions* of Veblen, Commons and Ayres, and the approach to institutional analysis developed within *organization theory* and the tradition of *organizational studies*, with an important anticipation in the work of Selznick. While constituting an obligate reference for any attempt of a renewal in institutional theory in the social sciences, Parson's contribution to institutionalism in a strict sense remains on the contrary confined within few albeit important methodological and metatheoretical remarks (Parsons 1990), however exerting a major indirect influence through the critical confrontation with his functionalist conception of socialization processes which marks the starting point of the 'cognitive turn' in social theory.

Despite differences in disciplinary matrices and research strategies, Commons' and Veblen's economic institutionalism shares with positions as far as Parson's and Selznick's a basic rejection of atomistic schemes of action and of equilibrium models typical of neoclassical approaches, which are contrasted with a multidimensional and voluntarist conception of social phenomena, framed into an evolutionistis and holistic inquiry of society. As has been noted (Wisman and Rozansky 1991; Harvey and Katovich

1992), questioning static perspectives of equilibrium and a 'logic of consequentiality' related to an instrumental assumption of ends introduces in classical institutional economics elements of a theory of cognitive phenomena and cultural dynamics which approximates it to the philosophy of pragmatism and to symbolic interactionism.[14] The theoretical operator of its institutionalist conception of social phenomena assumes the shape of a dualism between instrumental and ceremonial forms of action; the latter, intended as tendentially static, customary and 'backward-oriented' behavioral constructs, may influence lines of action, and their understanding becomes a key element of an holistic evolutionist analysis of society. Such an assumption of the symbolic and cultural dimension remains however attached to an essentially dichotomic understanding of this twin dimensions of action, tending to a kind of 'enlightened' view on the opposition between tradition and modernization.[15]

An heritage of institutional economics is directly recognizable mainly in two specific new-institutionalist lines of research: a rather minority position represented by the so-called *radical institutionalism*,[16] and an approach which is on the contrary generally acknowledged as one of the main strands of new-institutionalism in political science, defined as *historical institutionalism* (e.g. Steinmo, Thelen and Longstreth 1992).[17]

Historical institutionalism, a line of research developed mainly in a perspective of comparative politics, shares some of its basic assumptions with a two further new-institutionalist approaches within political science, a former assimilable to a *new-institutionalist theory of political regimes*, and a latter defined as *positive theory of institutions*, united by a critical reaction against atomistic behaviorism and goal-oriented rationalist understandings of action as well as against an abstract and hyposocialized conception of action situations. All of them in fact strongly point to an understanding of institutions as independent variables and to the role of institutional factors in determining the environment for the actors' rational and strategic choices. The conception of power and of social institutions pursued is tendentially opposed to the idea of that of their availability to the free interplay of social groups typical of the pluralist and of the structural-functionalist models. Emphasis is rather put on the importance of institutional factors in defining conditions for policy choices: institutional choice are themselves seen as sophisticated forms of policy action and of exercise of power. At the same time, however, a clear distance is marked from the formalist and substantially descriptive attitude of classical comparative constitutionalism. While institutions are seen as constitutive factors in the development of policy preferences and paradigms – and thus of the very aims of policies – the actors are not seen as passive 'receivers' of institutional choices. In fact, the former approach in particular has focused on the need for new interpretive frameworks in comparative politics which may account for the relative autonomy of state institutions in shaping trajectories of political change (Evans, Rueschemeyer and Skocpol 1985) – an issue emerging in particular in light of processes of transition from totalitarian to liberalist institutional regimes – but also around the proposal of new interpretive frameworks for understanding cycles of change within political regimes (Hall 1993).

An institutional analysis of politics amenable of explaining historical continuities and transnational variations in the nature of policies requires an enlarged notion of institutions, extended beyond formal determinations of the political world and encompassing the institutional dimensions of society and the economy. The definition of institutions adopted by Hall (1986) is an explicit attempt in an alternative to models developed in the theory of groups. Institutions are seen as formal rules, procedures of conformity and standardized operational practices which structure relationships between individuals across various units of the political sphere and of the economy: they bear as such a rather formal than cultural character, but are not necessarily derived from 'legal' rationales – traditionally intended as opposed to 'conventional' rationales. Emphasis in the definition of their role and nature is thus on their relational character, i.e. on the way institutions contribute to patterning interactions: as such, they bear organizational qualities, and are hence intended as synonymous to organizations. Pressure from societal groups thus does not constitute an exogenous factor of organizational dynamics, but is rather mediated by them, as far as to be affected by a form of institutional *imprinting* in both the definition of preferences and in outcomes. In a later contribution Hall (1993) has revised previous interpretations inclining to a theoretical distinction between cultural and institutional determinants in a strict sense, approaching a more complex interpretation of relationships between institutional regimes and political-cultural paradigms which has rather softened such a dualism, pointing to the construction of systems of beliefs strengthened by their insistence on systems of rules and practices routinized within institutional environments as a constitutive dimension of policy-making.

Positive approaches to a theory of institutions are characterized by an original reformulation of principles of rational choice which borrows important analytical tools from the new economics of organizations, as far as being identified with it as part of a common 'rational-choice school' of new-institutional thought (Hall and Taylor 1996). Their focus is on an original reformulation of issues of institutional equilibrium and persistence, based on an idea of political behaviors oriented towards rules and based on an enlarged and path-dependent assumption of rationality. The empirical terrain for these contributions is paradigmatically defined by an interest in the effects of stability in parliamentary systems based on formal majoritarian principles resulting in face of partisan policy-making rationales through the influence of decision-making rules and procedures (Shepsle and Weingast 1987), already central in Arrow's theory of 'preference disequilibrium' and of a whole tradition of political studies. Equilibrium assumptions typical of rationalist approaches to processes of political exchange, radically questioned within the framework of the rational-choice model (Arrow 1963), are reframed according to an idea of politics as an equilibrium of games in which, again, routines, rules and procedures, as factors for the reduction of transaction costs and of uncertainty, are attributed a key role in defining strategic preferences and relatively stable patterns of political behavior. The idea of an equilibrium of political games pursued is thus marked by a shift from a notion of *preference-induced* to that of a *structure-induced equilibrium*, which may account for the situationally and institutionally determined character of the constitution of political preferences, accounting for microanalytical factors of persistence and change in the evolution of institutional choices in a perspective of path-dependency (Shepsle 1986; 1989).

Within these assumptions, which have been summarized as a 'calculus approach' in opposition to a 'cultural approach' to institutionalist explanations of politics (Hall and Taylor 1996), we may also distinguish affinities and differences between these contributions and the main strand of 'sociological' institutionalism, the one emerging from the tradition of organization theory, and more recently associated with the names of March and Olsen – possibly the most paradigmatic, as well as contested, new-institutionalist approach to the study of politics.

As has been noted with reference to the difficult dialogue between the pluralist tradition of policy analysis, centered on the formation of coalitions and on the interplay among interest groups, and the emergent new-institutionalist paradigm,[18] approaches to institutional analysis inspired by organizational studies are influenced by a classical tradition of sociological theory, of Weberian ascendance, focussing on "the role of conditions of legitimacy and of processes of rationalization in structuring social action" (Lanzalaco 1995, p. 15). Their perspective of inquiry thus certainly builds on a conception of organizations as constructs directed to establishing valid and persistent lines of behavior, influenced by mechanisms for the distribution of rewards and sanctions, which is still present in Selznick's approach: organizational actors are 'kept in line' by systems of control –

such as hierarchies, rules, procedures, sanctions, and patterns of compensation – which respond to an overall organizational rationale. However, contributions inspired by behavioral and cognitive studies have introduced a decisive shift in attention towards the role of action within organizational contexts, i.e. towards implications of a cognitive and strategic order in the constitution of behaviors oriented towards rules and institutionalized procedures and in the reflexive definition of situations.

A common trait of these approaches is the meaning attributed to cognitive factors in the dynamics of organizational behavior and of processes of organizational adaptation and persistence, and the subversion of an instrumental and formalist conception of organizational rationality, in favor of a the idea of an 'organizational culture' as expression of the constitutive *sensemaking* dimension of organizational processes and of its role in shaping of the actors' preferences and strategies.

The Microanalytical Foundations of New-Institutionalism

In an important essay of 1934, which represents his most explicit but never followed-up contribution to institutional theory, Parsons claimed that "the theory of institutions must be concerned with the dynamics of institutional change", and pointed to the fundamental question of *structural change* intended as a dimension "concerned with the processes by which existing value-systems change and new elements come in" (Parsons 1990, p. 333). By this, Parsons aimed at developing a *subjective* approach to institutional analysis based on a theory of (rational) action and on "the point of view of the individual acting in relation to institutions", in contrast with an objective approach based on "the point of view of a sociological observer" (Parsons 1990, pp. 319-20).

Issues of institutional change, in the 'structural' sense pointed out by Parsons, are central to a sociological theory of institutions, and still represent a most critical aspect of new-institutionalist thinking. However, contributions of sociological theory to institutionalism may be traced back precisely to a profound criticism of Parsons's approach, introducing to a reflection on a path-dependent dimension of action which he had only been able to suggest.

A renewed interest for institutions in social theory arises in the wake of the 'cognitive turn' which marks since the 1960s the progressive shift from the influence of a Parsonsian action-paradigm, rooted in Freudian psychology of *ego*, to a theory of practical action based on the influence of ethnomethodology and of developments in cognitive psychology. While organizational studies represented to a certain extent an avant-garde in this direction, anticipating recourse to cognitivist interpretive frames, contributions inspired by the phenomenological sociology of Schütz and oriented towards a new analytical foundation of the sociology of knowledge, and particularly those by Garfinkel and Berger and Luckmann, for the first time focussed on the microanalytical foundations of social phenomena of institutionalization.

Compared to Garfinkel's work, Berger and Luckmann put emphasis on processes of institutionalization and institutional legitimation rather than on an exploration of practical reason as the foundation for the microconstitution of the social order. Their research marks therefore a significant difference from the 'radical realism' featured by the ethnomethodological approach, with its reduction of macrosociological concepts to mere epiphenomena, or mere descriptive constructs, of an 'essential' microsociological level. As a common basis of both lines of research stands an emphasis on the interactive foundations of the social order and on the centrality assumed by tacit and ritual forms of knowledge in governing action and communication, as well as a polemic rejection of the categorial apparatus of norms and values and of quasi-rational explanations typical of hyposocialized models of rational actors. However, the explicit treatment of institutionalization phenomena in a temporal dimension imparts on the latter a primary orientation towards a reconceptualization of institutional theory, exerting a pivotal role in an institutionalist reinterpretation of lessons from phenomenological sociology and ethnomethodology.

Commonsense knowledge, intended as the form of knowledge which is shared with others in the routinary and self-evident practices of everyday life, as the foundation of the intersubjective perception of reality, is the central issue of a sociology of knowledge: "[i]t is precisely this 'knowledge' that constitutes the fabric of meanings without which no society could exist" (Berger and Luckmann 1966, p. 15), as well as the premise for the understanding and for any theoretical activity on society. It is thus the very cognitive basis of its 'social construction' that constitutes the dual character of social reality, as a Durkheimian 'reality *sui generis*', caught in an irresoluble dialectics between *objective facticity* and *subjective meaning*. The essential task of the sociology of knowledge consists hence in exploring the processes through which the objectivation of subjective meaning is produced, i.e. in asking the question: "How is it possible that human activity (*Handeln*) should produce a world of things (*choses*)?" (ibidem, p. 18).

Processes of *institutionalization* are inquired into by Berger and Luckmann along the progressive path towards an *externalization* of known reality, enacted through its distancing in space-time from the 'here and now' of everyday experience. Institutionalization, as the primary dimension of the process of externalization and objectivation of knowledge, and of their "anthropological necessity" (ibidem, p. 52), is the key concept for the interpretation of the nexus between the subjective and interactive foundation of knowledge and the empirically perceived stability of the social order: the social construction of meanings is the perspective to which processes of apprehension and learning tend, moving from the knowledge of everyday life and from the prototypical intersubjective condition of face-to-face interactions – through ongoing and progressively anonymyzed typifications – towards the constitution of social structure as "the sum total of these typifications and of the recurrent patterns of interaction established by means of them" (ibidem, p. 33).

Institutionalization thus takes to the utmost consequence the processes of selective cumulation, constitution and transmission of social stocks of knowledge analyzed by Schütz. Knowledge constructed on the basis of the reality of everyday life is constitutive of the social structure, as well as the latter is an 'essential element' of everyday life: societal order is thus an ongoing social product, and is given only as an outcome of an ongoing process of externalization. At the same time, however, it becomes a parameter for processes of *habitualization*; the coextensive character of its features – the 'ontological security' (Giddens 1990; 1991) which emanates from its immanent effect of reduction of complexity – and of its institutionalization in form of a progressive *taken-for-grantedness* is an outcome of forms of reciprocity which tendentially nullify the reflexive dimension of knowledge:

> Institutionalization occurs whenever there is a reciprocal typification of habitualized actions by types of actors. Put differently, any such typification is an institution. (Berger and Luckmann 1966, p. 54)

The tendential outcome of institutionalization processes is thus paradoxical: it is equivalent to a process of reification and of subsequent internalization, where human phenomena tend to be seen as things, as an *opus alienum* on which no control is given, and where an extreme objectivation corresponds with an abolition of the subjective distance between the individuals and the role they bear, with a 'false conscious-ness': in other terms, "man is capable paradoxically of producing a reality that denies him" (ibidem, p. 89). The institutional world is thus experienced as an objective reality. As an external reality, however, it cannot be apprehended introspectively: individuals have to 'step out' in order to comprehend it. The objectivity of institutional reality remains a man-made objectivity, and it is objectivized as an external product: "despite the objectivity that marks the social world in human experience, it does not thereby acquire an ontological status apart from the human activity that produced it". The relationship remains a dialectical one, and "[t]he product acts back upon the producer. Externalization and objectivation are moments in a continuing dialectical process" (ibidem, p. 60-1).

What is taken-for-granted in the knowledge of society becomes thus coextensive to the knowable: "[k]nowledge about society is thus a realization in the double sense of the word, in the sense of apprehending the objectivated social reality, and in the sense of ongoingly producing this reality" (ibidem, p. 66).

What implications of this conception are relevant for a sociological theory of institutions? First, the focus on the process-based, dynamic, both ongoing and path-dependent dimension of institutionalization phenomena: institutionalization is seen as a property of relational structures of human agency which is produced in the framework of processes of an essentially symbolic-cognitive nature. Their outcome, what we call 'institutions', is thus to be intended as a relational pattern, and as a path-dependent configuration situated in a field of typifications and roles. As such, its rationale is not coincident with the rationale of single functional institutions or institutional spheres of social life and with their 'external' reality, with their own systems of typifications and roles: it is not, in other terms, a phenomenon 'overlapping' reflexive consciousness, but a phenomenon which is ongoingly enacted by it through its everyday embodiment in a structure of roles. The institutional order is 'real' only to the extent it is performatively realized within roles:

> It is thus possible to analyze the relationship between roles and knowledge from two vantage points. Looked at from a perspective of the institutional order, the roles appear as institutional representations and mediations of the institutionally objectivated aggregates of knowledge. Looked at from the perspective of the several roles, each role carries with it a socially defined appendage of knowledge. Both perspectives, of course, point to the same global phenomenon, which is the essential dialectic of society. The first perspective can be summed up in the proposition that society exists only as individuals are conscious of it, the second in the proposition that individual consciousness is socially determined. (Berger and Luckmann 1966, p. 78)

The problem of the logical and functional coherence of institutional orders is set within the perspective of legitimation needs, but legitimation is itself mainly a phenomenon of a cognitive order, as a form of 'objectivation of the second order' of relevant meanings. "Legitimation 'explains' the institutional order by ascribing cognitive validity to its objectivated meanings. Legitimation justifies the institutional order by giving a normative dignity to its practical imperatives" (p. 93). Legitimation therefore is not only a matter of values, as well as not only values are at stake in legitimation processes; rather, cognitive aspects are implied in it which constitute its logical *prius* and its social foundation. "In other words" – and, again, in Berger and Luckmann's terms (1966, p. 94) – "'knowledge' precedes 'values' in the legitimation of institutions".

Berger and Luckmann's contribution marks a theoretical advance in an institutionalist perspective of social thinking precisely for the explicitation of the 'institutional' dimension of symbolic-cognitive phenomena, through the link with a temporal-evolutive dimension and the identification of the nexus thus established between institutionalization and legitimation, intended as first- and second-order forms of objectivation and of progressive distancing from everyday life experience. Thus, "[i]f the integration of an institutional order can be understood only in terms of the 'knowledge' that its members have of it, it follows that the analysis of such 'knowledge' will be essential for an analysis of the institutional order in question": Berger and Luckmann however underline that this form of knowledge is a pretheoretical one, is 'commonsense', and that it is mainly this form of knowledge which "constitutes the motivating dynamics of institutionalized conduct" (ibidem, p. 65). Hardly a better expression than this could be given to the double hermeneutics in which any theory of institutions grounding on the phenomenological and interactive dimension of cognitive processes is involved.

A synchronic comprehension of an institution or of its constitution via *fiat* does not tell us anything in itself, hence, if separated from its diachronic-coevolutive dynamics and from the performative aspects which are jointly displayed in it: representations, symbolic apparatuses, and the constitution of social relationships, roles and networks. New-institutionalist approaches inspired by phenomenological sociology express thus a methodological rejection of static and conclusive interpretations of institutions, directing their inquiry rather to the intersubjectively constructed and intimately problematic character of phenomena of institutional persistence and change.

Among the contributions that have most explicitly addressed the issue of the dynamics of institutional innovation and change moving from phenomenological and ethnomethodological assumptions are those by L. Zucker (1977; 1987; 1988). Dealing with issues of *cultural persistence*,

Zucker (1977) has inquired into three fundamental levels of institutionalization, according to a conception of institutionalization as a process articulated along degrees which may influence its dimensions – the generational uniformity of *cultural understandings*, their stability, their resistance – and thus alter the dynamics of persistence and change. The higher the degree of institutionalization, intended in the terms by Berger and Luckmann, the higher the degree of these dimensions:

> internalization, self-reward, or other intervening processes need not be present to ensure cultural persistence because social knowledge one institutionalized exists as a fact, as part of objective reality, and can be transmitted directly on that basis. For highly institutionalized acts, it is sufficient for one person simply to tell another that this is how things are done. Each individual is motivated to comply because otherwise his actions and those of others in the system cannot be understood [...]; the fundamental process is one in which the moral becomes factual. Yet institutionalization is not simply present or absent; unlike many of the earlier approaches, institutionalization is defined here as a variable, with different degrees of institutionalization altering the cultural persistence which can be expected. (Zucker 1977, p. 726)

Reference to an ethnomethodological approach becomes explicit in the idea of reality as both socially constructed and intersubjectively experienced as a world 'known-or-knowable-in-common-with others', as historically preexisting the actors and providing "the resistant 'objective structures' which constrain action":

> To arrive at shared definitions of reality, individual actors transmit an exterior and objective reality, while at the same time this reality, through its qualities of exteriority and objectivity, defines what is real for these same actors. Macrolevel and microlevel are inextricably intertwined. Each actor fundamentally perceives and describes social reality by enacting it and, in this way, transmits it to the other actors in the social system. (Zucker 1977, p. 728)

Thus, institutionalization is seen both as a *process* and as a variable *property*. It is seen as the process of transmission of the meaning of an act among the actors and – at the same time and at any time of the process – as the definition of the degree to which it may be considered as a taken-for-granted part of reality. Institutionalized acts must therefore be perceived as both objective *and* external, often in a relationship of covariance (as in Berger and Luckmann's conception of the covariance between 'reification' and 'objectivation'). Along with this covariance, also the degree of institutionalization varies. Whenever acts are endowed with *ready-made accounts* they are completely institutionalized (Garfinkel 1967), and these accounts are social creations which, however, count for objective rules since their social origin is unknown (Schütz 1932). At the same time, they define what is possible – i.e.: institutionalization dictates what is 'rational' in an 'objective' sense – rendering other forms of action even unthinkable. Thus, social control through rewards or sanctions becomes tendentially unnecessary: rather, applying sanctions may unveil the effects of power, the feebleness of institutionalization, and therefore raise explicit demands for legitimation.

The degree of institutionalization influences the perception of the meaning of an act: thus, social settings may vary according to the degree of institutionalization which defines them. Because of being embedded in broader contexts in which some acts are seen as institutionalized, other acts in specific situations – incurring in what Zucker (1987) calls a 'contagion of legitimacy' – may become institutionalized.

Consequences of this conception emerge mainly in terms of *transmission, maintenance,* and *resistance to change*. Transmission is the mode of communication of *cultural understandings*: it may occur in a branched or sequential way, but anyway from actor to actor independent from any previous interact. This is a non-problematic process in case of highly institutionalized acts: communication occurs on acts intended as objective, and dealt with as such by whom is receiving them. This modality is variable, however, according to the degree of institutionalization: transmission will increase with its degree of objectivation and externalization. The same is valid for the continuity of transmission, which will increase institutionalization, thus tending to causally produce objectivation and externalization.

Transmission of acts is itself sufficient for ensuring maintenance. The degree of institutionalization therefore sensibly and directly affects the role and importance of direct social control: while the latter is necessary for acts at a low degree of institutionalization, it may be tendentially replaced by mere transmission where institutionalization is high.

Institutional resistance to change is hence high inasmuch as highly institutionalized acts are seen as external facts, as imposed by given settings while actually defining them. Attempts in change through individual influence can therefore hardly succeed, and may lead to a redefinition of the actor rather than of the institutionalized act. Persistence and institutional change are thus eminently *collective* phenomena.

The lessons of Schütz and of ethnomethodology constitute the basis on which, along with Zucker, scholars like Meyer and Rowan (1977) have grounded their new-institutionalist approach to organizational analysis, developing Berger's view of modernization processes (Berger, Berger and Kellner 1973) along a conception of institutions as cognitive constructs and as belief systems constituting a field of ongoingly evolving organizational relationships. Institutional rationales, intended as specialized and differentiated cognitive and normative systems, are seen as constitutive of modernly regulated human activities, but are at the same time components of a repertory of behavioral rationales disposable for differing interests and strategies (Swidler 1986). The relationship between institutionalization phenomena and their cognitive basis becomes thus problematic. The possible sources for beliefs at the basis of institutionalization processes are multiple but, throughout this multiplicity, institutionalization processes may induce conformation to legitimation and persistence criteria which tend to radicalize their process of externalization, progressively detaching criteria for organizational persistence from the definition of the organization's purposes and efficiency (Meyer and Zucker 1989), as well as from the cognitive roots of *taken-for-grantedness* criteria of individuals. The result of these observations is a focus on institutionalization as a process of the constitution of properties of distinct sets of institutional elements, which directs towards a multidimensional conception – potentially riven by intimate contradictions – of institutions as interrelationships between distinct social spheres (Friedland and Alford 1991) or as fields of interorganizational relationships, stressing the critical conditions for the insurgence of change in the face of internal pressures towards conformity (DiMaggio and Powell 1983).

Organizational Behaviors and Organizational Cultures

According to the synthetic definition given by two of its main interpreters,

> [t]he new-institutionalism in organization theory and sociology comprises a rejection of rational-actor models, an interest in institutions as independent variables, a turn toward cognitive and cultural explanations, and an interest in properties of supraindividual units of analysis that cannot be reduced to aggregations or direct consequences of individuals' attributes or motives. (DiMaggio and Powell 1991, p. 8)

By distancing itself from the approach of new-institutional economics, and particularly from its rational-choice matrix, the new-institutionalist line of research in sociological analysis thus identifies a primary line of ascendancy in a critical supersession of the behavioral paradigm in organizational studies, which may be traced back to heritage of the 'Carnegie School' and to the 'cognitive turn' in the social sciences.

A particular position between organizational and political studies is assumed by Selznick's work, which bears a pioneering role in attributing an autonomous analytical function to organizational dynamics in explaining political phenomena and in approaching the issue of implementation as an endemic decoupling of the intentionality and of the outcomes of processes. In his research on the activity of the Tennessee Valley Authority (Selznick 1949), the matter of inquiry is the process through which the conduct of a political-administrative agency, directed towards a technical-instrumental mission, extends its own aims beyond the technical requisites of its statutory task, autonomizing the definition of the own mission through what Selznick calls a process of *institutionalization*: through the dynamics of interactions among groups and coalitions of interests internal to the organizational environment, and through the progressive distancing from an exogenously determined definition of the institutional mission, the end of organizational action comes to be progressively subverted, as far as to be identified with the outcomes of the internal dynamics of the organization itself. The institutionalization process is thus intended as a process by which the organization becomes the object of a form of *value-infusion*, leading to an introjection of relevance criteria and to rendering them quasi-autonomous: in Selznick's classical definition, "'to institutionalize' is to *infuse with value* beyond technical requirements of the task at hand" (Selznick 1957, p. 17; orig. emphasis).

Organizations thus tend to be distinguished from bodies driven by an outcome-oriented rationality: they become a value in itself, as a sort of 'natural community' aimed at its own reproduction, along a path troubled by unintended consequences and non-foreseen outcomes; they thus tend to incorporate attitudes which determine their predisposition to ineffective performance seen from the viewpoint of the accomplishment of their statutory task. Processes in the evolution of institutional bodies hence express an endemic tendency towards forms of *goal displacement*, which are amenable of negative evaluation, as 'disfunctional' phenomena, particularly where the intended mission focuses on an accountable idea of the public benefit.[19]

Although Selznick's subsequent research (1957) has been more specifically directed towards an inquiry into particular roles and strategic behaviors in shaping organizational trajectories, starting a new style of research on leadership and approaching a conception of the enacted character of organizational identity which attributes broader scope to the possibilities of design and intervention, in his work a fundamental theoretical distinction is upheld between technical-instrumental bodies and organizational contexts of collective action which have undergone a process of institutionalization:[20]

> organizations are technical instruments, designed as means to definite goals. They are judged on engineering premises; they are expandable. Institutions, whether conceived as groups or practices, may be partly engineered, but they have also a 'natural' dimension. They are product of interaction and adaptation; they become the receptacles of group idealism; they are less readily expandable. (Selznick 1957, pp. 21-2)

Besides a relativization of explanatory criteria of an instrumental and intentional order, Selznick thus strongly underlines the importance of the evolutive path of organizations as well as the necessary focus on the context-dependence of analysis, adhering to an holistic interpretive approach based on a 'nested' conception of relationships between local processes and global conditions.

Developments in the behaviorist tradition in organizational studies stands under the sign of Simon's work and of the cross-sectional attention to structures of economic and administrative activity introduced by it, opening to a much broader influence in both theoretical and applicative terms.[21]

Simon's research moves from premises similar to those of theorist of economic organizations – an essentially individualistic (but radically anti-subjectivist: Simon 1973) orientation and an adhesion to economic models of choice in interpreting organizational phenomena – introducing however to a research tradition that leads to its radical internal subversion. The starting point of his research shares with economic theories of organizations the call for a deeper understanding of their internal dynamics, based on an endogenous explanation of their forms of rationality and efficiency. The reasons of structurally imperfect features of market exchange which are at the origin of forms of organizational internalization are seen in the cognitive constraints of individual actors. The notion of a *bounded rationality* (Simon 1947), which introduces a theoretical challenge to the rational-choice paradigm, aims at substituting the traditional model of rationally behaving individuals with an empirically adequate theory of choice, laying the foundations for a general theory of organization amenable to be extended to political-administrative structures and behaviors.

The observation of rational actors' constraints in gathering and in handling relevant knowledge, in disposing of informations and of computational capacity, favor the overcoming of perspectives of organizational equilibrium and optimization, and open to individualistic forms of explanation of the origin and function of organizational routines, rules and social norms. However, while the enrichment of the profile of individual rationality inherent in the discovery of the 'boundedness' of its orientation to economic efficiency and optimization allows for a dialogue with contributions from cognitive and behavioral psychology, underlining the role of memory, of the selective nature of knowledge, of processes of learning and of phenomena of adaptation and socialization, it also introduces to a reformulation of the intrinsical rationality of organizational structures in 'socialized' terms, as outcomes of recursive interactions between behaviors of rationally-bounded actors and managerial programming strategies. The process-based and dynamic interpretation of the organizational rationales which govern conformation to routines stems from an observation of the sequential and interactionally determined nature of decisions, which emphasizes the importance of adaptive forms of learning and programmed behavior, constitutive of a 'style' of organizational rationality highly determined by internal relationships (March and Simon 1958). The tendency of this line of studies is thus towards an interpretation of determinants of the organizational universe in *endogenous* terms, in radical opposition to a research tradition focussing on their assumption as exogenous and reducing organizational identity to an epiphenomenon of the aggregate of individual economic choices.

The herein implicit polemic against a neoclassical model of rationality becomes central in the contribution by Cyert and March (1963), in direct confrontation with the economic theory of the firm. Their empirical attitude of inquiry into phenomena of economic organization moves from a rejection of some of the theory's assumptions (i.e. optimization, profit maximization, perfection of information) and develops particularly in the direction of the dynamic and process-based dimension stressed in previous contributions (March and Simon 1958).

Cyert and March insist in particular on two interpretive aspects, in which the notion of the sequentiality and dependency of decisions from the evolution of organizational processes play a decisive role: the *strategic*, situationally and interactively determined nature of organizational objectives, intended as outcomes of internal processes of the constitution of reciprocity ties and of coalitions oriented towards specific policy objectives among members of the organization, and the intrinsical *inefficiency* of organizations, intended as an adaptive resource functional to facing external pressures, which are mediated by a stock of structurally non-exploited organizational capacities (the so-called *organizational slack*) constituting a buffer of resources for maintaining internal equilibria and for dumping the effects of change.

Cyert and March's work shares with Lindblom's notion of *disjointed incrementalism* and with Wildavsky's research on implementation an empirical orientation and a theoretical framework aiming at overcoming assumptions of consequentiality between the definition of problems and of objectives of action and the ways and means of concrete action, which are still present in Simon (1972) and in the tradition of problem-solving approaches. The actor's choices and the concrete forms assumed by their preferences are not defined clearly nor in an abstract *a priori* way, but rather influenced by the interactive conditions given within organizations: objectives may be multiple and conflictual, and decisions may be the result of moves suited to avoid conflicts among objectives. Decision-making behavior becomes thus an ongoing sequence of steps, an incremental path guided by a kind of feedback-loop between actual behaviors and the definitions of their motivations. Research on these aspects introduces to an understanding – for instance – of strategic selectivity in defining the context of choices, of systematic non-correspondence between ends and means, of the role of actions performed independently of specified ends (Wildavsky 1979), of inversions in the relationships between ends and means (Cohen, March and Olsen 1972), or even of the rationality of 'nondecisions' (Bachrach and Baratz 1962) or 'nonaction' (March and Olsen 1976).

The assumption of a sequential relational bond among decision-making processes is further developed, in the form of a *path-dependency* argument, in Cohen, March and Olsen (1972), March and Olsen (1976) and Weick (1979), implying a sensible semantic dilatation of the notion itself of decision-making. Proposed solutions may anticipate the formulation of problems: the traditional relationship between interpretation of an action situation and action itself, as it is assumed by the problem-solving approach, is thus inverted. The relevance of knowledge is determined not only with reference to an oriented and purposive sequence of objectives, goals and actions: it may also be defined regardless of its relationships to purposive action, as far as to render nonaction possible, but is at the same time mutually dependent on action situations, characterized by the copresence of competing strategies, the cognitive determinants of which represent a constitutive component of decision-making rather than its premise.

Apart from the different epistemological meaning attributed to problem-solving procedures, the organizational universe is hence interpreted by these authors in terms which imply a radical redefinition of the idea of the intentionality of action. March and Olsen (1976) interpret the cognitive challenge facing the actors as fundamentally defined by *ambiguity*, intended as the absence of an univocal definition of meaning, i.e. as a lack of meaning as well as a copresence of a plurality of meanings. Weick (1979) rather uses the term *equivocality*, intended as an endemic condition of overlapping of a plurality of different meanings.

Organizational choices are thus still seen by March and Olsen as outcomes of a activity of dealing with ambiguity: but this is not necessarily traceable to forms of purposive, intentional action, and may result in agreements not to act or to the pursuit of better knowledge. Thus, problem-solving intended as an orientation to action has to be considered as a quite partial explanation of organizational behavior: the assumption of problem-solving as the aim of an organization appears more as an 'ideological' claim of values than as an empirically observable practice.

While March and Olsen remain explicitly attached to a traditional vision of participants to organizations as problem-solvers and decision-makers, however looking at organizations from a perspective attributing primary importance to the question of how individuals and organizations make sense out of their experience and modify their behavior in relationship to their interpretation of this experience, Weick (1979) tends on the contrary to deny any autonomous relevance to problem-solving activities as such in organizational behavior. Organizations are rather seen as *loosely-coupled systems* (Weick 1976), which act mainly as tools for dealing collectively with equivocality. Weick is prevalently concerned "with ways in which organizations make sense out of the world and of the fact that they spend the majority of their time superimposing a variety of meanings on the world": equivocality is hence "a prominent component of an organization's existence" (Weick 1979, p. 175). The purpose of organizing, according to Weick, is thus that of *making sense of a situation*: it is precisely the potential richness and diversity of meanings applicable to a situation that makes what an organization is called to deal with.

Compared to Weick's, March and Olsen's position, at least before their new-institutionalist turn, keeps to a more atomistic and individualist foundation. Common to their positions however is the central meaning attributed to the activity of the *interpretation* of their own situation by members of organizational settings. The interpretive characterization of situations and contexts by organizational actors is seen as crucial for the definition of their action perspective.[22] The role of information, central in Simon's model, changes therefore according to the conception of decision-making which is assumed. According to a paradigm of bounded-rationality, the amount of information amenable to be treated offers possible alternatives for optimizing or satisficing choices, but remains functional to the active pursuit of specific objectives. Within incremental models, information is rather seen as a source of feed-back which remotivates towards action and, at the same time, contributes to direct towards solutions: information at hand is thus tendentially 'biased', or oriented, and is dependent on the procedure to which it is associated. In organizational behavior, accordingly, the research of solutions tends to cluster around areas where other solutions have been already found (Cyert and March 1963), producing partial effects of cognitive and especially organizational cumulation. In the *garbage can* model, which takes this conception to the extreme, information may in the course of the processes assume an active role in exploring problems and solutions and in discovering preferences, but may as well become a factor supporting the maintenance of intra- and interorganizational lines of communication – and thus of according strategies – or of forms of legitimation of particular decisions or courses of action (Cohen, March and Olsen 1972; Feldman and March 1981).

From Organizational Theory to Political Science and Policy Analysis

As previously noted, the new paradigm of organizational theory inspired by cognitive psychology represents a key source for the development of a new-institutionalist approach in political sciences. Starting with a line of research introduced by Simon and March's contributions around a theory of choice radically alternative to microeconomic models, the focus of research has progressively shifted towards issues of the allocation of cognitive resources: a growing emphasis has thus emerged on the role of *uncertainty* and of its reduction through organizational routines, on the central function of *sensemaking* and of the 'organization of attention' in the framework of decision-making processes, on the role of *ambiguity* in the interpretation of reality and in the definition of preferences. Decision-making comes to be intended as political process structurally determined by conflicts, differences and ambiguities in the definition of issues, by the involvement of multiple actors featuring incoherent preferences. Aspects related to customary behaviors and to commonsense forms of knowledge, as informal rules and routines, are thus no more seen as passive derivative features, but rather as aspects selectively influencing the structure of attention, producing effects of exclusion of different possible modes of decision. Organizational behavior depends more on the conformation to rules than on a calculus of the consequences of action, nonaction and decisions; and thus also innovation comes to be dependent on a sequential, evolutive and quasi-casual order of relationships between problems and solutions and of their perception and categorization rather than on a logical-functional order. The focus on organizational behavior shifts from a normative to a cognitive orientation towards action, emphasizing a dimension of sensemaking which moves from an assumption of values, motivations and commitments as premises for action towards an attention to the bond between action and the interpretation of action situations according to 'practical reason' which may account for rule- and routine-oriented behaviors.

Accounting for such phenomena according to a renewed commitment to empirical research and to a path-dependent interpretive perspective was to become a major stimulus for a new strand of sociologically-inspired new-institutionalism school of thought in political science.

March and Olsen's institutionalist turn (1984; 1989) is introduced by a plea for new theoretical approaches to institutional phenomena which moves from the empirical observation of the paradoxes in the 'social construction' of policies:

> The primary source of the institutionalist challenge is empirical. Observers of processes of decision making regularly discern features that are hard to relate to an outcome-oriented conception of collective choice. The pleasures are often in the process. Potential participants seem to care as much for the right to participate as for the fact of participation; participants recall features of the process more easily and vividly than they do its outcomes; heated argument leads to decisions without concern about their implementation; information relevant to a decision is requested but not considered; authority is demanded but not exercised [...]. The processes of politics may be more central than their outcomes. Politics and governance are important social rituals. (March and Olsen 1984, pp. 741-2)

March and Olsen's appeal for empirical realism is explicitly aimed at contrasting theoretical styles – polemically seen as dominating contemporary political sciences – defined by exogenist models of the interpretation of political processes.[23] Against these approaches, the need is stressed of models of inquiry allowing for "a more autonomous role for political institutions", and questioning the social determinism in understanding institutions which pervades political studies: "[p]olitical democracy depends not only on economic and social conditions but also on the design of political institutions" (March and Olsen 1984, p. 738). Accordingly, the attribution of "interests, expectations, and the other paraphernalia of coherent intelligence to an institution is neither more nor less problematic, a priori, than whether it makes sense to impute them to an individual"

(March and Olsen 1984, p. 739). The key question becomes therefore that of the acknowledgment of a behavioral 'rationale' of institutions which, similarly to that of individuals, may be understood as an outcome of processes of identification, such as "to justify viewing a collectivity as acting coherently" (ibidem).

The assertion of an autonomous role of institutions in the explanation of politics leads to reframing modes of attention on its internal processes. The central thesis is that of the necessity of considering processes and factors which are fundamental to the definition of the nature of decisions and of policy-making, as the formation and the nature of the actor's preferences, according to their assumption as endogenous to institutions. Their reduction to exogenously determined and tendentially stable factors is in fact recognized as the main shortcoming of current models of political analysis, with obvious consequences in terms of the stylization of contextual conditions (as typical in the case of studies on market conventions and of certain models of political exchange), tending to 'exogenize' and, *de facto*, to expunge cognitive and organizational-behavioral factors from the explanation of choices. On the contrary, it is stated that "preferences and meanings develop in politics, as in the rest of life, through a combination of education, indoctrination, and experience. They are neither stable nor exogenous" (March and Olsen 1984, p. 739). They are seen, as such, as outcomes of processes which are endogenous to political dynamics. The same is stated of the distribution of political resources, according however to a circular pattern of relationships: understanding the endogenous connotation of political resources implies a shift from a merely distributive and allocative assumption to an appraisal of the constitutive features of the process: "[p]olitical institutions affect the distribution of resources, which in turn affects the power of political actors, and thereby affects political institutions" (ibidem). Nor is the formulation of 'rules of the game' something which may be considered completely external to the political game: it is rather an outcome of its specific forms of interaction.

The primary consequences of such a vision of politics and governance, emphasizing their meaning as "important social rituals" (March and Olsen 1984, p. 742; see also 1983), are a challenge to the 'primacy of outcomes' and to its 'rationalized' assumption in the explanation of decision-making and organizational processes, as well as a reframing of the notion of intentionality which relativizes references to forms of purposive, goal-oriented rationality shaped by an ideal of ends-means consequentiality. The meaning of such aspects, already implied in the *garbage can* model, however, undergo a major revision according to an emphasis on the features of *path-dependency* of processes: the shift from a model of rational-consequential order to that of a temporal-coevolutive order of action is the key for interpreting the constitution of 'valid' behavioral patterns, made of a combination of routines, rules, values, which highlight the role of institutions as constitutive of an established order of reality. At issue in such processes are, of course, elements of order which go beyond the dictate of instrumental rationality, of free competition or of patterns of coercion: it is rather a new form of order, of an endogenous nature, linked to forms of obligation as, for instance, an orientation to rules and values of a collectivity. The institutional dimension of such forms of obligation is summed up in the idea of an alternative to the *logic of consequentiality*, represented by a *logic of appropriateness*:

> Politics is organized by a logic of appropriateness. Political institutions are collections of interrelated rules and routines that define appropriate actions in terms of relations between roles and situations. The process involves determining what the situation is, what role is being fulfilled, and what the obligations of that role in that situation are. When individuals enter an institution, they try to discover, and are taught, the rules. When they encounter a new situation, they try to associate it with a situation for which rules already exist. Through rules and a logic of appropriateness, political institutions realize both order, stability, and predictability, on the one hand, and flexibility and adaptiveness, on the other. [...]
>
> In a logic of appropriateness [...] behaviors (beliefs as well as actions) are intentional but not willful. They involve fulfilling the obligations of a role in a situation, and so of trying to determine the imperatives of holding a position. Action stems from a conception of necessity, rather than preference. Within a logic of appropriateness, a sane person is one who is 'in touch with identity' in the sense of maintaining consistency between behavior and a conception of self in a social role. Ambiguity or conflict in rules is typically resolved not by shifting to a logic of consequentiality and rational calculation, but by trying to clarify the rules, make distinctions, determine what the situation is and what definition 'fits'. [...]
>
> From this perspective, the polity embodies a political community and the identities and capabilities of individuals cannot be seen as established apart from, or prior to, their membership and position in the community. (March and Olsen 1989, pp. 160-1)

March and Olsen's new-institutionalist turn has been seriously criticized for its highly provocative, somehow rather pamphlet-like style of reasoning.[24] Their provocation has been however instrumental in stressing three crucial implications of a new attention to institutional factors: an understanding of political processes as practices of sensemaking, an emphasis on the rooting of political representations and choices in action and interaction, and an assumption of the transformative potential of purposive design made autonomous from assumptions of instrumental rationality, and rather put in relationship with the development of 'cultural' determinants of action. With reference to this position, which tendentially overcomes the dichotomy rationality-culture and which "breaks down the conceptual divide between 'institutions' and 'culture'", fostering a 'cultural approach' to understanding the link between institutions and individual action (Hall and Taylor 1996, p. 947), March and Olsen's arguments remain nonetheless prevailingly tied to a formal, objectivating conception of institutional phenomena (Lanzalaco 1995), which marks their selective reception of influences from new strands of sociological institutionalism. The main shortcoming of this attitude may be seen – as has been noted by Zucker (1987) – in an interpretive strategy which equates the 'behavior' of institutions to that of autonomous actors, while devoting scarce attention to aspects connected to the origin and to the evolutive dynamics of institutionalization processes.

Institutional innovation and change, intended as an analytical-interpretive-theoretical as well as a normative issue, has been more explicitly thematized in the framework of inquiries which have approached the rationale and the organizational dynamics of institutional factors from the perspective of a broader notion of institutions: not only rules, procedures, organizational standards and governance structures, but also conventions and habits. Criticism of the individualism and the functionalist assumptions on institutions typical of rational choice models has led representatives of this line of research (Meyer and Rowan 1977; DiMaggio and Powell 1983; 1991; Meyer and Scott 1983; Scott and Meyer 1994) to reject a conception of individual freedom in the choice of norms, rules and procedures which is akin to the notion of a logic of appropriateness; their interpretation however stresses in particular the role of symbolic-cognitive aspects and of their concretion into cultural phenomena as rituals, ceremonials and symbols. The interpretation of the cultural dimension of institutional phenomena is mediated through the thesis of their embeddedness within interorganizational environments, which underlines inertial aspects related to the role of legitimation processes dependent on forms of isomorphic adaptation and mimetism, but also transformative potentials related to the constitution of new relational or discursive clusters. Research on the 'irrational' character of institutions (Sjöstrand 1992) has thus focussed in the framework of interactionist approaches mainly on the influence played on persistence and change of institutional forms by aspects of a cultural order (Meyer and Rowan 1977; Scott and

Meyer 1983) and on the internal relational dynamics of institutional environments conceived as *interorganizational fields* (DiMaggio and Powell 1983). Isomorphic change and mimetism as adaptive legitimation mechanism are linked to conditions of uncertainty and ambiguity which are seen as pervasive of the whole interorganizational field: they are related to aspects of a strict cognitive order (e.g. definition of the own preferences and of means-ends relationships) and with aspects of a strategic order (e.g. knowledge of the other's preferences, positions and moves).

A significant analogy in interpretive approach is displayed in the study of phenomena of survival of *permanently failing organizations*, i.e. of organizations stabilized along a performative path marked by a structural displacement from their statutory goals. Meyer and Zucker (1989) have inquired in particular into the effects of 'disfunctional' features on the dynamics of organizational structures in relationship to their persistence over time. Survival of a 'failing' organization, despite functionalist-instrumentalist conceptions of organizations, is possible and empirically given; however, it implies a 'reinstitutionalization' of ends which implies a decisive shift away from the possibility of interpreting processes of organizational adaptation and conversion according to a 'logic of consequentiality'. The issue of innovation is thus itself to be understood in terms of a process which implies – more or less explicitly – a reframing of organizational preferences and ends. The centrality attributed in Meyer and Zucker's scheme of organizational innovation and change to argumentation may be seen as equivalent to the shift in modes of legitimation found in the mechanisms of institutional isomorphic change analyzed by DiMaggio and Powell, particularly to the shift from a prevailingly normative-coercive dimension to a cognitive and exploratory dimension, based on the mutual reframing through interactive procedures of the construction of agreements on shared redefined ends of the organizational entities.

Notes

1   Reference goes in the following to reviews of new-institutionalist theories presented by: Scott (1987), DiMaggio and Powell (1991), Ostrom (1991; 1995), Thelen and Steinmo (1992), Koelble (1995), Lanzalaco (1995), Hall and Taylor (1996).

2   The logical development of this attitude in a 'realist' perspective is represented by the evolution of the theory of the firm in relationship to the development of complex forms of organization of markets and to the introduction of variables of a superior order (theories of networks and of the ecological determinations of organizational change and persistence). For a critical review of the application of the theory of the firm and of the principal-agent model in bureaucratic environments and in public policy-making, see: Moe (1984).

3   Reference in this case goes mainly to the attitude of ethnomethodological research.

4   The notion of *field*, recently reevaluated by authors like Bourdieu (e.g. 1994) and DiMaggio and Powell (1983), reveals here its derivation from the meaning proposed by Lewin (1951), according to which behaviors of members of interacting groups are interpreted as 'forces' interacting in a 'field', within which all the groups' members are mutually influenced by these forces and all behavior is hence mutually *contingent*. Thus, the action of individual members must be considered in the overall context of the groups' interactions. A field is therefore conceived as non-coincident with the horizons of bounded rationality of a single group or individuals, but rather as overarching different groups' or individuals' rationales in situationally defined terms, and is as such amenable to the contingent introduction of integrative and innovative definitions and solutions for policies.

5   The arguments followed here are based on Jepperson (1991) and Friedland and Alford (1991), who have in particular emphasized the theoretical and methodological consequences of a process-based and multidimensional conception of institutionalization phenomena.

6   DiMaggio and Powell (1983, in: Powell and DiMaggio 1991, p. 64) define as *organizational field* "those organizations that, in the aggregate, constitute a recognized area of institutional life: key suppliers, resource and product consumers, regulatory agencies, and other organizations that produce similar services or products": the notion thus accounts also for 'weak ties' among organizations, i.e. not only for actors directly connected by forms of interaction or competition but for the totality of relevant actors (evidently amenable of alterations according to the dynamics of the field). Attention to the structuration of organizational fields, i.e. to their institutionalization, intended as a contingent outcome of a dynamic process, allows for an understanding of relative characteristics such as the adaptability and consolidation of their ends, their aperture and closure to the access of other organizational formations, the commutation and hypostatization of practices within their domain, in relationship to some relevant phenomena of their evolution (ibidem, p. 65):
- the increase of interactions among organizations comprised in a field;
- the emergence of dominant interorganizational structures as well as modes of coalition;
- the growth of information loads organizations in the field are confronted with in their action;
- the development of forms of mutual acknowledgment of the sharing of a common endeavor among participants to an aggregate of organizations.

7   The role of aspects of legitimation in orienting organizations towards adaptive behavioral patterns has been analyzed particularly by DiMaggio and Powell (1983) and is central to authors inquiring into the nexus between the degree of institutionalization and of environmental isomorphism and the persistence of organizations (e.g. Meyer and Zucker 1989; Baum and Oliver 1991).

8   Moving from the lesson of Simon (1947), the 'positive' role of informal structures and procedures for organizational performance has been highlighted by the studies of Cyert and March (1963), Weick (1979), Scharpf (1978; 1994b), Chisholm (1989), to recall only some of the most influential.

9   Ritual and ceremonial components of action and procedures in the framework of formal organizations have been analyzed by authors like R. Scott and J. Meyer (Meyer and Rowan 1977; Meyer and Scott 1983).

10   In line with the above, DiMaggio and Powell (1991) stress the change in the critical attitude of new-institutionalism towards some central questions of rational theories of

action: in particular, the criteria are re-formulated for a criticism of utilitarianism, shifting from the centrality of a positive theory of the mechanisms for the aggregation of interests to that of the symbolic-cognitive dimension of action.

11 Reference here is of course to the contributions by March and Olsen (1976, 1983, 1984; Cohen, March and Olsen 1972). Here conceptual change in the consideration of structural determinants of action implies a shift from a problematization of the *efficiency* to that of the *effectiveness* of collective action, and this under the sign of the problematic role of the cognitive dimension for innovation and change.

12 Readers who are well-acquainted with new-institutionalist research may therefore like to skip this review and jump directly to ch. 12, were some relevant dimensions of institutionalization processes are discussed in more detail.

13 The origins of "new-institutionalism" in sociology and political science in a strict sense may be traced back to the late 1970s and to the pioneering essays by Meyer and Rowan (1977), Zucker (1977), Scott and Meyer (1983), and March and Olsen (1983; 1984).

14 The main element joining symbolic interactionism and new-institutionalism consists in a rejection of atomistic images of action and in the adoption of a dynamic conception of conjoint action, intended as a process set in a constitutive relationship with social reality, defined by its relationship with the situation and with the dynamic of social structures as well as itself element for their definition. Social actors are seen as agents of change of the social context as well as objects of such a change. The image of the individual is thus that of a creative mind endowed with purposive consciousness, structurally embedded in an institutional context intended in a relativist and evolutionist cultural perspective (Harvey and Katovich 1992).

15 In general, within contributions to the renewal of institutional theory, a pioneering role is acknowledged to these authors in pointing to an autonomous theoretical dimension of institutional phenomena and to an endogenist perspective of inquiry on their dynamics. Their limits are conversely seen in the descriptive and definitory rather than explanatory role of their theoretical models, leaving substantially unquestioned the 'how' of the processes analyzed.

16 Radical institutionalism, mainly represented by authors like Dugger (1989), Bromley (1989) and Hodgson (1988; 1993), is an approach to institutional analysis defined by its alternative position to its neo-classical counterpart, and explicitly inspired by an anti-capitalist political philosophy and by an emancipatory and egalitarian political economy.

17 "Historical institutionalism [...] focuses on the impact of political struggles on institutional outcomes and the way institutional outcomes in turn shape further rounds of political struggles over policy and institutional rules. The approach points to the important interaction between intended and unintended consequences of political exchange and to the impact of this interaction on the determination of strategic goals of individuals. It thereby alerts us to the interplay between structure and agency, with a greater emphasis on the role of structure than agency" (Koelble 1995, p. 242).

18 A vivid example of this difficulty is given by the polemical stance adopted by Jordan (1990) who, in reply to March and Olsen (1989), while pointing to the vagueness and abstraction of new-institutionalist notions, claims their irrelevance and uselessness for an understanding of political networks, themselves intended as 'working institutions', as pursued by pluralist models.

19 The implicit normative orientation assumed by such an appraisal in Selznick's analysis, along with aspects as the underscoring of the role of commitment and of the

sharing of values in the actors' adhesion to the objectives of complex courses of policy, represents a significant anticipation of key issues of implementation research.

20 Zucker (1987) has underlined the mainly heuristic meaning of this distinction, since environments of a technical and of an institutional character overlap in reality within single organizations, and are normally interrelated within single organizational fields.

21 As Moe (1984) has noted, the behavioral paradigm remains in fact historically in a marginal position within the body of works on the theory of organizations, however fostering a variety of multidisciplinary contributions and developments, in particular in the field of empirical research on decision-making processes and policy analysis (e.g. works by Allison, Lindblom and Wildavsky).

22 This is true with the exception of the possibility of the presence of an 'indifference zone', which allows to understand the role assumed within organizations by a certain amount of tolerance towards hierarchical impositions: Simon 1947).

23 Let us remind here of their definitions: *contextualism* as a form of exogenist determinism; *reductionism* as a form of microsociological reductionism of macroorganizational phenomena; *utilitarianism* as dominance of an instrumental and intentional conception of decision-making processes, *functionalism* as a form of historical determinism of social equilibrium; and *instrumentalism* as a form of primacy of outcomes in the interpretation of decision-making processes (March and Olsen 1984, pp. 735-8).

24 For sharp criticism of March and Olsen's version of the 'new-institutionalism', see: Jordan (1990), Pedersen (1991), Sjoblom (1993).

12 Three Relevant Dimensions of Institutionalization Processes

Introduction

A conception of policy-making and planning as a form of generative action, caught up in a tension between the aim of effectiveness, innovation and discovery, requires a revision of the conception of some fundamental aspects of the institutionalization of social phenomena. Attention in this chapter is devoted to three selected but paramount dimensions of institutionalization processes, identified in the definition and sharing of rules, in the development of networks, and in the interplay of social roles. The aim is to highlight the nature of the social mechanisms involved in their construction and to understand their meaning for a critical approach to interactive planning and policy-making processes, introducing to a discussion of issues collective action as a paradigmatic field for in rethinking their institutional dimension.

Rules

Conformity to Rules and Orientation towards Rules

The first dimension of institutionalization processes to be addressed is the character of *rules*. This does not simply happen to be a central aspect in the attempt to confer rational grounding to institutional theory, according to a line of research trying to combine research on the dynamics of strategic behavior with the criticism of rationality emerging from a cognitivist-behavioral focus in the sociology of knowledge and in organizational studies. Indeed it represents a fertile and, to some extent, indispensable attempt to overcome the dichotomy between a conception of rules as 'rules of the game', as constraints of an exogenous nature to the definition of values, and a conception which reduces them to behavioral regularities, i.e. to an aggregate outcome of individual strategies.

Traditional attempts at overcoming this dichotomy build upon theories of non-cooperative games elaborated in the framework of rational-choice assumptions, in a common theoretical pursuit of avoiding a hypersocialized conception of the normative functions of institutional settings through an enlarged and situated, non-hyposocialized conception of individual rationality. Two main positions may be distinguished in this direction.

On the one hand we may recognize positions more or less tied to a paradigm of exchange, which highlight the possibility of cooperation in the context of non-cooperative strategic interactions among maximizing individuals, and which understand the emergence of rules, norms, and institutions as a non-intended outcome tending to a stabilization of relations among actors and to minimize transaction costs. This 'functionalist' conception of the insurgence of rules is expressed by part of new institutional economics, seeing institutions as functional outcomes of transactions among rational self-interested actors, i.e. as "properties of the equilibrium of games and not properties of the game's description" (Schotter 1981, p. 155, quoted in: Ostrom 1986, p. 4). The search for a coherent explanation of the emergence and persistence of cooperative behavioral patterns and of effective forms of collective action, traced back to strategic patterns emerging from the pursuit of maximizing behaviors in the context of iterated interactions, thus becomes an explanatory factor for the emergence of rules and social and cultural norms as well as of institutions themselves.

The aim of safeguarding the explanatory role of a notion of maximizing rationality, however, apparently turns into a peculiar form of 'deductivism' of rules. These positions remain thus dependent on 'weak' functionalist assumptions which, while deemphasizing the role of authority and opening to an understanding of the potential order embedded in situations of disorder and anarchy, nonetheless imply a peculiar scale- and system-determinism, which seemingly assumes a unilateral relationship between social structure and forms of action.

On the other hand we may observe explanatory models which, moving from a criticism of idealtypical assumptions of game theory concerning constraints to individual rationality, address the need to introduce a voluntarist component of social regulation as an explanation for the minimization of defection and egoistic, non-cooperative behavior. These conceptions, according to which rules and norms, in a perspective of socialization and collective behavior, emerge in the form of social sanctions, run into explanatory problems typical of models based on principles of authority and coercion, while proving unable to account for the relational, precontractual aspects, i.e. for the non-rational, *symbolic-cognitive* dimension of influence on behavior endowed by rules and sanctions (Giglioli 1989).[1]

An interpretation pursuing a principle of rationality of purposive action, i.e. keeping to a conception of social actors as *knowledgeable* (Giddens 1984) without running into hyperrationalistic temptations, must evidently face the above mentioned contradictions. The reasons for them lie in an underestimation of the role played by the symbolic-cognitive component of action in relating to rules, as well as of the function of dialectic mediation this element plays between a *regulatory*[2] and a *conditional* consideration of the nexus between *rules* and *action*. This calls for a reevaluation of *conditional* over *normative* determinants of action, according to a notion of rules which underlines effects of *attraction towards action* instead of *compulsion* or *determination towards action*; and this translates into a conception of behavior towards rules which is not grounded on explanatory principles which are exogenous to action, and which at the same time may be able to incorporate rule-oriented action in the understanding of the actors' intentionality.

The notion of rules deriving from this effort tends to be redefined around a conception of action comprising the active role played in it by symbolic-cognitive and interpretive components, and to reframe the notion of *rule-bounded* action into that of *rule-oriented* action.[3]

This conceptual shift expresses the explanatory tension between the constitutive condition of *embeddedness* of the subject and of its acting in the forms of structuration of the institutional context, and the normative need to make sense of a capability for intentional and purposive expression by the subject of action.[4]

At the basis of this conception is the idea that rules do not only represent constraints, i.e. that they do not respond to a linear relationship between structure and action. Rules assume a function in the codetermination of action which is also understandable as a decisive explanatory factor of the *strategic behavior* of actors. Building on discussions on the role of *uncertainty* as a factor of the actors' orientation towards rules and as an implicit presupposition for the possibility of a strategic orientation to cooperation against neo-classical paradigms of rationality (Heiner 1983; 1985), the switch from a conception of boundedness to rules to a pragmatic orientation towards rules is interpreted as the outcome of a kind of subjective 'rationalization', in the form of a strategic principle guided by bounded rationality, in the pursuit of self-interest (and in the intentionality) of actors.

Rowe (1989) has expressed the conceptual scope of this reinterpretation of the strategic determinants of individual rationality by introducing a distinction between the notions of *act-individualism* and *rule-individualism*:

> An agent who follows a rule acts as if he had precommitted his own future actions. By rejecting discretion and committing himself to a rule, the agent sacrifices his freedom to act rationally, given other agents' expectations of his actions, in order to gain the ability to influence those same expectations. (Rowe 1989, p. 6)

According to these principles,[5] the observation of interactions among rule-oriented actors highlights the dissimilarities from behaviors as they would be expected according to a perspective of act-individualism. Institutions are in this sense understandable as particular forms of behavior by some actors who, by these actions, are considered members of a social institution; these are constituted by actors who follow rules of action and who think other actors do the same, and this behavior is seen as 'rational' since, through this behavior, an actor may influence the others' expectations of his own future actions, and therefore influence their behavior to his own benefit (Rowe 1989, p. 227).[6] Social institutions thus appear as a 'theoretical construct' which allows for a reinterpretation of act-individualism without running into empirical falsification.

The contradiction between the *teleological,* forward-oriented perspective typical of act-individualism, and the *deontological,* backward-oriented, and *path-dependent* perspective typical of social institutions appears amenable of conciliation within such a perspective centered on the *constitutive duality* of rules. While act-individualism only follows a teleological principle, rule-individualism in fact encompasses both a teleological and a deontological dimension: the justification of acts according to a rule is in fact of a deontological order, while justification of the rule itself is teleological. For this reason, an institutional perspective appears to be amenable to conciliation with principles of economic rationality only if able to comprehend the specific form of rationality on which the constitution of rules which give orientation to action grounds.

Focussing on rule-oriented action highlights the inadequacy of methodological-individualist theories of the institutional meaning of rules in explaining *generative* aspects of action: determinants of an institutional order in the emergence of rules are assumed there as essentially *exogenous,* implying a substantial inability to account for an *endogenous* generative dimension of rules which may range beyond their assumption as mere constraints. However, a conception of rule-oriented action also bears the risks of running into a peculiar form of tautology, as expressed by the tendency to a self-referential conception of institutions implicit in several cognitivist interpretations of new-institutionalism. While highlighting the symbolic-cognitive bases of institutions and their constitutive role for individual action and social interaction (Lanzalaco 1995, p. 59), this conception is in fact threatened with failing to identify the nexus between the intentionality of action and its institutional determinations.

The key to overcoming this contradiction may be found, again, only outside equilibrium assumptions, and in the framework of a radical assumption of the dimension of reproduction of institutional factors, i.e. in a dimension of *institutionalization*. This calls for a consideration of the *multidimensional configuration* and of the *disposability* of the regulatory universe pertaining to action, and of the impossibility of defining a meaning of the intentionality of action outside these dimensions.

The Constitutive Dimension of Rules

Key to addressing these dimensions is a discussion of the nexus between structure and action and of its role in a conception of rules.

Normative determinism, "starting from the outcome of action, retrospectively imputes it to antecedent norms conceived as necessary and sufficient causes, bypassing both the cognitive processes through which the new instances are submitted to rules and the fact that the development of action modifies the circumstances in which the acting subject finds himself" (Giglioli 1989, p. 124). On the contrary, the reflexivity through which actors, through action and interaction, reframe the circumstances of action itself is fundamental in reproducing their knowledge of the social world: the process of production is coincident with the process of understanding, in the form of an *accounting-for-action* and of an *accounting-for-themselves-in-action*.

Rules stand therefore in circular relationship with the knowledge that grounds them and with the action that constitutes their object and *raison d'être* of being. The relationship between rules and action, and between normative and symbolic-cognitive components of action, is not dominated by the normative aspect, i.e. it is not unidirectional, but rather circular and 'constitutive': the definition of rules is non-separable from the practices which constitute their application and which, by applying them, redefine them.[7] Using phenomenological terminology, *accounts* on facts are a constitutive part of the facts themselves: "it follows that the fundamental relationship between normative rules and socially organized action is not of a *regulatory* kind, but of a *constitutive* kind: norms are evoked by actors in order to give sense to conduct, to reflexively constitute the circumstances and the activities to which they apply" (Giglioli 1989, p. 124).

Interpreting the meaning of the expression 'following a rule', of the dynamic occurring between rules and pragmatic action, in the framework of social behaviors situated into concrete contexts of action, is therefore an exercise which cannot be limited to identifying the observance of certain rules, but must extend to an understanding of the relationship which in any context is instituted time and again with rules (Crespi 1986, p. 399).

The pattern of relationships under which the nexus between structural-regulatory and individual-voluntaristic determinations of action may be understood in the *longe durée* of structuration processes must thus be represented in a circular form, as for instance in the following scheme:

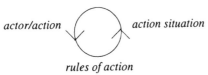

actor/action   *action situation*

rules of action

Rules are thus intended as *constitutive* elements of a relationship of duality which characterizes processes of structuration, and the form of co-determination which rules exert is characterized simultaneously by effects of *constraining* and of *enabling* action. The relationship between rules and action thus appears similar to a process in which "prescriptions presuppose interpretation, but consequent acting results conversely in reinterpretation" (Ferrari 1986, p. 371).[8]

The consequences of reference to a common sphere of sense may hence be understood even in terms of the relationship between rules and action. The notion of 'meaningful action' entails reference to rules as constitutive elements of action: an intimate connection is thus instituted between meaningful acting, the social context and rules. The possibility of analytically distinguishing action from rules is tied exclusively to the co-presence of different rules and of different ways of adhering to rules or of breaking rules, i.e. to a constitutively *multidimensional* conception of rules:

> By the moment a rule is identified, either through the interpretive process of the observer or by means of the self-understanding of actors, the rule acquires in fact a *definite* character which, as such, constitutes a *reduction of complexity of the lived experience*. It is therefore only as complexity that action may be distinguished from the specific rule which gives orientation to it; but, based on the assumption that there is no meaningful acting deprived of rules, it is apparent that the complexity of lived experience may not be caught directly, but only as the possibility of a shift from a rule to another rule or between different modalities of using the same rule. (Crespi 1986, p. 403)

Rules, in their relationship to action, are therefore themselves subject to forms of institutionalization. *"The discursive formulation of a rule is already an interpretation of it*" which "may in and of itself alter the form of its application" (Giddens 1984, p. 23, orig. emphasis): rules are placed on a continuum along which each form of discretization, starting from their discursive formulation, constitutes an institutional innovation.

Rules thus assume a primary role among factors of institutionalization of social practices; their degree of formalization is however itself the outcome of a stratification of practices placed at different levels of institutionalization. The degree of formalization of a rule thus appears not deducible *a priori*: in other terms, there is no phenomenological uniformity of rules and of their normative role; under the label 'rule', normative phenomena of a very different nature may thus be found.

Acknowledging this is key to a conception of the possibility of the interpretation of normative systems, i.e. to a conception of *manipulability* as *interpretive manipulation*. Interpretation is intended as a form of modification intervening on the sense itself of a normative proposition, on its semantics as well as on its syntax, but which does not constitute its suspension: it represents neither a momentary interruption of its effectiveness nor a moment of normative undefinition, but rather the possibility of acting in conformity to a rule and, at the same moment, interpretively affecting its coevolution.[9] Once again, however, the level of amenability of a normative system to interpretation is not homogeneous throughout the different rules which constitute it and throughout their different levels of formalization. In this sense, the study of rules pertains not only to their 'structural' connotations but to their differentiated and changing 'structural properties' (Giddens 1984, p. 23).[10]

This conception of rules, while contrasting with a conception of the taxonomy of regulatory enunciates, implies an attitude towards normative experimentation.[11] This in turn implies acknowledging the internal articulation of the universe of rules in terms which blur the classic distinction between *constitutive rules* and *regulatory rules*, rather interpreting it along a *continuum* of determinants.[12]

Rules of a constitutive character and rules of a regulatory character never appear to be mutually non-related: if the definitional connotation of the former is the function of conferring meaning to actions, "then regulatory rules must be based on constitutive rules in order to define the terms of the relationship in the formula that expresses the enunciative structure of regulatory rules [...]. Regulatory rules imply constitutive rules, which therefore are not independent. Their dependence is further set by the relationships which, in a complex normative system, constitutive rules establish among themselves" (Ferrari 1986, p. 394). But a further factor which weakens such a distinction is the incomplete, *ambiguous* nature itself of normative systems, and their ideographic semantic dimension, which involves the actors of their interpretation in a relationship of hermeneutic circularity with their very formulation. Rules display themselves along a multidimensional *continuum* throughout different degrees of explicitation and formalization, of coherence and completeness, of internal articulation, differentiation and hierarchization. The clarity of a regulatory system depends therefore on the always relative completeness and coherence of these configurations; its clarity however is never completely reducible to those requisites, it cannot be abstracted from the complexity of the communicative medium and of the interpretive processes enacted through social interactions. In other terms, rules must be assumed as *hermeneutic models of action*, rather than as models of *imprinting*.

This position highlights, on the one hand, the instability and the endogenous potential for alteration of rule systems and, on the other, the structural and political-organizational constraints to the manipulability of rules, which render illusory a voluntaristic, ahistorical and decontextualized principles of disposition of rules.[13] It also stresses, however, the opportunities given for strategies working within rules and acting on their margins of contradiction and conflict in the framework of communicative practices, and their potential capability of transforming situations of instability into dynamic games, into 'games about rules' (Mutti 1986).

The demise of aprioristic, ontological conceptions of rules, detached from the concrete context of lived experience and action (Crespi 1986), and the radical relativization of distinctions between different regulatory components, introduces a conception of rules defined in relational terms, allowing internal degrees of institutionalization and the development of forms of action interacting with their configuration, and potentially inducing endogenous paths of change.

The Configurational Character of Rules

Implications of such a conception of rules become apparent in the framework of interpretations assuming the 'constitutive' dimension of rules as a factor for the definition of institutional phenomena, i.e. as a strategically determined social construct. The assumption of the *configurational character* of rules is also the starting point for an important reinterpretation of the conditions for collective action (Ostrom 1990).

Relevant for a definition of rules, intended as "prescriptions commonly known and used by a set of participants [to an action situation] to order repetitive, interdependent relationships", is not so much their alleged self-sufficient character of formal prescriptions,[14] but their character as "artifacts that are subject to human intervention and change" (Ostrom 1986, p. 5), i.e. their situated and contingent character, as well as a redefinition of the requisite of enforcement in terms of its relational connotations:

> Rules are the result of implicit or explicit efforts by a set of individuals to achieve order and predictability within defined situations, by (1) creating positions (e.g., member, convener, agent, etc.); (2) stating how participants enter or leave positions; (3) stating which actions participants in these positions are required, permitted, or forbidden to take; and (4) stating which outcome participants are required, permitted, or forbidden to affect. (Ostrom 1986, p. 5)

The meaning of rules hence lies not so much in their orientation towards the prescription of single actions or effects, as in their being elements of the definition of clusters of actions or effects, which are conferred a combination of normative values through the specific, complex and mutually interrelated configuration of different regulatory enunciates.

This conception of rules introduces a shift from the observation of *rule-bounded* towards that of *rule-oriented* behavior. Observing rules as configurations of social orientations towards action, and abandoning determinist temptations, weakens their orientation towards the definition of *effects*, emphasizing rather their orientation to the definition of *action situations*. Rules are accordingly seen as "the means by which we intervene to change the structure of incentives in situations" which define action (Ostrom 1986, p. 6), as elements of the very definition of action situations themselves.

An important implication is the possibility of interpreting in strategic terms – i.e. in mutually constitutive rather than in deterministic causal terms – the behavioral components of action. While in fact, from the perspective of actors, rules intervene in forms of complex configurations which manifest themselves through complex chains of behaviors and actions, i.e. in terms which influence action in strategic and indirect ways, from the perspective of an analyst, as well as of a hypothetical 'normative strategist',

> rules can be viewed as relations operating on the structure of a situation. Rules can be formally represented as relations, whose domain are the set of physically possible variables and their values, and whose range are the values of the variables, in the situation under analysis. [...] Viewing rules as directly affecting the structure of a situation, rather than directly producing behavior, is a subtle but extremely important distinction. (Ostrom 1986, p. 7)

The analysis of action situations characterized by multiple interactions and high interdependence needs therefore a description of a situational configuration which is given by a correlation of *structural variables* and *regulatory variables*. While variables of a structural order of an action situation may be defined as constrains of a physical and behavioral nature, constituting the 'surface structure' of its representation, "[t]he rules are part of the underlying structure that shapes the representations we use" (Ostrom 1986, p. 19).

Two main analytical consequences may be drawn from this. First, attention towards rules is redefined as a strategy of inquiry into such rules which may directly influence an action situation. Redefining the composition of a regulatory configuration thus becomes thinkable in relationship to the structural conditions with which rules interfere, through a correlation between structural variables and regulatory variables of a situation.[15]

The significance of such a correlation however implies assuming a *multidimensional* conception of rules, excluding their confinement into a formal-juridical dimension, and emphasizes the amenability to coevolution of their configuration. Regulatory systems are thus defined as systems of *rules-in-use*, i.e. as configurations of non-static nor explicitly formalized rules, but of rules defined by their operational and contingent linkage to an action situation; as such, their operational character results from the contingent sharing of their symbolic-cognitive implications. It is therefore not by chance that such a reinterpretation of the notion of *working rules*[16] emphasizes the central role of forms of *shared knowledge* and of the dimension of experience in defining the operational character of rules:

> In other words, working rules are *common knowledge* and are monitored and enforced. Common knowledge implies that every participant knows the rules, and knows that others know the rules, and knows that they also know that the participant knows the rules. Working rules are always monitored and enforced, to some extent at least, by those directly involved. In any repetitive situation, one can assume that individuals come to know, *through experience*, good approximations of the levels of monitoring and enforcing involved. (Ostrom 1990, p. 51; emphasis added)

Secondly, attention to rules is redefined as a multilevel strategy of inquiry. In this direction a three-tier distinction has been proposed (Kiser and Ostrom 1982) between *operational, collective choice*, and *constitutional choice* levels. Such an articulation of rules does not coincide *in toto* with statically defined settings of actors and situations; rather, the need for articulation derives from methodological questions concerning the understanding of *institutional change*. Rules may be represented in this sense as systems 'nested' into further systems of rules, in a relationship which defines their possible paths towards innovation.

As such, the relationship between *operational rules* (rules directly influencing everyday action in situations of conjoint action), *collective-choice rules* (rules determining the definition of operational rules), and *constitutional-choice rules* (rules defining the conditions of choice itself) may be analytically conceived as an ordered sequence of regulatory levels, activities and processes. Rules are accordingly articulated at different levels related to their different relative degrees of institutionalization.

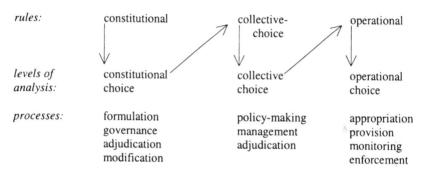

Figure 12.1 Linkages among levels of rules and levels of analysis
Source: Ostrom (1990, p. 53)

Isolating these levels, however, is an analytical stylization (fig. 12.1). Not only are rules not fixed in relationship to their outcomes; they are not even fixed in relationship to knowledge, and *vice versa:*[17] the internal articulation of their implications of a symbolic-cognitive order bears evolutionary features which require further levels of analysis:[18]

> The rules affecting operational choice are made within a set of collective-choice rules that are themselves made within a set of constitutional-choice rules. The constitutional-choice rules for a micro-setting are affected by collective-choice and constitutional-choice rules for larger jurisdictions. Individuals who have self-organizing capabilities switch back and forth between operational-, collective-, and constitutional-choice arenas, just as managers of production firms switch back and forth between producing products within a set technology, introducing a new technology, and investing resources in technology development. (Ostrom 1990, p. 50)

The definitional condition of rules-in-use itself, as well as of their application and of their monitoring in the course of action, makes them therefore an outcome of a reflexive process: but this process is not limited (if not in the framework of needs for a space-time reduction of analytical dimensions) to the sole operational dimension.

Let us exemplify this, in terms of a 'real world' context, by observing the interaction between different dimensions in the framework of concrete action situations (fig. 12.2).

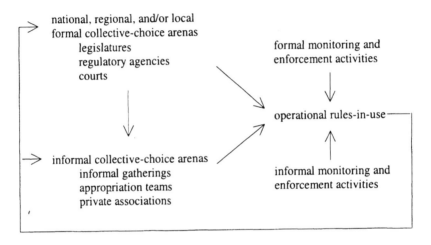

Figure 12.2 Relationships between formal/informal collective choice arenas and operational rules
Source: adapted from: Ostrom (1990, p. 53)

Focussing on a pragmatic dimension of action, i.e. on contexts of an operational type, the environment defined by operational-choice rules may be seen as an environment of operational-rules-in-use, subject to influence from a level of definition of collective-choice rules. In this practice-centered perspective, the constitutional dimension of choice and of rules may not be explicit; it is however present, i.e. not only influent, but also acknowledgeable, at different degrees of formalization and institutionalization, in the articulation of the possible forms assumed by collective choice: the field of collective action is in fact tendentially articulated in a continuum of arenas characterized by different levels of formality or informality, with different degrees of relationship with arenas of constitutional choice. The internal composition of the field of collective action and choice (schematized here as a duality between formal and informal) is itself a 'nested' system of relations with the constitutional dimension of choice, which is articulated in diverse forms and through which different forms of potential relations are instituted between rules of constitutional choice and the level of rules-in-use which pertain to operational choice. In the framework of these typically non-linear relations, differential opportunities are given for a retroaction between different forms of rules and choices.

A distinction of rules in a perspective of action is thus an analytical construct, rather than a connotation of distinct ontological spheres of regulation, and their relative rigidity or flexibility of relations is partially a matter of choice:[19] it is a phenomenon correlated to the qualitative connotations of the relational forms which are constituted, or acknowledged, between different levels of rules.

A specific constructive and project-like dimension of the relationship to rules thus becomes apparent. The condition however for rules to reciprocally express both this constructivist and project-like dimension is a methodical suspension of the reified existence of rules; and this presupposes systematic and interactive forms of disposition of rules and of learning about rules by the actors through their very enactment.

Networks

The Networking Potential of Weak Ties

Network analysis represents a crucial perspective in connecting micro- and macroanalytical strategies of inquiry in the context of a theoretical paradigm tending to reframe relations between structure and action beyond symmetrical forms of reductionism, whether in the form of a hypo- or of a hypersocialization, in, considering the institutional dimension of social processes; at the same time, it represents a fundamental intermediate level of inquiry for the understating of processes of individual mobilization towards collective action (Melucci 1987).

In addressing the role of social networks in constraining as well as in enabling action, the research tradition of network analysis has focussed on two main conceptual strategies, which may be defined respectively as a *relational* and *positional* strategy (Emirbayer and Goodwin 1994).

The *relational* strategy, grounded on a paradigm of social cohesion or integration, focuses principally on direct relationships between actors, tending to an explanation of behaviors and social processes related to aspects of social connectivity taken *per se*; among its key notions are therefore the density, strength, symmetry, and relative degree of ties. From this perspective, dense, strong and highly insulated networks are interpreted as factors which tend to develop subcultures as well as 'strong' forms of social identity; at the same time, however, the role of weak ties in bridging structural holes between groups within networks is highlighted as well as the crucial influence of their configuration on the trajectory of interaction processes. With respect to the nature of relations into which their action is inscribed, this research strategy thus interprets the identity of and the opportunities for actors as a function of their specific forms of embeddedness into existing patterns of social bonds.

The *positional* strategy on the contrary focuses principally on the nature of ties of actors not among themselves, but with respect to third parties; this strategy tends to account for the nature of behaviors and social processes in terms of patterns of relations which define the actors in their relation to all other actors in a social system. The prevailing perspective is therefore that of systemic integration; however, in its framework emphasis is placed mainly on aspects of structural equivalence, i.e. on symmetry and sharing of equivalent relations among various actors with respect of third parties, to which a decisive role is attributed for the understanding of individual and collective behavior. A relevant aspect of analysis thus focuses on the identification of the specific *position* or *role* occupied by a patterned group of actors (i.e. by a *block*) in a system in its whole (*blockmodeling*).

The work by Granovetter (1973; 1985), originally situated in the former field of approaches, but progressively developing 'positional' implications through the exploration of the notions of *institutional fields* or *spheres*, represents one of the founding contributions to the development of the notion of *embeddedness* introduced by Polanyi (1944; 1957) towards an analysis of social networks. Moving from the analysis of interactions at the microlevel, his research explicitly aims at extending beyond application to relationships of a dyadic type and to small groups confined into defined organizational or institutional settings, and at offering indications concerning the dynamics of connections and relations between microlevels and broader phenomena like diffusion, social mobility, political-ideological organization, and social cohesion, thus constituting a bridge between microanalytical and macrosociological levels of inquiry.

Granovetter's contribution is relevant to our scope mainly in a perspective of analysis of processes of transscalar diffusion, through different forms of social relations, as well as for its implications in a perspective of mutual learning and coevolution among social networks.

The microanalytical unit of inquiry is represented by the nature of interaction, analyzed in its function and potential in constituting complex and integrated networking relations according to a basic distinction between *strong* and *weak* ties.[20] The role of weak ties is strategic as, through weak ties, a much broader potential of connection among members of different small or restricted groups is enacted than through strong ties, which for their part tend to self-referential concentration inside the particularity of small groups.[21]

A key function is attributed to the concept of *bridges*, intended as lines of a network which define the only possible relational path (functional, in the simplest possible example, to the exchange of a unit of information) between two points or nodes, i.e. between the units of a dyadic relation. The extended application of the notion of bridges to triadic forms of relationship, i.e. to the constitution of multiple networks and of their modes of correlation, highlights the central role of weak ties in understanding *diffusive* and *evolutive* aspects of complex forms of network relations. While both weak and strong ties may assume the role of bridges, the relational connotation they assume here is in fact different: strong ties, according to their definition, may constitute bridges exclusively under the condition of the absence of further strong ties on the side of the individuals constituting this relation; this phenomenon, typical of small groups, and an expression of their relative insulation as well as of the differential, quasi-oppositional form of their identity, is however much less frequent compared to the weak relational forms characterizing complex societies. All bridges, under these conditions, assume the relative shape of weak ties. The strategic potential of weak ties lies in their ability to substitute for strong ties through the constitution of 'local bridges', i.e. of network-like trajectories substituting for absent features of vicinity and intensity of relationships with shorter and more efficient alternative paths.

Three main consequences follow. The first and most obvious is that any form of diffusion may reach a greater number of individuals and groups, and cross greater 'social distances', through weak ties; the second is that the nature of social relations tends to become less and less dependent on a (hypothetical) explicit definition of the identity of individuals and groups, while more and more connected to its relational definition, in strict connection to the structural configuration of networks; the third, referred to the intertemporal dimension of relations, is that the possibility of interactions and of their iteration represented by the substitution of strong ties by a system of weak ties enables the development of different morphological paths, and introduces multiple possible configurations of networks.

The access of individuals or groups to an enlarged system of opportunities, defined by different forms of material or immaterial, power- or knowledge-based resources, appears possible only through the networking connection to other individuals or groups situated at different positions and endowed with different resources; this kind of connection, crossing different social circles, is by definition only possible in the form of a 'weak' tie, i.e. independently of forms of affiliation, belonging and internal dominance of statuses and roles, whilst the prevalence of ties of a strong kind, depending on forms of obligation within distinct social circles, tends to close relationships inside their particularistic criteria for defining networks, roles and rules of relation, thus limiting access to aggregate social opportunities.[22] The strategic character and the pervasive diffusion of information in the context of extended network patterns of relations thus attributes a crucial importance to the ability to build bridge-like relations among different groups. Networks based on weak ties therefore represent relatively open systems of relationships, amenable to the constitution of 'networks of networks' connected through ties based on low levels of institutionalization. This relative degree of openness defines the field of alternatives and opportunities for their development, while highlighting their relative elusiveness with respect to a temporally and situationally bound definition of intentionality.

Networks of Interaction and Processes of the Constitution of Identity

Interconnecting networks enacted through weak ties, i.e. through weak forms of institutionalization of interactions, implies both enlarging relational networks among individuals and groups and entering interactional contexts characterized by multidimensional forms of institutional determination.

A multidimensional perspective on network analysis draws attention to the role of flows and patterns of relations in defining the nature of interactional processes: it highlights the role of the multiplicity of networks in defining complexity and evolutive potentials of interactions. In this, it also highlights the mutual determination through which networks and role structures affect the intertemporal dimension of the definition of the actors' volitions and identities (Pizzorno 1986; Melucci 1991).

Research on dynamics of recruitment and participation in social movements represents a significant combination of relational and positional elements of network analysis with the consideration of processes of reframing and of identity formation. Authors like Snow (Snow *et al.* 1980; 1986), Melucci (1987) and Tarrow (1998) in particular adopt a polemic stance with respect to approaches to social movements based on understandings of individual participation which underestimate the role of the symbolic-cognitive and interpretive components of social experience. In the first instance, these positions may be traced back to 'structuralist' theories of resource mobilization, and to "the meta-assumption that this [ubiquity and constancy of mobilization grievances] exhausts the important social psychological issues and that analysis can therefore concentrate on organizational and macromobilization considerations" (Snow *et al.* 1986, pp. 465-6), ignoring or taking for granted questions of alignment and framing. Symmetrically, a comparable attitude is to be found in culturalist and sociopsychological approaches, putting excessive emphasis on subjectivist and voluntaristic variables, as a static assumption of individual motivations, and excluding the consideration of patterns of the embeddedness of actors in networks of social ties. In fact, a clear motivation to adhere *per se* to a context of interaction, defined *a priori* of interaction itself, is rarely to be found; conversely, with the varying of activities and of modes of involvement, interests for participation and modes of participation also tend to vary:[23] "the 'motives' for joining or continued participation are generally emergent and interactional rather than pre-structured" (Snow *et al.* 1980, p. 795). Motivation to participation can neither be seen exclusively as an individual nor as a discrete variable: it is rather constructed and consolidated through interactions and in the framework of their process-like dimension (Melucci 1987), whereas "rationales for participation are both collective and ongoing phenomena" (Snow *et al.* 1986, p. 467).[24]

In this sense, the attributes of relational networks as microstructural elements of interaction constitute at least as important factors in defining paths of participation and cooperation as dispositions do. Preferences, dispositions, and strategies must thus be interpreted according to a network perspective in terms of the ties of a normative, instrumental and affective order between actual and potential participants.

Contrary to conceptions which see the outcomes of social movements as functions of the definition of objectives and of the group's ideology (thus binding them to the sharing of defined 'values'), this conception understands processes of collective mobilization as phenomena subject to significant variability according to differences in relational structures. Research on the networking conditions for collective engagement is furthermore important for the relevance it attributes to phenomena of a symbolic-cognitive order. The correlation between identity and pre-structured cognitive statuses of actors and the structural conditions of interaction is interpreted in terms of a dynamic and coevolutive process of *frame alignment*, i.e. of the alignment between individuals' or groups' frames and *collective frames*. In its complex nature as a learning process, "frame alignment [...] is a necessary condition for movement participation, whatever its nature or intensity, and that is typically an interaction accomplishment" (Snow *et al.* 1986, p. 467). At the same time, it is a condition for the reproduction and coevolution of interaction forms.[25]

Networks of Interaction and Networks of Meanings

Network analysis has become a key contribution to understanding the essentially process-like character of social relations, as well as the influence exerted on their forms of structuration by preanalytical and meta-intentional elements of agency and social communication, like patterns of discourse and cognitive maps. This contribution bears some important implications. The first is a rejection of temptations towards a *reification* of its subject of analysis, implicit in structuralist models of the relationships between social structure, the cognitive universe, and action.[26]

While rejecting an identification between social ties and their concrete entities, as well as an identification between their logic of constitution and the logic of constitution of their subjects, network analysis has moreover shifted towards a multidetermined conception of social action. In this perspective, human agency implies the ability of knowledgeable actors placed into socially embedded structures to appropriate, reproduce, and innovate the cognitive determinations and the material conditions of action which are given in a certain situation with regard to their personal and collective ideals, interests and commitments. But this does not make the concept of human agency coincident with that of social action: social actors are much more and at the same time much less than mere 'agents', for their empirical action is determined in multiple ways, i.e. it is not driven only by intentionality, but simultaneously structured by its belonging to various 'environments of action', such as cultural environments and social networks (Emirbayer and Goodwin 1994). And this simultaneous and never definite reciprocity of determinants between action and the dynamics of social structures is the reason, as Crespi (1989) would say, for the 'unfindable' character of action.

Network analysis in this sense may constitute a useful explanatory framework for social action only if connected to models of action which comprise metaintentional and metarational factors of collective identity, such as cultural factors, among the determinants of individual action. Along this line, transcending structuralism towards a theory of structuration, a growing importance is attributed in the framework of network theory to the interpretation of the active role and of the potential transformative impact of the constitution and evolution of cultural idioms, representations, and forms of learning.

The overcoming of residual conceptions of structural determinism thus favors a deep transformation in the conception of networks and of their relations to action: the latter no longer appears conceivable as a product of the networks' parameters and constraints, but is rather conceivable as the 'dynamic face' of networks. The notion of network is thus detached from a reified conception of structural relations, and defined rather as a field of opportunities in a permanent tension between the multiple elements defining an action situation and the action which reproduces and modifies them.[27]

At the same time, the notion of social network transcends its interpretive assumption as a factor of the determination of action and as a structural principle of the cultural universe, and may be intended as a 'network of meanings', in the framework of which its own conditions of existence are coevolutively reproduced.

The circular tension between structure, symbolic-cognitive universe, and action thus becomes a central interpretive element. In this perspective, the mutually constitutive character of culture and social structure may be interpreted as an outcome of the concrete interrelation between cultural formations and social structures on one side, and actors which reproduce and transform them, on the other. Reference is made here primarily to a conception of culture as a *category of action*, no longer conceived as a constraining framework of values, but as an enabling and constitutive factor of strategies of action, through the very *enactment* of which, in the framework of concrete action situations, their opportunities for persistence and their evolutionary path are defined (Swidler 1986; Douglas 1986; 1992).

Roles

Identity of Actors, Plurality of Roles, and Practices of Positioning

The fertility of a reflection on the dimension of identity and on the actors' roles in the constitution of interaction processes is strictly dependent on the conception of social roles which is assumed.

In order to be relevant for our analysis, reference to role theory must in the first place overcome structural determinist conceptions of social roles typical of most traditional approaches. According to hypersocializing macroanalytical premises, in fact, role theory preceding the 'cognitivist revolution' in sociology depended on a rather static and tendentially hypostatized understanding of the social status of individuals or groups. The notion of roles was accordingly intended as a form of expected behavior related to the social position defined by the concept of *status*, and is assumed in an essentially *conformative* sense, Apparently, this was at the expense of an understanding of the actual dynamics of behaviors and of an acknowledgement of the flexibility and of the negotiated nature of roles which may be empirically found in the everyday experience of social life.

At the same time, a conception must be rejected which, while acknowledging the microanalytical foundations of behavior as well as an articulation of social dispositions which defines the notions of role and status along a *continuum* related to the dialectics between the dimensions of individuality and the dimension of socialization of experience, interprets their degrees in the form of a tendential dualism between an *institutionalized* and a *non-institutionalized* dimension of social relations.

This aspect is well illustrated by the ambiguity of the notions of status and role in Parsons' work, where the duality of terms seems to correspond to the tension between a static, structural-functional, positionally determined setting of expected behaviors, and a dynamic system of intentional, purposive behaviors, without however an analysis of their mediation through concrete forms of interaction. It is similarly to be found, albeit on a quite different level, in the perhaps richest and most fertile contribution to the understanding of relations between roles and forms of interaction, in the tragic declination of the dramaturgical metaphor of the scene through which Goffman interprets concrete forms of everyday life.

In his studies, Goffman emphasizes the dimension of ritual activity and of a ceremonial of affirmation of the self which characterizes interactions, weakening the premise of intersubjectivity and of consensus on values, and rather equating interaction systems to self-referential games, based on the acquisition of credits and on the application of sanctions. The actors' commitment is thus essentially directed towards the self, rather than towards values, or towards explicit, concrete matters around which interactions develop, i.e. towards contingent aspects of performing a role (DiMaggio and Powell 1991). The radical relativization of the context of interactions and the centrality of situational aspects expresses an attitude based on a rejection of the idea of a *continuum* between social statuses and social roles, as well as of a notion of status itself distinct from the actual reality of interaction, which is typical of symbolic interactionism:[28] "the plurality and the hierarchy of roles in which the position of the single actor is inscribed, who plays contemporarily on different levels and scenes" is the expression of a "distance from the role [as] condition for the autonomy of the subject from social roles, allowing it to 'play' between different manifestations of the self, satisfying the requisites of stability of its identity" (Cremaschi 1994, p. 79).

The continuously renewed scene of interaction processes and of their situational conditions, in the framework of Goffman's microanalytical attitude, becomes however a measure of the void and of the distance between individuals and the institutional determinations of their action, which are experienced as a radical and complete form of externalization.[29] Such an interpretation has been put under revision by sociological approaches which aim at rejoining the microanalytical dimension of the dynamics of interaction with an interpretation of processes of social structuration.[30]

Giddens' critique of the concept of role reacts to exogenous predefinition of conditions for interaction which is implied, in its double version, in both Parsons' and Goffman's work:

> The notion is connected with two apparently opposite views [...]. One is that of Parsons, in whose theory role is fundamental as the point of connection between motivation, normative expectations and 'values'. This version of the role concept is much too closely bound up with the Parsonsian theorem of the dependence of societal integration upon 'value consensus' to be acceptable. The other is the dramaturgical viewpoint fostered by Goffman [...]. The two conceptions might seem to be contrary to one another but actually have a definite affinity. Each tends to emphasize the 'given' character of roles, thereby serving to express the dualism of action and structure characteristic of so many areas of social theory. The script is written, the stage set, and actors do the best they can with the parts prepared for them. Rejecting such standpoints does not mean dispensing with the concept of role entirely, but it does imply regarding the 'positioning' of actors as a more important idea. (Giddens 1984, p. 84)

According to Giddens, 'positions' must thus be understood as a result of 'positioning' and, similarly, the term 'role' as an element of a *stratified* system of definitions of roles, i.e. as a product of the interaction of social actors with the multiple contexts of structuration of their action, in order to understand the socially constructed nature of the 'prescriptions of roles associated to positions' displayed in processes of structuration.

The intersection between the immediateness of practices of positioning in everyday life and phenomena of regionalization constitutes the first element of stratification and differentiation of practices of positioning:

The positioning of actors in the regions of their daily time-space paths, of course, is their simultaneous positioning within the broader regionalization of societal totalities and within intersocietal systems whose broadcast span is convergent with the geopolitical distribution of social systems on a global scale. The significance of positioning in this most rudimentary sense is obviously closely bound up with the level of time-space distanciation of societal totalities. In those societies in which social and system integration are more or less equivalent, positioning is only thinly 'layered'. But in contemporary societies individuals are positioned within a widening range of zones, in home, workplace, neighbourhood, city, nation-state and a worldwide system, all displaying features of system integration which increasingly relates the minor details of daily life to social phenomena of massive time-space extension. (Giddens 1984, p. 84-5)

It is however mainly the institutional dimension of forms of positioning which, far from justifying their hypostatization, constitutes the conditions in the framework of which a monistic type of reference to social roles loses its meaning; in fact,

it is the intersection between these forms of positioning and that within the *longue durée* of institutions which creates the overall framework of social positioning. Only in the context of such intersection within institutionalized practices can modes of time-space positioning, in relation to the duality of structure, be properly grasped. (Giddens 1984, p. 85)

In that sense, the stratification of forms of behavior tends to an articulation which renders reference to defined sets of social roles, particularly in an interactional and experiential perspective, rather useless:

As its dramaturgical origins indicate, it is useful to speak of role only when there are definite settings of interaction in which the normative definition of 'expected' modes of conduct is particularly strongly pronounced. Such settings of interaction are virtually always provided by a specific locale or type of locale in which regularized encounters in conditions of co-presence take place. Settings of this sort tend to be associated with a more clear-cut closure of relationships than is found in social systems as a whole. (Giddens 1984, p. 86)

This *excursus* on Giddens' notion of positioning illustrates the tension between assumptions about social roles and the experience of structured processes of social interaction, a tension in which the role of involved actors appears to be stretched between that of a precondition for effectiveness and that of an issue at stake, whereas the reconstitution of roles potentially involves its promoter to the same degree as its players.[31] The definition of roles, for this reason, necessarily stands in a tension with the definition of forms of interaction. Definitions of roles are part of the very definition of identities and preferences of strategic actors which is – more or less explicitly – at stake in the game which develops as conditions for structured forms of interaction are displayed.

Multiple Role Structures and Multiple Networks of Interaction

Friedland and Alford (1991) have criticised the assumptions of 'classical' sociological theories of roles on the basis of the adherence to a principle of duality between individual action and institutional structures. Significantly, and with analogies to Giddens' conception, their position refers to a conception of multidimensionality and of an intimate contradictory character of the institutional universe, as well as to the idea that the cognitive universe and the symbolic constructs which define an institutional context (and, along with it, pertaining structures of roles) are continuously subject to redefinition through the belonging of social actors to different domains of institutions and roles. In this sense, the bi-dimensional conception of roles typical of a certain tradition of social theory may – and not by chance – be critically compared in its explanatory shortcomings with another dominant paradigm of relations between individuals and society, the *rational-actor model*:

> Role theory abstracts the role from the person and the institutional memberships that he or she must manage. Because humans live across institutions, it is necessary, unlike in role theory, to specify the institutional conditions under which individual behavior can be explained by a persons' role as a worker, voter, or lover. Conversely, rational-actor models which generalize microeconomic theory to all institutional arenas abstract person from role, assuming an egoistic, calculating actor that can be specified independent of the multiple roles that constitute the self. Because the meanings and relevance of individuality and rationality depend upon the specific institutional context, it is also necessary to specify the institutional conditions under which it makes sense to analyze individual behavior in these terms. (Friedland and Alford 1991, p. 25)

The multidimensional conception advocated by Friedland and Alford is one of the main contributions introduced into role theory by Merton. Its innovative aspect consists in the broadening of the conception of role itself. Based on the eminently interactional conception proposed by symbolic interactionism, Merton extends the definition of the nature of a role to the ensemble of individuals with whom the bearer of a role is involved: the concept of *role-set* thus derived indicates "that complement of role relationships which persons have by virtue of occupying a particular social status" (Merton 1957, p. 369), whereas not a single role, but multiple roles may be connected to a single social status, and different role relations may be constituted with respect to other roles which are situated at different places (statuses) of the social structure.

A role-set, according to Merton, is eminently a combination of role relations proper to a single social status. It is therefore not by chance that contributions to role theory inspired by Merton have stressed the need to overcome a conception of role which, after all, still appears defined by an essential limitation of the typology of possible interactions. This conception remains basically restricted to forms of relations between defined statuses (or, in other terms, to bilateral relations between different social circles), ignoring for example aspects of interdependence and interaction between members of the same social circle: the relation which defines the role is thus reduced to a simple form of reciprocity, and the definition of status remains immune to an examination of its potential articulation and internal diversification (Znaniecka Lopata 1991).

As has been noted, the tendency to underestimate aspects like strategic behavior and personal freedom, which would broaden the analysis of the social function of roles in a pluralistic and multidimensional sense, descends from an interpretive perspective emphasizing issues of social order and disorder (L. Coser 1991). According to Merton, individuals tend to find themselves in structural situations which force them to confront a variety of role partners: these partners of the role-set of an individual usually reflect different positions of status, and therefore also incorporate different and often incompatible expectations about the dispositions and behaviors of the others. In the absence of a social mechanism for the treatment of contradictions and conflicts which may thus arise, society therefore faces a perspective of disfunctionality.

However, if a multidimensional institutional perspective is introduced, paying attention to the contextuality of fields in which relations are defined, the notion of status intended as a univocal principle of social identification loses meaning, opening to a possible interpretation of its plural dimension and of a different kind of tension with the notion of role.

Roles, on the one hand, may be at least in part culturalized, i.e. 'institutionalized', in relation to specific contexts, and may thus constitute stable and durable – albeit always partial – factors of anticipatory socialization. The roles assumed by an individual at a given moment, however, are always set in a multidimensional universe: i.e. they may be multiple, not necessarily completely unconnected, but also not perfectly connected; accordingly, the concept of *role-cluster* introduces a principle of identification of roles which are in the domain of a common institution or institutional field, and therefore present greater affinities than others in the domain of different institutions or institutional fields. The complexity of social life consists precisely in involvement in multiple roles, a conception which recalls the notion of *life space* introduced by Lewin (1948), intended as a multidimensional condition constituted by the assumption of multiple social roles in the framework of diverse institutional dimensions. The complexity of life space may be understood in this sense as a condition for the development of considerable forms of complexity of personality and of intellectual flexibility: it may in fact lead occasionally to peaks of individual involvement in more than one single set of role relations. The ability to assume diverse roles proper to other participants of other social circles thus strengthens the capacity for empathic comprehension and mutual recognition, and becomes the condition for social practices of *framing* and *collective learning*.

If assumed in terms of systems of relations between social individuals and participants of different social circles, rather than in terms of expectations of conformance to static settings of positions, role theory may therefore bear a potential in exploring social life (Znaniecka Lopata 1991) as well as in orienting action.

Particularly by taking distance from the 'defensive' aspect of Merton's conception, and by introducing a 'creative' perspective, contributions of authors like Gouldner and R. Coser have stressed the relevance of symbolic-cognitive implications and of processes of the constitution of identity and personality connected to the complexity and segmentation of social roles:[32] in these terms, the complexity of role-sets in modern societies is seen as a potentially liberating factor rather than as a factor of alienation.[33]

R. Coser (1975, 1991) has proposed an interpretation of the attachment to multiple roles along a line of research on the conditions for individual autonomy and creativity which helps highlight the connection between opportunities for learning and the development of identity and their structural conditions.

According to this interpretation, in analogy to Merton, "rather than having the individual confronted with ready-made social norms that are external, [...] individuals have to find their own orientations among multiple, incompatible, and contradictory norms" (R. Coser, p. 239, cited in: J. Blau 1991, p. 133). Settings of social roles and statuses are again seen as relatively independent but complementary structures; however, the ideal condition for mediating individualism and the social complexity of role relationships is identified in contexts in which no institutionalized status and no role relation exert an exclusive influence on the individual: in the language of network analysis, social ties prevailing in interactions have to be weak in order to allow the development of intellectual and behavioral independence.

A Provisional Synthesis: Multiple Role Settings as Structures of Weak Ties

This perspective, understanding the development of individual autonomy as a reflexive process in identity construction, embedded into its social and institutional conditions of possibility, is of great interest in relationship to the analysis of the possibilities of forms of structured interaction.

The development of conflictual expectations, the presence of segmented roles and of complex role-sets, the coexistence of contrasting rules in everyday confrontations, constitute, according to R. Coser, conditions through which the development of an autonomy of individuals in modern societies becomes possible; the multiplicity of roles, however potentially contradictory and incompatible these may be, is in fact a potential factor for articulation of the individual consciousness which is constituted around them.[34]

Such a perspective, which, in an explicitly optimistic way, identifies in the pluralism of social roles an opportunity for tolerance and for the development of flexible forms of democracy, nonetheless points to some of its preconditions, and to their problematic reflexes in terms of the institutionalization of social practices: among them, the freedom of choice of the individuals which bear roles and statuses, but also the flexibility of positions in the framework of complex role-sets, as premises for the capability of entering different forms of relation in respect to different role partners. For these reasons, aspects of the complexity of role-sets and aspects of the differentiation of roles in which social actors are involved throughout processes of interaction constitute opportunities for the development of individuality, but at the same time constitute structural preconditions for learning processes which sustain it.

Under these conditions, the nature of interaction processes is strongly influenced by the connotations of communication processes and forms of argumentation: the ability to make use of forms of abstract thought and of translation codes, i.e. of elaborate and multiple linguistic codes, amenable to being shared through interaction and not exclusively belonging to different social circles, becomes a necessity for the intersubjective opening of processes and for the ability to conduct interactions among a variety of different role partners, adapting (in both communicative and strategic terms) to different domains of reciprocal expectations.

This theoretical approach points to the question of which structural and organizational conditions may grant the sustainment of such connotations of interaction forms, as defined in opposition to organizational forms based on hierarchy, functional-departmental fragmentation and sectoral segmentation, while safeguarding requisites of effectiveness and accountability which are expected to be granted in the public domain of democratic societies: i.e. expressing an orientation towards coordination, self-evaluation, and accountability of choices, but also towards persistence and the constitution of institutional identity and memory.

Another decisive factor for the constitution of forms of action based on complex settings of roles is the nature of correlations and interactions between different social components or, in the terminology of policy analysis dealing with multiactor interactional settings, between the different actors of multiple interconnected decision-making arenas. It is possible in the first instance to refer those conditions to the notion of weak ties proposed by Granovetter (1973). The effectiveness of the tie in constituting and reproducing complex role systems is inversely correlated to its 'strength', which in this sense represents a negative, inhibitory factor for diversification and for forms of integrative learning: whereas strong ties indeed tend to favor homophily among individuals and groups, thus enhancing *de facto* the dominance of internal factors of identity and the autarchy and relative insulation of social circles, weak ties allow for the development of differences among individuals and groups inside the same line of relations, therefore enhancing the intersection of social circles and promoting the constitution of complex role-sets.

It is therefore particularly in relationship to the nature of interactional ties that high levels of social differences (i.e. of the actors' preferences and identities) in the context of systems of structured interaction may be interpreted in their potentially integrative rather than disintegrative function.

Such arguments however call for an account of structuration processes in their evolutive dimension, interpreting their opportunities as outcomes of a dynamic process, self-reproducing its constituting practices and forms of knowledge. The risk of a determinist, 'structuralist' emphasis on structural components must accordingly be avoided: structural factors cannot be reduced to 'context' variables, to factors of an ecological order, but imply an endogenous and phylogenetic conception of institutionalization processes, intended as outcomes of collective action.

J. Blau (1991) has exemplarily addressed this question through an analysis of forms of structuration of weak ties, intended as overlappings of independent subnetworks, characterized by different statuses and roles of actors, in the framework of complex systems of iterated interactions. The nature of this model of analysis makes it possible to sketch a provisional interpretive synthesis of the notions which have been addressed in the previous pages.

The starting question is how weak ties may be sustained in the framework of an organization, at the same time granting satisfaction of the criteria of coordination, accountability, and efficiency proper to a functional structure. Complex networks constituted by overlapping subnetworks, organized into independent operational or thematic areas, while tendentially obliterating the formal structure of an organization, maintain a dimension of integration and coordination through the copresence of multiple actors' roles. The belonging of actors to diverse subnetworks, constituting independent areas and rationales of activity, is the condition for the relative independence of the statuses and roles which they occupy in each of them; at the same time, the overlapping of network memberships constitutes a complex pattern of relationships which enables integration among the different categories of the actors' positions. Throughout this complex web of relations, expressions of conflict may arise frequently and explicitly, but never produce forms of fragmentation and fracture: the multiple commitments of actors produce continuous forms of mutual realignment of cognitive frames of reference through practices of self-monitoring; and "[t]hrough this process whereby alignments continuously form and reform through exchange and co-optation, institutional life derives its tenacious integrity" (J. Blau 1991, p. 141).

It is interesting to link this representation, directed towards exploring the conditions for endogenous forms of dynamic reorientation, to discussions on the potential of cooperative networks in contributing to the effectiveness of forms of non-hierarchical coordination. The linkage with the notion of roles consists in the nexus between the constitution of *embedded* forms of *positive* and *negative coordination* and the role played by *trust* in the relationship between the degree of inclusiveness of decisional contexts and the overlapping of multiple roles and networks. According to Scharpf's (1994b) interpretation, positive coordination appears highly facilitated when relationships (in his argument: of a negotiated character) occur among parties characterized by cooperative attitudes: generalized trust and commitment to be worthy of trust indeed do reduce opportunistic strategies and negotiators' dilemmas, facilitating the possibility of reducing distributive conflicts through general agreements on rules of distributive justice. But these beneficial effects are only given in situations of coincidence between (predefined) needs of coordination and (predefined) structures of cooperative networks. This element thus introduces a factor of structural determination and constraint: the volatility of interactions in fact generally hinders perfect adaptation between actual constellations of problems and permanent networking structures. In conditions of the emergence or presence of new actors, higher transaction costs and higher difficulties in reaching agreements tend to limit the dimension of the group in which negotiated coordination may be successful: in this sense, positive forms of coordination face the limits of the effectiveness of trust.

Negative coordination may gain importance and effectiveness precisely in connection with conditions of the intersection and overlapping of cooperative networks and of patterns of factual interdependence: i.e. in conditions which would call for multilateral forms of positive coordination, and where a formal order based on the demise of hierarchical structures offers limited defense to the threats of individual actions.

This aspect may be exemplified by the frequent case of non-coincidence between structured contexts of negotiation and the structure of cooperative networks, whereas several of the actors involved have cooperative ties with some third party not directly involved in negotiations, and therefore suit multiple roles. When this occurs, according to Scharpf, possible manifestations of collective egoism connected to positive coordination are compensated for by the individual self-interest of negotiators, who must safeguard relations of high-level trust with external partners. Motivations of these actors towards maintaining a double orientation are therefore strong: they take the shape of *mixed motivations* oriented both towards the best negotiated outcome and towards the consideration of the impact of potential agreements on parties not directly involved. In this sense, multiple relations of roles connected to embedded forms of negotiation appear to be both a constraint exogenous to the situation, which limits the direct scope of action of the actors, and a potentially endogenous creative opportunity for integrative solutions and for a patterned evolution of the actors' preferences.

Notes

1   Examples of these positions may be found in the work of Taylor (1976; 1987) and Hechter (1987); very interesting in this connection is the confrontation on cooperation and social regulation between Taylor and Coleman on one side and Hechter on the other, which may be found in Cook and Levi (1990). See also the discussion

presented by Lanzalaco (1995, pp. 48-52).

2   'Regulatory' should be understood here in particular in the sense of the French term *règlementation*. On the meaning of this conceptual distinction in defining the notion of regulation, see ch. 1, note 4.

3   This revision of the role of rules in the constitution of social behaviors apparently also pertains to the definition of 'logic of appropriateness', which constitutes a paradigmatic reference of March and Olsen's (1989) new-institutionalist turn.

4   An interest in this aspect is central in the work of several further authors, e.g. Granovetter, Scharpf and Shepsle, or like Elster and Boudon, who try to reconcile the extremes of *atomistic* and *holistic* interpretations of social phenomena, and point to the need to de-dramatize dichotomies between *rational* and *irrational* behaviors as well as between *intentionality* and *unintentionality* (Rositi 1986); and it is central in Ostrom's contributions to new-institutionalist debates (1991; 1995).

5   Rowe in fact understands this as a positive distinction, contrary to the normative distinction between *act-utilitarianism* and *rule-utilitarianism* proposed by Hodgson in the framework of his conception of economic ethics; like Rowe, Hodgson moves from the effects of *time inconsistency* of action directed by utilitarian principles and of its outcomes: "certain good consequences depend on the existence of *expectations* of actions, and that, under certain conditions as to knowledge and rationality, an agent's avowedly acting upon the act-utilitarian principle could preclude such expectations and such good consequences" (Hodgson 1967, p. 85, cited in: Rowe 1989, p. ix). However, the explanatory limits of principles of methodological individualism are addressed by Rowe from a different point of view than the postulation of an ethical principle of utilitarianism, tending to show the economic rationality of perfectly altruistic behavior; instead, his adherence to principles of individual self-interested rationality of action tends to bring back the normative question ('why should rules be followed') to an essentially positive level ('why are rules followed').

6   Rowe builds in particular on discussions of the paradox represented by commitment to forms of altruistic behavior and by its strategic foundation as proposed by Schelling (1960). In game theoretical terms, the issue has been addressed in the so-called 'Stackelberg-equilibrium' model, where a *leader* A acting first, and who reveals through this action elements of his or her strategy, benefits from the advantage of potentially influencing through his or her *precommitment* the subsequent action of B, who on the contrary assumes the action of A as exogenous, being therefore bound to choose his or her action on the basis of expectations of a reaction by A. The *pre-commitment* by A thus constitutes a *strategic constraint* to B although B's action is actually subsequent to A's; the iteration of interactions however produces a form of dynamic equilibrium of an endogenous kind among the strategies of actors. The paradox of self-constraint turning into a resource recalls arguments by Heiner (1983) and the moral of *tit-for-tat* discussed by Axelrod (1984). It is however interesting to draw attention on the analogy with the interpretive problem of the role of the *institutional entrepreneur*: in this perspective, identifying the *institutional entrepreneur* with the leader A, the 'third party' or *enforcer* is itself involved in a learning process, and an equilibrium of games is conceivable only in endogenous and constitutively coevolutive terms.

7   In discussing the dialectic relationship between normative and conditional aspects, Giglioli (1989) introduces a criticism of theories of action based on an overemphasis of *Weil-Motive* at the expense of *Wozu-Motive*, i.e. of 'determining' over 'attracting' factors of action, which is shared by Crespi (1989). As is familiar, an embryonic conception of the constitutive character of rules, as a cause as well as a premise for

the knowledgeability of action, i.e. as a condition for the actualization as well as for the conceptualization of action, has been identified in the work of Weber; the meaning of rules in the understanding of action is in fact triple, according to Weber, i.e. that of a *causal determinant* of action, of a *heuristic principle* of causal knowledge of action, and of a *cognitive principle* of classification and conceptual comprehension of action (Ferrari 1986). This implicit conception of the constitutive character of rules and of the duality of motivations for action is the point of departure of the distinction between *Weil-* and *Wozu-Motive* developed by Schütz (1932) and is central to recent proposals for the reconstitution of a form of rationality around the idea of 'good reasons' (Boudon 1987) or around rational argumentation (e.g.: Elster 1991).

8   The importance should be underlined which in this conception is granted to the inter-subjective ackowledgeability of rules, as well as to the centrality which, in this sense, the concept of rules as communicative codes assumes in an ethic of discourse and in a theory of communicative action. This aspect furthermore introduces a fundamental correlation between following a rule (i.e. the effectiveness of a rule and its deontical content in terms of effective action) and the linguistic context in which a form of communication is perceived as a prescription and a behavior is perceived as governed by a rule. These, like many others, are aspects of a theory of rules which pertains to the *configurational* conception of rules to be addressed below.

9   Rositi (1986) underlines the centrality attributed to the concept of *manipulability* or *interpretive manipulation* of normative systems by phenomenology and by symbolic interactionism and ethnomethodology in terms of a research strategy making it possible:

a. to reintroduce certain types of strategic action into the framework of normative action: the literal meaning of a normative system does not represent only a constraint, but also a resource for actors in maximizing their private advantages, even in contexts where acceptance of norms is granted, or 'institutionalized': this, for example, weakens the opposition between rhetoric on rules and following a rule, recognizing on the contrary an intimate constitutive solidarity between the two;

b. to understand phenomena like rigorism (i.e. a rigid and/or literal interpretation of a normative system) and severity (i.e. overly expeditious application of sanctions) as mechanisms for the reduction of the interpretive *alea* of a normative system;

c. to understand the possibility of becoming 'deviant': in manipulable systems, the mechanism of exclusion does not intervene so early as to hinder the growth of deviant behaviors, as far as they are admissible as interpretive phenomena internal to the normative system.

10   According to this line of reasoning, Giddens (1984, p. 17-8) underlines the need for a revision of the characterizing elements of current notions of rules:

- the need to overcome a prevailingly formalized conception of rules, intending them as 'rules of the game'; if it is possible to understand them as 'rules of the game', so Giddens, this should be in the sense of the children's games as Wittgenstein discussed them, and to the extent these games may be assumed as a metaphor of social life;
- the need to overcome a monistic and isolated, 'singular' view of rules, as well as the attribution of univocal forms of correspondence of rules to behaviors or situations;
- the need always to pay attention to their adherence to a principle of power or, in other terms, to a strategic principle which ties them indissolubly to resources and concrete forms of practice;
- the need to conceive rules in the light of their constitutive correlation to methodical procedures of social interaction;
- the need to understand the coexistence in rules of aspects of the constitution of

meanings and of aspects of the sanction of social conduct: these in fact are not two distinct typologies of rules, but two aspects of rules, which refer to the constitutive relationship enacted in them and through them.

11   Considerations about the fundamental *constitutive* character of rules in the context of social action do not imply the loss of the fundamental heuristic and methodological function of a distinction between types of rules (Ferrari 1986). Rather, an analytical turn towards a *configurational* understanding of rules is a necessary path towards an experimental and normative perspective.

12   The concept of *inviolability*, i.e. of non-disposability to interpretive manipulation, is conversely one of the main criteria of distinction of constitutive rules proposed by Searle. The typical elements of a distinction between regulatory and constitutive rules may be summarized as follows (Rositi 1986, pp. 352-3):
      a. constitutive rules are those which define, create or 'constitute' new forms of behavior, which may be defined as "practical" in order to distinguish them from other types of intentional action; regulatory rules are those which intervene in regulating, organizing already existing behaviors;
      b. 'practices' cannot be described and/or enacted in the absence of knowledge and full understanding of the rules which organize them;
      c. constitutive rules are paradoxically inviolable, in the sense that a non-conforming practice, if ever thinkable, would fall outside their domain;
      d. in some cases constitutive rules provide for further rules for the continuation of a game, or of any practice characterized by them, after a breach of the rule (i.e. inviolability is applied to a system rather than to single rules);
      e. constitutive rules allow for the transformation of material fact into meaningful facts in the framework of practices: they bear a semantic and definitional dimension (i.e. in the strong version by Searle they have analytical-descriptive and quasi-tautological form), against the imperative or hypothetical-imperative of regulatory rules;
      f. the commitment to practices typical of constitutive rules is intentional and voluntary;
      g. a commitment to practices typical of constitutive rules implies acceptance of the entire system of rules which constitute them as that system (sometimes called a 'normative system of conjoint validity') is an ensemble of strictly interdependent and coherent rules.
      As has been noted (Rositi 1986), it is however difficult not to think of rules as constitutive and *at the same time* regulatory, i.e. connotated by both of these elements: there is in fact an 'intrinsical solidarity' between the concepts of norm and interpretation which leads to a questioning of the usefulness of such a distinction.

13   Mutti (1986) points to the illusory and (it might be added) irresponsible character of calls for 'stable' and 'certain' rules, intended as outcomes of a voluntarist or decisionist principle, in conditions of *uncertainty* and *ambiguity*.

14   While it is not possible here to discuss thoroughly the complex conceptual field spawning between the notions of *norm* and *rule*, the importance has to be underlined of a distinction between the notion of rule and a formalized system of prescription and sanctions in Ostrom's work: a logic inversion is, on the contrary, highlighted there in the relationship between the regulatory value of a prescription and its formal connotation assumed by 'naive' conceptions of social regulation: "[t]he term 'rules' should not be equated with formal laws', since 'enforcement is necessary for a law to become a rule" (Ostrom 1986, p. 6). Moreover, emphasis on the configurational dimension of rules in the framework of institutional phenomena, as Jepperson (1991) underlines, highlights the non-equivalence between institutions and norms. However,

the logic inversion thus introduced evidently bears a very different meaning from that form of discretization which, on the contrary, uses the notion of norm in order to underline the formal-juridical aspect of institutionalization: an example of this is given by Parsons (1951, p. 20), who contends that a norm is institutionalized only under the condition of its enforcement, i.e. if rewards and sanctions are attached to it. The relevant and problematic consequence which is drawn by some authors (Schotter 1981), which Ostrom tends to reject, is that norms constitute a component of institutions not subject to processes of self-policing. The solution to that question is seen on the contrary by Ostrom in the continuous and multidimensional configuration of the regulatory field and in the redefinition of the nature of *enforcement*.

15 Ostrom (1986, pp. 17, 19) proposes the following typology of correlations between structural variables and regulatory variables:
1. the set of positions to be held by participants / *position rules*;
2. the set of participants (including a random actor where relevant) in each position / *boundary rules*;
3. the set of actions that participants in positions can take at different nodes in a decision tree / *authority rules*;
4. the set of outcomes that participants jointly affect through their action / *scope rules*;
5. a set of functions that map participant and random actions at decision nodes into intermediate or final outcomes / *aggregation rules*;
6. the amount of information available at a decision node / *information rules*;
7. the benefits and costs to be assigned to actions and outcomes / *payoff rules*.
As we have seen, Scharpf (1989, p. 152) refers in particular to *boundary rules* and *decision rules* (here: *aggregation*) in his discussion of the interrelationship between decision rules and styles in defining the nature and effectiveness of public policy.

16 Commons (1957, p. 6) conceives institutions as defined by working rules which "tell[.] what individuals *must* or *must not* do (compulsion or duty), what they *may* do without interference from other individuals (permission or liberty), what they *can* do with the aid of the collective power (capacity or right), and what they *cannot* expect the collective power to do on their behalf (incapacity or exposure)".

17 The assumption of 'common knowledge' in this perspective bears a very different sense than in game theory (where it is functional to the analysis of equilibrium conditions, and where it is assumed as a context condition, *quasi* as a *ceteris paribus* clause) or in the idealtype of rational choice theory, where it is a corollary of the assumption of perfect availability of information: on the contrary, its emergence is intended here as a possible outcome of an experiential process, through which the coevolution of preferences and strategies of actors is produced. This assumption bears great importance for the notion of *self-policing convention*.

18 This is also the direction in which the contribution by Axelrod (1986) moves, which tends to a behavioral interpretation of the origin, sustainment and evolution of norms not on the basis of criteria of rational calculation, but of the operationalization of effective strategies, of which he underlines situational conditions as well as phenomena of reproduction and adaptive behavior, mediated through processes of learning which recall the symbolic-cognitive role assumed by metanormative aspects.

19 It is, in this sense, a matter of collective choice, but, as such, it is also a matter of dilemmas of collective action: the question, which is obviously decisive, recalls classical problems concerning the endogenous introduction and self-sustainment of forms of *self-organization* to be addressed in the next chapter.

20 The definition of *strong ties* given by Granovetter (1973) is associated with a combination (of a tendentially linear nature) of devoted time, intensity, emotivity,

intimacy and reciprocity; it is therefore correlated to relational factors which may be summarized by spatial (e.g. 'vicinity') and temporal metaphors (e.g. 'frequency'), but also connected to (non-necessary) voluntaristic aspects, e.g. 'reciprocal choice'.

21   Among the arguments of a 'positional' kind offered by Granovetter, I would like to stress the question of trust in the framework of social relations with an 'egocentric' connotation. While *leadership* is constituted mainly through personal forms of contact and transmission of reputation, and trust is a function of reputation, i.e. of the predictability of behavior of the leader, in the framework of networks based on 'high density', i.e. on strong interpersonal relations tending towards intensity and 'effectiveness' rather than towards 'extension', as in the case of *cliques*, motivations to the diffusion of the influence of trust are limited, and leadership tends to self-reproduce on the basis of a relative insulation of the members of the group from external forms of knowledge. The fragmentation of networks which derives from it therefore acquires a determinant role in inhibiting the diffusion of trust, and the nature of prevailing ties institutes an inverse relationship with its robustness as a factor of systemic integration. An analogy may be found in this with the formulation of second-order dilemmas of collective action (see next chapter).

22   According to Granovetter, the personal experience of individuals is therefore strictly connected to aspects of the macrolevel of social structure, which stand outside of their control, and the capacity of connecting aspects of the micro- and of the macro-analytical scale appear to be fundamental not only in a theoretical perspective of an analytical-descriptive kind, but also in a perspective of social intervention. This aspect is highlighted by the paradox by which the weakness of social ties, in sociological common sense, is frequently associated to alienation, while on the contrary it appears fundamental for enhancing social bonds and integration.

23   Development, persistence and reproduction of participation to collective action in the framework of a movement or a group, in Snow's terms, are thus contingent to the influence of network alternatives present in the field and to the intensity and frequency of interaction with members of the movement or group: "whereas the first factor determines whether one is structurally available for participation, the second factor gives rise to the rationale for participation" (Snow *et al.* 1980, p. 795).

24   Snow *et al.* (1986) underline how movements often demonstrate a great 'practical consciousness' of this aspect: participants spend a great amount of time in accounting collectively for modalities and motivations for participation, in a sensemaking process of a collective and evolutive rather than individual and discrete nature.

25   In the framework of this circularity of relationships, however, none of the possible forms of processes of collective framing is bound to succeed nor exempted from risks. Their development, persistence, and reproduction with respect to intentional or motivational types of reasons cannot be taken for granted: as variable in time, processes of framing are subject to ongoing processes of reassessment and re-negotiation, as "the reasons that prompt participation in one set of activities at one point in time may be irrelevant or insufficient to prompt subsequent participation" (Snow *et al.* 1986, p. 476). Interaction processes centered on practices of frame alignment appear in this sense as a hazard; and collective control on their vulnerable interpretive dimension becomes as, if not more, important than control of material and concrete types of resources for the processes. Once again we face the centrality of symbolic-cognitive aspects in sustaining collective forms of action.

26   Emirbayer and Goodwin (1994) talk in this respect of *structuralist determinism*, of *structuralist instrumentalism*, and of *structuralist constructionism*, as the three dominant forms of an implicit theory of action in the tradition of network analysis.

27   This apparently at the same time implies that the notion of network itself is endowed with a duality of meaning which hermeneutically encompasses the observer: it is a phenomenic reality, and at the same time a principle of evaluation and interpretation. A discourse on networks, in the framework of an inquiry into forms of social interaction as well as on other terrains, bears therefore in itself the principles of a normative orientation, in the form of an 'implicit project' (Dematteis 1995).

28   The conception of roles inspired by symbolic interactionism tends to understand them as interpersonal units of relations taken *in se* and *per se* for their internal relational nature, thus stressing the complexity of relations implicit in every single role relationship; situating social roles thus intended in a broader relational context, of an organizational or institutional kind, is seen as a subsequent analytical step. This form of reduction of roles to their essence as relations of an interpersonal kind makes it possible to understand the dynamic and process-like, constitutive rather than conformative aspects of the construction of roles through interaction, pointing to the production of meaning rather than to the constitution of social expectations of behavior; at the same time, it tends to overshadow processes in the institutionalization of roles, which is often hypostatized into a notion of *status* as an extreme degree of institutionalization of a role relationship.

29   The mask donned by Goffman's actors, while instituting a distance from the social role as a precondition for safeguarding the subject from alienation, represents an acknowledgement of a form of alienation: it is indeed in this dimension that Goffman's analysis, while emphasizing references to categories of an existentialist type, tends to distance itself from the comprehension of institutional factors as constitutive elements of behavior (see: Crespi 1989).

30   Thus, shifting from Goffman to Geertz, the metaphor of the theatre scene, expression of the hollowing-out of subjectivity and of the reduction of the individual to a mask and of its behavior to the mere dramaturgical function of 'indicality' of a pregiven structure of roles becomes an explicit expression of the social process of the constitution of subjectivity: as in Geertz's example, "art forms generate and regenerate the very subjectivity they pretend only to display" (Geertz 1973, p. 451; see also: Geertz 1980). Giddens (1984) himself takes over the metaphor in his conception of regionalisation processes and of 'regions' as 'theatres of interaction'.

31   This, by the way, is precisely something which almost never occurs, almost by definition, within models of mediated dispute negotiation.

32   Gouldner (1984) expresses his conception in the framework of a theory of intellectual creativity; this may however be extended to diffuse social contexts with a multi-cultural connotation. Based on a discussion of creative potentials related to the transcending of single linguistic codes and to the recursive translation among different codes, Gouldner indicates the condition for superior learning ability and intellectual productivity in the multiple cognitive perspective derived from this, and in an orientation towards trespassing the boundaries of a single tradition through confrontation with diverse traditions, according to a methodic intellectual principle of 'suspension of commonsense', of a phenomenological *coupure,* or *rupture.* A relevant aspect of this perspective is particularly its sociological rather than psychological rooting: the capabilities, it is argued, do not depend on qualities of the individual, "but rather on his critical assimilation of and unusual relation to established intellectual traditions, upon his access to *multiple* traditions, as well as his capacity to *switch* or translate from one to another" (Gouldner 1984, p. 205, orig. emphasis).

33   This conception bears a significant analogy with the analysis of the role of the intersection of social circles in the development of forms of freedom from pressure

and dominance of single societal groups in the work of Simmel.

34   This interpretive perspective of course does not fail to acknowledge that such condition may also provide for the insurgence of role conflicts and of forms of stress. However, considered as a dialectic construct, conflict is not seen here as a constraint but as a factor for the development of identity.

13 Institutionalization Processes and Collective Action

From the Dilemmas to the Opportunities of Collective Action

Research on self-organized forms in the management, production and distribution of common goods is a field in which some very interesting attempts have been made towards overcoming explanatory alternatives between macrostructural and microindividual variables in understanding institutional change. The possibility and conditions of emergence of new institutional forms of action are inquired into through an analysis of the constitution and evolution of structures of collective action in the concrete processes of their making (Melucci 1987). As a result of this strategy of research, renewed theoretical approaches to collective action have joined recent contributions to action theory in fostering a radical redefinition of the rationality of individual as well as of institutional action.

As a critical starting point of this line of research stands the mismatch between the axiomatic methodological function of rational actor models and their shortcomings in accounting theoretically for the multi-dimensional determinants of the actors' rationality, behavior and volitions which define action situations, i.e. in adopting a sufficiently 'thick' conception of rationality to be useful for understanding their situational connotations. In this sense, approaches to the question of collective action leaning on new-institutionalist theories (McCay and Acheson 1987; Ostrom 1990; 1994a; Ostrom, Feeny and Picht 1988; Keohane and Ostrom 1994) aim at reconciling assumptions of the individual rationality of actors with a demise of hypostatized assumptions of instrumental means-ends rationality, which appear unable to account for aspects like the evolutive commutation of preferences central to the explanation of stable forms of individual commitment to self-governed forms of collective action.

The necessary starting point of any discussion on the possibility and limits of forms of action coordination towards supraindividual ends is the classic definition of 'paradoxes of collective action' proposed by Olson (1965), along with its metaphorical or model-like variants, such as the image of the 'tragedy of the commons'[1] and its game-theoretical applications, particularly the 'prisoner's dilemma'.[2] Olson's famous thesis constitutes the most radical confutation of neoclassical assumptions of rationality (typically assumed to govern over the self-regulating mechanisms of markets) concerning the possibility of achieving quasi-natural social equilibrium and efficiency conditions as an aggregate outcome of individual choices based on principles of self-interested action rationality. Consequently, it has also represented the theoretical stimulus for an array of game-theoretical models inquiring into the possibility of the insurgence of social cooperation on the background of individualist rational choice assumptions and into its conditions and constraints (imperfection of information, strategic environment of action, and the like).

In its essence, Olson's thesis may be summarized in quite simple terms: in conditions defined by the presence of small numbers (i.e. scarcity of goods or of resources for their production) and of large groups, there are no reasons, in terms of economic rationality, for an individual orientation to contributing to a collective good: under such conditions, collectivities tend to assume the problem-laden feature of 'latent groups', subject to the opportunistic behavior of free-riders and to structurally unbalanced and suboptimal distributions of costs and benefits, such as to render in principle impossible the achievement of an efficient level of public goods.

From the point of view of rational choice theory, the sum of individual behaviors of rational actors leads therefore to 'irrational' collective outcomes. Olson's argument points to a fundamental shortcoming of methodologically individualist conceptions of rationality in understanding social phenomena, rooted in the very definition of such form of rationality: according to its assumptions, in fact, a self-interested rational actor should not have rational motivations to cooperation in achieving collective ends.

The 'wicked' logic of collective action delineated by Olson, and the dilemmas which have dominated reflections on the rationality of public action along with Arrow's theorem (Arrow 1963), is embedded *in toto* in an individualistic paradigm of the economic rationality of action. Along with it, however, it shares serious explanatory and heuristic shortcomings, which have been highlighted by various contributions in economic and organizational sociology. As R. Hardin (1982) summarizes it, the logic of collective action described by Olson is based on a rigorously static calculus of benefits and costs of action, which does not account for forms of exchange which cannot be reduced to the framework of utilitarian relations that defines individual rationality as distinct, or independent, from the features of groups and of group processes.[3] The consequence is that its explanatory value applies exclusively to situations defined by discrete actions, as typically in the case of single specific issues defined in relationship to the interest of large groups: in these cases, in the first instance, a strictly defined 'rational' motivation to collective action may be found only locally, on a small scale or at the level typical for the generation of conventional (e.g. 'communitarian') forms of action. However, issues of collective action are frequently constituted in different terms: they cluster around broad, vaguely defined issues, the very definition of which extends over ample timeframes and complex interactional settings, developing 'nested', incremental – i.e. local and partial – forms of collective action.

The classical alternatives to collective action defined from the point of view of rational choice theory, i.e. hierarchical models based on coercion and the exercise of authority (the state, the Leviathan) or contractual models based on quasi- or imperfectly hierarchical relationships of inter-dependence (the principal-agent model and the firm in the framework of 'institutionalized' markets), may be sensibly redefined if placed in a temporalized and coevolutive dimension based on the empirical observation of concrete practices of collective action. Such a perspective implies – assuming as a matter of inquiry, and as the main problem of collective action – the concrete processes of intersubjective and inter-temporal constitution and coevolution of preferences. Understanding the principles of rationality which underlie the pursuit of collective aims implies understanding the microsociological mechanisms on which the strategic behavior and the intentionality of actors are built.

This perspective is pursued by new-institutionalist research on forms of collective action, and it implies rethinking assumptions of the actors' rationality in terms of its social-constructive determinants. The question of the 'rational' foundations of collective action is reframed as an inquiry into the possibility of the production of collective activities which represent the low-cost subproduct of actions directed towards the pursuit of individual interests and which may produce selective sanctions and incentives, thus constituting action situations which retroact on the very form of rationality which guides them.

Endogenous Dynamics of Institutional and Organizational Change

In referring to the specific connotations of this approach it may be useful to recall one of the basic formulations of Olson's thesis:

> when a number of individuals have a common or collective interest – when they share a single purpose or objective – individual, unorganized action [either will] not be able to advance that common interest at all, or will not be able to advance that interest adequately. (Olson 1965, p. 7)

The issue of collective action, in line with this conception, is framed by authors like E. Ostrom in the first instance as an organizational issue. It is intended as such, however, in its process-like dimension, as an activity not amenable of model-like reductions: 'organizing' is seen as the outcome of a human ability which stands at the origin of different possible experiences of self-governed collective action.[4]

Whilst the issue is not new, traditional solutions to it do not appear to be satisfactory in a perspective of reconciling assumptions of both individual and collective rationality. A common feature of models of collective action proposed in political philosophy (central authority, contractualism) and in political economy (welfare economics, privatization or nationalization policies) is their tendency to address issues which constitute the core of 'second' and 'third' order dilemmas of collective action (i.e. the cost of monitoring, sanctioning and supply of incentives, and the cost of relevant and shared information) in forms which presuppose some kind of exogenization, by this impairing their explanatory potential.

The path taken by new-institutionalist approaches, on the contrary, moves from an essentially process-based conceptualization of the organizational issue of collective action. At the core of their inquiry is a question of *organizational action*: the question of how to introduce "changes that order activities so that sequential, contingent, and frequency-dependent decisions are introduced where simultaneous, non-contingent, and frequency-independent actions had prevailed" (Ostrom 1990, p. 39) or, essentially, the question of how to create different organizational settings, of how to innovate them, and of how to allow for their self-sustainment, in a perspective of path-dependency.

The key organizational problems for the management of common goods are classically defined by a triad of 'dilemmas' of collective action: the problem of the introduction and rooting of new institutional settings (i.e. of *institutional change*), the problem of *credible commitments*, and the problem of *mutual monitoring*.[5] These are problems strictly linked to each other, and endowed with cumulative effects, as real *nested dilemmas*. The limits to collective action in the face of these dilemmas may be summarized as the restrictions in access to the form of situational knowledge which is necessary in order to allow the merging of individual orientations into a collective rationality of a shared and coevolutive kind.

In general, under such conditions, apparently reproducing the ideal-typical dualism between centralized and decentralized action common to models of organizational coordination, the tradition of classical institutionalism has had trouble thematizing the origin of new institutional settings, tending to explain institutional innovation essentially as a discrete process of a foundational nature, and to stress the constraints to its emergence represented by the costs in terms of transactions and information needed for introducing new institutional action rules. The interpretive question concerning the endogenous or exogenous nature of institutional innovation tends thus in this conception to be radicalized, as does the question of the relationships between institutional origins and institutional change.

New-institutional strategies of inquiry propose on the contrary an inversion of the question of institutional change: they interpret innovative forms of collective action in a pragmatic perspective which points to the incremental and coevolutive dimension of cooperative settings and to an endogenous explanation of their institutional conditions of emergence.

What is in general questioned by this is the overall adequacy of traditional institutional theories in explaining the key problems of collective action, as summarized by the dilemmas of constituting apt institutional forms, credible commitments, and mutual monitoring in an intertemporal dimension.

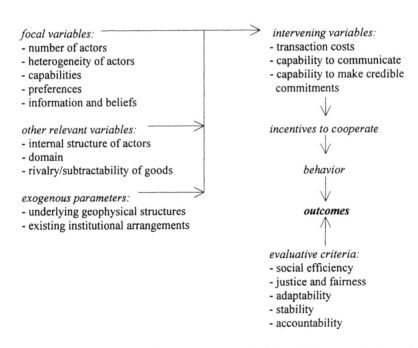

Figure 13.1 Relationships between the variables of forms of collective action at a given time
Source: Keohane and Ostrom (1994, p. 418)

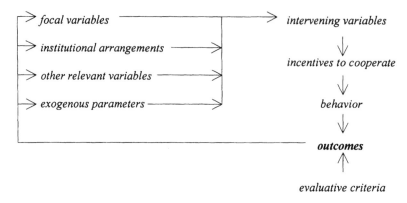

Figure 13.2 Relationships between the variables of forms of collective action over time
Source: Keohane and Ostrom (1994, p. 419)

This aspect may be represented idealtypically through a comparison of synchronic and diachronic sets of relationships between variables of cooperative forms of collective action (figs. 13.1 and 13.2).

The key for such an approach is the nature of individual rationality which is assumed to be enacted in the context of group processes. The form of rationality guiding actors is intended as essentially pragmatic, situational and contingent, i.e. oriented towards the adoption of *contingent strategies* in action situations, which are complex, highly differentiated and endowed with uncertainty and ambiguity.[6]

Rationality is thus assumed according to a 'broad' conception, which interprets the strategic horizon of the individual ('the internal world of individual choice') as a result of the intersection of four endogenous variables (expected benefits, expected costs, internal norms, discount rates) and which, through the mutable outcomes of such intersection and their influence on expectations on future action, establishes a retroactive relationship of interdependence with the external world. At the same time, the form assumed by rationality in individual strategic choice is marked by its belonging to a social context: "[w]hat types of internal norms an individual possesses are affected by the shared norms held by others in regard to particular types of situations. Similarly, internal discount rates are affected by the range of opportunities that an individual has outside any particular situation" (Ostrom 1990, p. 37).

This order of assumptions introduces a shift in the explanation of institutional change from variables of an exogenous kind to the internal features of interactions through which the constitution of preferences and agreements among actors is realized. The pertinent dimension of observation is thus redefined as an *intertemporal dynamics of institutionalization*, of a necessarily multidimensional nature.

Coordination and Institutional Innovation as Incremental Outcomes of 'Nested' Enterprises

A fundamental conceptual role in reinterpreting the dynamics of institutional change is attributed to the nature of rules in the domain of institutional phenomena. In the previous chapter we have discussed this issue with reference to new-institutionalist contributions. Research on collective action shares with them an understanding of the multidimensional domain of rules as well as of the constitutive connotation of their configurational profile. Both assumptions allow us to conceive institutional innovation and change as an incremental process, capable of arising at different levels of the regulatory definition of action:

> Both origins and changes in institutions can be analyzed using the same theory when both are viewed as alterations of at least one status quo rule. A change in any rule affecting the set of participants, the set of strategies available to participants, the control they have over outcomes, the information they have, or the payoffs [...] is an institutional change. The costs of changing the rules vary substantially from one rule to another, from one political regime to another, and from one level of analysis to another, and they also vary over time as participants and conditions change. Whether or not it will be costly to achieve any institutional change will depend on many variables [...], not simply on whether or not a new institutional arrangement is being created. (Ostrom 1990, pp. 140-1)

The plurality of levels for the definition of rules is thus backed by a plurality of levels of action: "self-organizing and self-governing individuals trying to cope with problems in field settings go back and forth across levels as a key strategy for solving problems", and this is the specific difference of their situation from that of "[i]ndividuals who have no self-organizing and self-governing authority" and who are therefore "stuck in a single-tier world", where "[t]he structure of their problems is given to them" and "[t]he best they can do is to adopt strategies within the bounds that are given" (Ostrom 1990, p. 54).

The consequences of this theoretical attitude are multiple. First, the burden in interpreting and explaining institutional innovation as a condition for collective action is no more set on a single factor. Similarly, it is no more reduced to a single, discrete institutive step or to a single unitary will, be it individual or collective. Much more than as a voluntarist phenomenon, institutional change is seen as an incremental, coevolutive and self-transforming process, of an emergent nature, which draws attention to its complex conditions of emergence and to the social investments which render it possible, rather than to singular actions or conditions. Second, the domain of possibility and the domain of pertinence of institutional innovation and change are not unique: innovation may occur and become effective within different arenas of action, and these may always find forms of connection between different but not distinct spheres (operational, collective, constitutional) of choice which provide orientation for action. Each level of institutional reality contains, at least potentially, elements which make the emergence and recognition of the constitutional connotation of collective choice possible. Third, in a positive as well as normative perspective on institutional and organizational innovation and change towards forms of collective action, attention is shifted decisively from an emphasis on *summary variables* to an inquiry on *situational variables*: the condition for this is a multidimensional assumption of the institutional determinants of action.

The Dynamics of Institutional Self-Transformation

The emergence of practices of self-regulation in contexts of interaction characterized by mutual acknowledgement of the actors, i.e. by the minimal conditions for the constitution of forms of social capital, is a decisive factor in introducing a dynamics of change. Through small-scale initiatives, a change in the incentives for action based on shared rules and cooperation may develop which bears cumulative innovative potentials:

> Success in starting small-scale initial institutions enables a group of individuals to build on the social capital thus created to solve larger problems with larger and more complex institutional arrangements. Current theories of collective action do not stress the process of accretion of institutional capital. Thus, one problem in using them as foundations for policy analysis is that they do not focus on the incremental self-transformations that frequently are involved in the process of supplying institutions. Learning is an incremental, self-transforming process. (Ostrom 1990, p. 190)

The primary condition for realizing such potentials resides, as we have seen, in a perspective of iteration and frequency of interactions among the groups' members. Another condition is given by the endogenous opportunities for evolutive responses in the intertemporal dimension of experience. Both aspects are linked in an essential way to the power of self-disposition of rules.

Intermediate benefits from initial, limited and partial forms of investment in coordination and cooperation are realized without necessarily high and generalized amounts of individual resources. Each change in action situations which is thus determined realizes a change in the structure of incentives according to which strategic decisions about future actions are taken. The enactment of new forms of communication and the development of processes of mutual adaptation or exchange connected to the initial investment (in the plurality of their possible forms of institutionalization) define an institutional setting which modifies the nature of the situation: from a situation in which decisions are assumed relatively independently to a situation in which options are mutually correlated through ongoing formal and informal practices of mutual monitoring of action. The conditions are thus given for forms of micro-constitutional choice which connect diverse arenas of choice and diverse levels of rules, extending in a potentially transscalar sense the effects of the building of an initial, 'local' social capital.

The potentially nested nature of change, supported by evidence of the evolution of self-governing experiences, stresses the importance of the ability of the actors to define and modify rules as a condition for the effectiveness, sustainment and the generative potential of initiatives in collective action. This is true relative to both the scope and the nature of group processes. Self-regulating capacity is in fact primarily an expression of the need for an adaptive attitude towards the nature of problems, and of the inherent inclination to *discovery* and learning implied in their collective reframing: it is, in this sense, an expression of the evolutive attitude needed in order to effectively pursue an orientation to ends. But self-regulation is also a condition for the linkage between the constitution of credible commitments among the actors and the enactment of mutual monitoring and self-evaluation. The adoption of contingent strategies by self-interested actors and the probability of forms of strategic self-monitoring among the actors are in fact two aspects set in a reciprocal relationship: in a perspective of iterated interactions, they represent the condition for the constitution of reputation and trust (Axelrod 1984) and of graded, low-cost forms of social sanctions, in which the collective cost of sanctioning is exceeded by the individual cost of being sanctioned and is lowered by the pursuit of individual strategies oriented to consistency with the rules of collective action.

Towards a Generalization: Challenges and Conditions for Self-Policing Forms of Collective Action

Public Goods and Collective Identity

The empirical observation of cases of "common property arrangement in which the rules have been devised and modified by the participants themselves and are monitored and enforced by them" (Ostrom 1990, p. 20) allows to advance some careful interpretive generalizations on the conditions which may render the rise of self-policing forms of collective action possible:

- relevant knowledge is shared by the actors, and is both practical and based on interaction: knowledge which directs efforts in the maximization of individual benefits is at the same time the knowledge relevant for the adoption of individual strategies and the knowledge relevant for the collective distribution of benefits. The ability to safeguard the reproduction of goods is inherent to the social use of this form of knowledge (it is not, for example, dependent on exogenous definitions of the nature of goods);
- the construction of rules is realized within the group of actors whose behaviors is regulated by them: the power of disposition on rules is internal to the group and to the nature of knowledge it develops and shares;
- recourse to external authority is limited to an arbitral function, and does not imply forms of externalization of enforcement and monitoring costs;
- enforcement and monitoring of rules are realized by the actors themselves as a byproduct of the pursuit of their own interest in the framework of a rule-oriented behavior, where the self-imposition of a regulatory constraint becomes an element of the redefinition of individual preferences, and where its enforcement in the form of a rule of collective action determines a substantive (albeit not substantial) constraint to free-riding;
- the solution to a collective-action problem is defined by an attitude of 'institutional realism' characterized by a decisive action-orientation and by a minimization of interferences with the prerogatives of existing formal institutions.

A preliminary question to be raised is however whether reasons related to the scale or to the subject of inquiry may be a constraint to the generalization of the lessons drawn from these experiences. [7]

Given some idealtypical assumption about the endogenous character of their resolution, research on the management of common-pool resources highlights a positive correlation between the adoption of incremental strategies for the redefinition of operational rules in a group and some internal features of the group itself:

1. Most appropriators share a common judgement that they will be harmed if they do not adopt an alternative rule.
2. Most appropriators will be affected in similar ways by the proposed rule changes.
3. Most appropriators highly value the continuation activities from this CPR [common-pool resource]; in other words, they have low discount rates.
4. Appropriators face relatively low information, transformation, and enforcement costs.
5. Most appropriators share generalized norms of reciprocity and trust that can be used as initial social capital.
6. The group appropriating from the CPR is relatively small and stable. (Ostrom 1990, p. 211)

The conditions mentioned present similarities, but not an isomorphic coincidence, with the attributes of communitarian forms which are sometimes highlighted as a condition in literature on collective action.[8] They certainly represent relevant contributions to the constitution of the social capital which is necessary to "address problems of mutual vulnerability":

Individuals who share similar beliefs are more likely to be able to communicate effectively about the problems they face. If the group is stable, can communicate directly, and will interact over a long period of time, the likelihood that the group will find solutions to many of the problems they face is indeed higher (assuming other variables are the same) than for those groups lacking these characteristics. (Ostrom 1992, pp. 343-4)

This line of reasoning, however, by pointing to a dimension of collective behavior defined through coevolutive processes of mutual adjustment and thorugh iterative interaction and communication, clearly differs from approaches inspired by contractual and exchange paradigms, as well as from distinctions among forms of collective action based on substantialist assumptions, based on some kind of quasi-natural 'factual order' of what is at stake in their formation. The latter is typical of positions which assume the nature of the good itself towards which action is oriented as an evaluation criterion, and which accordingly distinguish forms of action according to properties inherent to the good itself (as aspects of scale, accessibility, rivalry or subtractability). Typically, according to such a logic, based on neoclassical assumptions of rationality, dilemmas of collective action are defined according to a quasi-natural distinction between public or semi-public and private goods,[9] which presents substantial similarities with arguments on the communitarian preconditions for cooperation.[10] Rather than pointing to the alleged primacy of thier communitarian features, as has been observed by Douglas (1986), experiences in the constitution of self-policing forms of collective action highlight the fundamental character of social and emergent constructs of local societies and of stable mutual commitments, and to the symbolic-cognitive dimension of processes of interaction and exchange among reflective, knowledgeable actors.

Institutional Regimes and Institutional Change

The theoretical interpretation of effective experiences in collective action emphasizes the decisive role of the ability to promote interactional settings of a stable and ongoing nature, based on *self-monitoring* and *self-enforcement* by localized actors, which may be summarized into the notion of *self-policing conventions*.

The interactional foundation of the constitution of commitments is also a reason for rejecting a determinist understanding of the role of factors like the groups' size, the heterogeneity of their members, the degree of openness of organizational and negotiating arenas, traditionally at the core of analyses on the emergence of forms of cooperation. Although decisive for an understanding of actual experiences, the role of these factors cannot be generalized in abstract terms.[11] The meaning of abstract formal models of collective action is accordingly relativized: each of these factors is in fact located in a multidimensional system of relationships with others, and its connotation tends to modify other parameters of the situation which may exert influence on behaviors, thus departing from a linear causal nexus (Hardin 1982). Interpretation must therefore be *situated*. Whilst, for example, in some cases a negative correlation exists between plurality and heterogeneity of positions (due to differences in the nature and intensity of values) and the probability of achieving cooperative solutions, this correlation does not only point to a possible different definition of objectives, i.e. to an intrinsic opportunity for solution, but (as highlighted by the theory and practice of alternative dispute resolution) to the possibility that differences may constitute a fundamental resource for redefining a problem in integrative terms, based on the amplification of what is at stake and on a non-competitive diversification of distributed benefits (Fischer and Ury 1981; Sullivan 1984; Moore 1986; Susskind and Cruickshank 1987).

An analogous potential role in generating self-organizing schemes is attributed to the cross-cutting combination of interests among actors endowed with different levels of abilities and power, provided they are able to creatively define a vision of mutual benefits of cooperation (Ostrom 1994b). In summary, attention to determinants of the context is important in particular as it contributes to an interpretation of the features of action situations, and as far as these are interpreted not so much in terms of independent variables as in terms of the contribution of their specific combination in determining such aspects as transaction costs and discount rates for the future. The mutual acknowledgment of preferences and the generation of new collective abilities, which are indispensable for the emergence of forms of cooperation, depend above all on the constitution of effective forms of communication and on the frequency of interactions, an aspect strongly underlined by research on the evolution of strategic behaviors towards modes of cooperation in the framework of iterative models of non-cooperative games (Schelling 1978; Axelrod 1984) as well as by network analysis.

Some consequences may be drawn from these observations. A first aspect, implicit in the emphasis on the self-organizing and self-regulating features of the processes, is a demise of hierarchical forms of enforcement, which is extended to the overall structural features of cooperative settings. Determinant for the self-sustaining ability of forms of collective action is the ability to constitute credible commitments among the actors: "a remarkable convergence seems evident between two independent streams of literature in political science, economics, anthropology, sociology and related disciplines. At both local and global levels, researchers have found that when individuals or organizations (such as states) can make credible commitments, they are frequently able to devise new constraints (institutions, or sets of rules) that change the basic structure of incentives that they face" (Keohane and Ostrom 1994, p. 404).

Emphasis on this aspect builds on positions from other social disciplines which have pointed to the role of the temporal dimension and of symbolic mediation in constituting forms of social order, and, on the other hand, is a major reason for extending the lessons from local experiences to other levels of policy-making.[12]

The importance of local institutions and of the rooting of self-organization and self-regulation of collective action within *locales*, intended as units of recognition and definition of the environment of interactions (Giddens 1984), is thus endowed with multiple meanings. It responds to a principle of 'institutional realism', strictly correlated to the incremental nature of institutional innovation: it refers to the 'stratified' nature of opportunities for innovation and to the need for adequate margins of flexibility relative to the formal-juridical constraints of the institutional framework.

The incremental interpretation of change on a formal organizational level is, secondly, tied to a consideration of its cognitive implications, i.e. of the fundamental role of the sharing of problems among actors and of the relevance of action relative to a shared definition of a problem, which refer both to the situational and localized connotation of knowledge relevant for action, but also to the internal dynamic of its constitution, calling for a rooting in time and space in order to ensure effective coupling to practices of mutual adjustment and to evolutive adaptations of rules.

Adherence to a principle of locality is finally also an element for a possible strategy of innovation confronting the problems of scale related to organizational and cognitive factors of institutional change: innovation 'from below' may represent a condition for the constitution of nested enterprises, not reducible to general institutive rationales, but capable of generalizing styles of cooperative behavior, starting from small-scale problems and from the constitution of local forms of social capital, towards problems and contexts of a superior order of institutional innovation.

The attention of research has focussed however also on the external conditions for effective forms of collective action. A first level of considerations points of course to the relationships between different levels of the institutionalization of social practices.[13] As we have underlined, any attempt or inquiry in the experimentation of institutional self-reform necessarily comes across different levels of institutional determinants of the context. The local microconstitutional level of choice is always embedded into a framework of complex relationships with the macro-constitutional determinants of social institutions.

Considering political factors thus becomes key to a critical contextualization of experiences as well as to action. The relationship between actors at various institutional levels is in fact never linear, but (albeit problematically) circular.

This aspect becomes apparent in the way new-institutionalist approaches conceptualize the facilitating and enabling role of the state and, in general, of supralocal levels of government, as well as the function of the enforcer as a third party: its identity is independent from forms of hierarchical positioning, and is rather defined in terms which are inherent to the self-policing convention within which it exerts its function of arbitration.[14]

The role of the enforcer is not, and cannot be, alien to a sharing of values on which the self-governing collectivity is based, but must be set with it in a bilateral relationship, defined by an ability to sustain common understandings and mutual commitments as well as to act as a warrant of self-government.[15]

The Role of Information, Communication and Transaction Costs

The main reasons for the functionality of self-promoted, non-hierarchical forms of the constitution of collective commitments are often seen in the reduction of transaction costs and in the constitution of conditions facilitating the exchange, circulation and handling of information. While the former aspect reveals an adherence to a conceptual repertory tied to an interpretation of social phenomena in terms of aggregates of individual economic choices under contractual conditions (as typically found in public choice),[16] the latter points to the socioanthropological dimension of identity and to the processes through which actors 'discover' and change their preferences and volitions and the representation of their self-interest.

Information and communication, as well as their active and passive constraints, thus assume a key role. As has been highlighted by studies on the dynamics of social networks and of the role of 'civic culture' in political behavior, networks of communication and the contribution they offer in constituting mutual trust and in preventing a conflictual splitting of positions bear a great importance in generating and sustaining forms of collective action. Again, the role played by information must be understood in relational terms: while in some situations the sharing of uncertainty may lend to imperfect information a potential role in generating cooperation (Heiner 1983; 1985), and suspending routinary knowledge-based practices (as indicated by cognitive psychology and phenomenological sociology) may generate a potential in innovation, a relevant factor impairing orientations to cooperation and collective rules (as shown by applications of the principal-agent model in policy-analysis as well as by game theory and by experiences in dispute resolution) is represented by asymmetries in the ability to deal with information, which may alter the field of alternatives to individual strategies in a sense divergent from a perspective of the sharing of knowledge, possibly in the long run deleterious for the sustainment or emergence of cooperation. Even in this sense, the nature of social institutions exerts a considerable influence on the communicative competence of groups and local societies.

Conclusion: Institutional Settings for Collective Action as Self-Policing Conventions

Assuming emergent forms of cooperation under the category of agreements-in-action that are called *conventions* apparently implies a shift from the classic definition of dilemmas of collective action.

In fact, conventions are defined essentially as 'tacit contracts', not sanctioned by an explicit agreement: in Lewis's words, "[m]any social contracts will be sustained by the moral obligation of tacit consent or fair play, as recognized by the agents involved" (Lewis 1968, p. 94, cited in: Hardin 1982, p. 155).

In summarizing the conditions for the emergence of a convention listed by Lewis,[17] of this 'minimal' form of institution, i.e. the presence of a common interest towards a rule granting coordination of action, the absence of interests conflictual to such a rule, and the absence of deviations from the rule but as in presence of a manifest failure of pursued coordination, Douglas (1986, p. 46) observes that, under such definitory conditions, a convention is, by its own nature, *self-policing*. The explanatory challenge of its self-policing character, therefore, lies in its very generative and coevolutive conditions. Let us address them in brief.[18]

1. A major aspect defining a convention is that, once it is established, it becomes an intrinsical interest of the participants to hold to it. The existence of a convention becomes in fact a crucial aspect of the situation and of the definition of the actors' preferences in relationship to the situation; it thus becomes a factor in the definition of the strategic environment of action, generating mutual expectations of behavior and cumulative systems of rewards to conformance and sanctions for non-conformance. Conventions thus assume the meaning of social institutions which exert a power of influence on individual choices in that they embed expectations which entail high costs for the actor infringing them.[19]

The application of sanctions is itself, in principle, 'tacit', and thus reverts assumptions on their cost which are at the origin of their dilemmatic nature in classic theoretical models of collective action: the cost of sanctions is high for the individual who incurs to them and low for the group inflicting them.

The role of sanctioning is therefore circular with respect to the definition of a convention: the foundation of sanctions is the fact that it is supported by a voluntary, conventional recognition by the members of a group, but the foundation of a convention is the fact that it constitutes an authority for sanctioning.

What is, however, the nature of the sanction in this seeming paradox? It appears that its definition can be substantive only in secondary terms.[20] Its definition is in fact translated into social practices along a continuum which ranges from its tacit emergence – through progressive degrees of formalization – to its possible translation into legal or regulatory constraints (as, for instance, selective forms of access to benefits and resources); at the basis of its conventional character, however, is its essentially non-intentional and non-intended nature, grounded on the constitution of principles of obligation, of commitments. The paradox of the conventional nature of sanctions is thus reformulated in terms of its principle of power: in Schelling's expression (1960, p. 43), "the power to be sued [or sanctioned] is the right to accept a commitment". The nature of sanctions is hence – at least potentially – always amenable of purposive-intentional interpretations, but is still always the outcome of ongoing processes of reinterpretation: the degree of its objectivation depends on the nature of institutionalization processes to which its definition is liable.

2. The conditions of emergence and persistence of a convention, as we have noted, place a considerable weight on aspects of the *sharing of knowledge*. While a convention has been defined as a process-based, 'emergent' and essentially non-intentional outcome, the nature of knowledge implied in the assumption of its shared and mutually acknowledged nature is still to be discussed. Assuming that the constitution of a convention is not primarily an intentional process also entails that the knowledge-base on which it is grounded need not be referred to substantive knowledge on preferences of the group's members. The knowledge pertaining to the constitution of conventions is rather defined as an outcome of strategic relationships among the actors and of the influence of their patterns of interaction on the constitution of their preferences.[21] Assumptions of *common knowledge* in Lewis's formula thus relate to a broad conception of the knowledge involved, of an essentially pragmatic foundation, but largely unconscious, which may be referred to the kind of knowledge Giddens (1984) calls the 'practical consciousness' of the actors. It is, again, a form of tacit knowledge, of a relational nature, built through interaction and through the reciprocity of relationships, constituting the premise and foundation for the sharing of relevance criteria on further knowledge-creating activity directed towards action and the regulation of conducts.

A self-policing convention may thus be considered as an institutional archetype of what, according to the definition proposed by Blau (1964, p. 312), may be called an 'intrinsically rewarding' organizations, wherein rewards and sanctions are not distinguishable from the associative phenomenon itself. Such an interpretation highlights the role taken by processes of identification through which a collectivity constitutes and acknowledges itself in the formation of what has been defined as 'common knowledge'.

However, this should not overshadow the key role of issues related to the production and diffusion of knowledge. No 'organic' knowledge form can be assumed at the origin of self-policing forms of collective action.

Between the domains of 'practical' and 'discursive consciousness' of the actors involved in a conventional situation, on the contrary, the nature of its generative and evolutive path is at stake. On the one hand, limits to the cognitive base are constitutive of conditions of uncertainty in the relationships between costs and benefits of agreements, which highlight the shortcomings of contractual approaches to collective action, and rather stress the exploratory and evolutive potential of conventional forms. On the other hand, the vagueness of objectives of a conventional nature and the always possible ambiguity or equivocality of their sharing may become constraints to the pursuit of an outcome-orientation which may reproduce and enhance the legitimation of the very conventional construct, as far as depriving it potentially of any sense and of any possibility of self-sustainment and evolution. All these aspects thus point to the critical role of constraints and opportunities for the constitution and reproduction of shared forms of knowledge given within and around social settings of a conventional kind.

3. The issue of the strategic and relational character of knowledge at the origin of conventional forms of collective action, and of the connection of this character with the nature of the actors' identity and preferences, are strictly connected to the *intertemporal dimension of interactions*. Research on cooperative behaviors based on the sharing of explicit knowledge and of prominence solutions (Schelling 1960) as well as research on strategic behaviors based on the constitution of tacit forms of knowledge (Axelrod 1984) have stressed the dependence of the possibility of the emergence of forms of cooperation on the actual iteration and on the perspective of ongoing iteration of interactions. Aspects as those previously considered cannot thus be seen as randomly patterned, but rather as path-dependent. However, as Hardin (1982) notes, the constitution of sequential iterations of an interactive situation is in itself no more probable, from the point of view of a goal-oriented rationality, than the possibility of an effective sanctioning of non-cooperative behaviors: both are 'second-order dilemmas' of collective action.

Again, the constitution of conventional forms of collective action faces a dilemma if intended as an intentional and univocal process, i.e. as a result of an organizational or institutional 'design'. The same takes a different shape, however, if seen in a networking and coevolutive perspective, as outcome of forms of overlapping of decentralized relationships, whereby the constitution of common knowledge is defined by the networking of 'local' forms of knowledge. Several contributions to the study of strategic behaviors (Hardin 1982; Axelrod 1984) point to the fact that, even moving from assumptions of individual rationality, the constitution of spontaneous or voluntarist forms of cooperation within small groups and their networking integration through effective forms of communication may constitute a solution, of an emergent and coevolutive kind rather by design, to the problem of 'latent groups' discussed by Olson (1965): dilemmas of collective action within large groups (i.e. at the macrolevel) may be solved in form of outcomes of iterated interactions within small subgroups (i.e. at the microlevel of interaction) in the framework of conventions of a 'local' kind, oriented by forms of 'local knowledge' (Geertz 1983). Their networking connection may hence define – through the availability of weak ties and through the definition of bridges (Granovetter 1973) – an extension of the 'common knowledge' relevant for forms of action directed towards shared ends, based on multilocational and constitutively relational patterns.[22]

Such an interpretive perspective of the emergence of collective action, nonetheless, points to the necessity of considering that the locales where the definition of preferences and the constitution of relevant knowledge occurs are hardly anymore univocal, in conditions of 'late modernity', neither in space nor in time (which, again, speaks against any 'organicist' assumption of these notions): both are increasingly connoted in a multidimensional sense, in connection to multiple arenas of action and choice. This stresses the role played by the structuration of such arenas in making the reproduction of the premises for conventional patterns of action and their self-sustainment possible.

Notes

1   The definition of the 'tragedy' inherent in the inability to perceive the long-term threats for collective well-being deriving from the short-term pursuit of individual interests is based on a famous image by Hume and is summarized by G. Hardin in his metaphor of the common pasture, bound to be exploited by excessive appropriations in the absence of regulation: "Therein is the tragedy. Each man is locked into a system that compels him to increase his herd without limit – in a world that is limited. Ruin is the destination toward which all men rush, each pursuing his own best interest in a society that believes in the freedom of the commons. Freedom in a commons brings ruin to all" (Hardin 1968, p. 1244, cit. in: Ostrom 1990, p. 2).

2   Here only a few essential references to this literature: Schelling (1960; 1978), Hardin (1982), Taylor (1976; 1987), Axelrod (1984), Hechter (1987). Our interest is essentially in contributions which share an application of strategic models of game theory to 'real life' situations.

3   Many critiques of Olson's thesis stress aspects of collective identity and of the intertemporal dimension of its constitution (Pizzorno 1986). Also comments inspired by communitarian and organicist positions have pointed to the existence of a collective identity as a precondition for the 'rational' calculus of costs and benefits of action

presupposed by Olson. We agree however with Melucci (1987, p. 47) that the conception of collective identity proposed by these positions is too static and uni-dimensional as to be able to clarify "the process of the constitution of a collective actor through interactions, negotiations and relationships with the environment", i.e. through its concrete process of acting and organizing.

4   The aim of this line of research is very similar to that of studies on self-organized and self-governed forms of local development, as represented in planning and development theory, e.g. by Sachs and Friedmann. The differences in theoretical matrices, however, apparently do not help mutual communication, as the absence of reciprocal references shows.

5   The dilemma in these issues resides in their mutual implication. The introduction of forms of coordination of action in a field dominated by individualist action rationales is the classical problem of collective action, solvable only through the constitution of forms of reciprocity or community, i.e. through the creation of the 'non-contractual preconditions of contracting' (*dilemma of the first order*); these preconditions take the shape of the credibility of mutual commitments of the actors, i.e. per definition of the outcome of an endogenous process of the actors' self-motivation, without the intervention of an enforcer or of external sanctions, but rather based on an effective mutual monitoring of the actors (*dilemma of the second order*); this is again essential for building a commitment, but without commitment its introduction represents a non-reasonable cost (*dilemma of the third order*). The problem thus becomes circular.

6   The notion of *contingent strategy* derives from game-theoretical contributions which have inquired into the insurgence of cooperative strategies in conditions of incomplete information; its adoption however is extensible in diverse situations, which "vary in terms of the level of initial cooperation extended and the action of others required for switching behavioral patterns" (Ostrom 1990, pp. 36-7).

7   The identification between communitarian connotations of the social structure and solutions of a conventional nature to problems of collective action is proposed, for example, by M. Taylor as a sort of exception: "In many small-scale communities no 'selective incentives' or controls are needed: it is rational to cooperate voluntarily in a production of the public good of the social order" (Taylor 1982, p. 94, cit. in: Douglas 1986, p. 26). In introducing her research, Ostrom on the contrary defines her strategy "as that of a 'new institutionalist' who has picked small-scale CPR [common-pool resource] situations to study because the processes of self-organization and self-governance are easier to observe in this type of situation than in many others" (Ostrom 1990, p. 29): it is, in this sense, an overtly instrumental choice. For a debate on this issue, see: Ostrom (1992).

8   Ostrom (1992) refers in particular to a definition by M. Taylor, intending community as an ensemble of persons who 1. share a system of beliefs, including preferences and beliefs endowed with normative value, which go beyond those directly constituting the problem of collective action, 2. are stable, 3. have expectations of continuation of reciprocal interactions in time, and 4. are internally characterized by direct, non-mediated and multiple reciprocal relationships (Taylor 1982).

9   The thesis of the cultural-political construction and of the inherent normative content of such forms of distinction has been shared with Douglas by the late Wildavsky (Malkin and Wildavsky 1991; Douglas and Wildavsky 1982).

10   Again Taylor, in discussing a distinction between public or collective goods and semi-public goods (which Ostrom calls common pool resources): "There is, in particular, a very important class of collective action problems which arise in connection with the use of resources to which there is open access – resources, that is, which nobody is

prevented from using. These resources need not be public goods" (Taylor 1987, p. 3, cit. in: Ostrom 1990, p. 32); or again: Hechter (1987), on the basis of considerations not substantially different from those of Olson (1965) as regards their assumptions of rationality.

11   For an example of symmetrical conclusions on the role of heterogeneity of actors in different contexts of collective action, see: Libecap (1994) and Martin (1994).

12   "Not surprisingly, many of the 'design principles' underlying successful self-organized solutions to CPR [common-pool resources] problems appear relevant to the design of institutions to resolve problems of international cooperation as well as those at a strictly local level" (Keohane and Ostrom 1994, p. 404).

13   We are less interested here in other dimensions of the relationship between context variables and perspectives of the success of self-policing conventions, as, for example, aspects tied to the features of resources (intrinsical variability, external flow dependence, specific distributions of costs and benefits) or external factors (e.g. changes in market parameters).

14   It must however be noticed that the enforcer retains some form of functional if not hierarchical definition, dependent on the specific conditions of cooperation: for example, its central position may assume a potentially determinant role in relationship to specific features of the group, such as its size, and to the consequent difficulties in terms of coordination and monitoring costs (Snidal 1994; Keohane and Ostrom 1994). In this sense, as previously stated, these approaches never completely reject a functionalist interpretation of social roles within cooperative settings.

15   Although fundamental in her theoretical approach, this aspect is unfortunately discussed only to a lesser degree by Ostrom in terms of its consequences. We may however notice an analogy of this problem, to a certain extent a dilemma, with the definition of the social position of the planner given by Friedmann (1987).

16   In her work, E. Ostrom (1990, p. 40; Keohane and Ostrom 1994) emphasizes her theoretical adhesion to the theory of the firm and to transaction-cost theory as developed in the framework of new institutional economics; whilst however the principal-agent model and the theory of the firm may be considered metaphorically coherent with her approach, we are less convinced of the significance of transaction-cost theory in the framework of the role attributed by Ostrom to considerations of a behavioral and symbolic-cognitive kind.

17   It is useful to present here the original definition by (Lewis 1968, p. 78, cit. in: Hardin 1982, p. 158) underlining – and adding emphasis on – the assumption of *common knowledge*: "A regularity R in the behavior of members of a population P when they are agents in a recurrent situation S is a *convention* if and only if it is true that, and it is *common knowledge* in P that, in almost any instance of S among members of P, 1. almost everyone conforms to R; 2. almost everyone expects almost everyone else to conform to R; 3. almost everyone has approximately the same preferences regarding all combinations of action; 4. almost everyone perefers that any one more [person] conform to R', on condition that almost any instance of S among members of P could conform both to R' and to R".

18   The following discussion owes much to: Hardin (1982).

19   The most typical example is given by conventions of social customs and by their constitution into semantic codes. "Conformity to the convention, in the ordinary sense, of a dress code generally becomes a convention, in Lewis's sense, because violation of the code provokes censure, which is commonly not in one's self-interest if one prefers approbation to disapprobation from the relevant community".

20   The substantive definition of incentives and sanctions lies at the core of organizational and institutional models based on contractual assumptions, as the principal-agent model, which have been criticized from a phenomenological perspective for their assumptions of 'institutionalization of trust' (Giglioli 1989).

21   The assumption of 'common knowledge' which is central to the notion of convention may thus intended in terms of a mutual understanding of the own and the other actors' preferences, which is tied to the strategic understanding of the situation, and is combined with conditions of iterative reproduction of the action situation and with the production of low-cost forms of incentives and sanctions through this iteration.

22   In other terms: an action-convention within large groups, in situations defined by limitations in direct interaction (i.e. in the potential for the development of cognitive resources and for the generation of alternatives) and, as such, subject to restraints in definition, may on the contrary develop on the ground of an overlap of localized forms of interaction constituted in the framework of a broader class of situations (i.e. in aggregate conditions favoring the development of cognitive resources and the generation of alternatives) (Hardin 1982).

14 Conclusion

Introduction

In the course of this research, an actor-, action- and process-oriented perspective on planning practice has been privileged. The centrality attributed to interaction has led us to inquire into a conception of planning and of its aims as process-based and as socially constructed: a conception not grounded on the presumption – or on the objective – of a substantial consensus on values, but rather directed towards exploring the conditions for the resolution of actual problems of collective action.

The rejection of an idea of planning as a form of 'institutionalization of ends' has hence led us to inquire into the emergent and contingent character of the definition of its aims and forms of conduct in the course of action, as well as into a coevolutive conception of its processes of institutionalization. The forms taken by the institutionalization of social practices through the conduct of patterned forms of interaction in the framework of planning practice have been accordingly explored – from a normative viewpoint – in their ability to favor the exploration of conditions for meaningful and effective action. A conception of the institutional dimension of planning has been upheld, in other terms, which explicitly aims at the 'exploration' rather than at the 'exploitation' of social resources for the development of potentials for collective action.

An important focus of reflection, as previously stated, has been placed on the possibility of the emergence and sustainment of patterns of mutual commitment and cooperation through the conduct of interactive planning approaches. While innovative forms of collective action have been intended as possible, but non-necessary, and rather emergent and contingent outcomes of interactive processes, the proactive and transformative potential inherent in *self policing-conventions* has been inquired into in light of a new understanding of the nature of processes of institutionalization involved in their development.

Reference to a 'conventional' dimension of institutionalization, allowing for coevolutive processes – as we have called them – of *institution-building*, appears to be problematic, however, whenever an explicitly normative and proactive perspective is adopted, i.e. when processes of institution-building are related to ambitions of *institutional design* and to defined assumptions on the effectiveness of institutional action. A crucial issue which has been inquired into is therefore that of the possibility of conceiving the emergence of forms of collective action – according to the previous assumptions – as contingent and emergent, self-organizing and self-policing interactional constructs, while developing an institutional design of interactive process which may favor their emergence and coevolution.

A pragmatic but non-reductionist answer to this question is perhaps possible – again – only through a critique of such an ambiguous and contradictory notion as that of 'consensus'. Let us therefore dwell, once more, on the meaning we have attached to such an elusive concept, before drawing some conclusions from our critical itinerary.

From Consensus-Building to the Challenges of Collective Action

The Everyday Life of Consensus-Building

In discussing the dilemmas and the possible contrasting interpretations of consensus-building approaches to policy-making, we have nonetheless taken seriously the innovative potential being developed by their experimentation within the changing landscape of territorial governance. Through the display of forms of joint deliberation, negotiated agreement, cooperation, networking, and of the devolution of powers, a radical evolution is underway in the design of territorial policy-making as well as in its forms of institutionalization. As far as this evolution expresses a normative orientation towards the effectiveness and innovativeness of policy-making and towards a reform of its institutional settings, the design rationality of consensus-building strategies represents both a relevant task and a challenge. The consciousness of their *design rationality*, however, must be backed by an ongoing commitment to *practical reasons*. It is the pursuit of such a commitment, in fact, that confers meaning to any reference to 'consensus' in the framework of the design of interactive planning and policy-making approaches.

'Consensus' in itself, yet, appears as anything but an elusive concept. It refers to a status never to be reached, to an ideal horizon looming on a concrete background of social practices. Far from being irrelevant, or even misleading, any assumption of consensus as a reference for planning and policy-making must hence bear consciousness of referring to a dimension of social life which only takes shape through actual practices of strategic-communicative action and through a reality of conflict, and which directs attention to the social processes through which the quest for consensus comes at issue and to the conditions and claims through which it is continuously reframed.

The idea of 'consensus' – to use a metaphor by Berger and Luckmann (1967, p. 45) – is thus only "a narrow cone of light" thrown on our path through the experience of everyday life. Consensus-building highlights the constructivist dimension of social and institutional life, the 'everyday life of the public sphere' (De Leonardis 1997).

Consensus-Building and the Metapolitics of Institutions

A parallel between a notion of 'consensus' thus intended and the philosophical debate on the ethics of communicative action may be mind-opening in this regard. A 'discourse ethics' (Habermas 1981; 1983) that admits a weakening of its foundation on idealtypical assumptions of 'inter-subjective universalizability' enables to understand and to experience the situational and constitutive character of its own principles. This bears consequences for a conception of consensus – or, as in the following citation, 'free consent' – as a reference for communicative practices:

> Consent alone can never be a criterion of anything, neither of truth nor of moral validity; rather, it is always the rationality of the procedure for attaining agreement which is of philosophical interest. We must interpret consent not as an end-goal but as a process for the cooperative generation of truth or validity. The core intuition behind modern universalizability procedures is not that everybody could or would agree to the same set of principles, but that these principles have been adopted as a result of a procedure, whether of moral reasoning or of public debate, which we are ready to deem 'reasonable and fair'. It is not the *result* of the process of moral judgement alone that counts but the process for the attainment of such judgement which plays a role in its validity and [...] moral worth. Consent is a misleading term for capturing the core idea behind communicative ethics: namely, the processual generation of reasonable agreement about moral principles via an open-ended moral conversation. (Benhabib 1990, p. 345)

As it is true that "[w]e are not born rational but we acquire rationality through contingent processes of socialization and identity formation" (Benhabib 1990, p. 356), the object of an ethics of discourse is necessarily *constructive*. Power relations – differently than as they are assumed at the core of a discourse-ethical critique of institutions – do not bear the feature of merely distributive patterns, suiting given social resources to given patterns, or 'structures', of social relationships: they are always embedded in socio-cultural frames of knowledge, interpretation and communication which are themselves constitutive – and reproductive – of connotations of power. Public dialogue, hence, is a dimension that is never exogenous, or alternative, to 'power', but rather constitutive of power relations: it constitutes a 'metapolitics of institutions'. A different critical perspective may accordingly be envisioned on the relationships between communicative action and institutionalization processes: a perspective which implies shifting away from a radical critique of institutionalization processes as constraints to an idealized normative conception of intersubjective universalizability, in favor of a constructivist understanding of their generative and enabling potential:

> There is an interesting consequence here: when we shift the burden of the moral test in communicative ethics from consensus to the idea of an ongoing moral conversation, we begin to ask not what all would or could agree to as a result of practical discourses to be morally permissible or impermissible, but what would be allowed and perhaps even necessary from the standpoint of continuing and sustaining the practice of the moral conversation among us. The emphasis now is less on *rational agreement*, but more on sustaining those normative practices and moral relationships within which reasoned agreement *as a way of life* can flourish and continue. (Benhabib 1990, p. 346)

The construction of this 'reasoned agreement' is not conceivable, however, as a mere proceduralism: discourse always unfolds in connection to concrete forms of strategic-communicative action and through symbolic-cognitive constructs. The dimensions of *agency* and of *symbolic-cognitive mediation* are indissoluble in understanding processes of structuration, and none of these dimensions taken alone may exhaust the understanding of their actual enactment (Giddens 1984; Crespi 1982; 1989; Giglioli 1989).

Adhering to a conception of communicative practices which does not conceive its intersubjective foundation according to idealtypical universalizable principles of discourse, but as an interactional and ongoingly negotiated construction of meanings, implies acknowledging that the nature of social action is always *conventional*, a phenomenon emergent from the complex intersection of 'practical reasons', performative acts, and symbolic-cognitive frames which define concrete action situations.

Assuming a non-reductionist focus on interactive planning practices implies therefore an assumption of the plural and generative dimensions of social practices.

But, again, what is the possible socio-ethical foundation of a normative commitment to interactive planning practices? What conception of the 'common good', of 'collective' ends and identities, of the 'public', is implied in making reference to interactive, experimental consensus-building approaches as they have been previously discussed?

Consensus-Building and the Constitution of the Public Sphere

In considering an interactionist and proactive approach to consensus-building, an attitude towards processes of collective learning and sensemaking is meant which rejects both subjectivist or reified conceptions of social identity and of its formation. Reference for such a conception is not given by either communitarian identities or instrumental definitions of social practices. Attention is rather drawn towards the character of social and emergent constructs of local societies and of commitments to collective action, and towards the cognitive dimensions of processes of interaction and exchange among reflexive, knowledgeable actors.

In an interactionist perspective, the problem of the constitution of identity has to be reframed in terms of the processes of constitution of concrete systems of action: 'collective identity', far from being given, as well as from constituting a precondition for action, is a concept which points to "an interactive and shared definition which groups of individuals develop about the directions of action and the field of opportunities and constrains in which it is situated", where "'interactive and shared' means negotiated through an ongoing process of enactment of relations which reciprocally bind the actors" (Melucci 1987, p. 46). Local systems of action, as well as local societies, appear in this perspective as "never static, nor self-stabilising, but being built continuously by a process of rational bargaining and negotiating", through which "the categories of political discourse, the cognitive rules of the social order, are being negotiated". In this perspective, "the whole system of knowledge is seen to be a collective good that the community is jointly constructing" (Douglas 1986, p. 29); and social behavior may thus be interpreted only in conjointly transactional and cognitive terms, aiming at understanding "the role of cognition in forming the social bond" (p. 19), i.e. in constituting that peculiar 'public good' which is represented by the shared social acknowledgement of the nature of goods around which action develops (Giglioli 1989). In other words, such an interpretive attitude holds to the conviction that "[h]ow a system of knowledge gets off the ground is the same as the problem of how any collective good is created" (Douglas 1986, p. 45).

Assuming an interactionist stance thus leads to rethinking 'public goods' in terms of the processes of constituting a sense of what is 'public'. The 'public' character of choices depends in this sense on the process of situating policy issues "in the sphere of social visibility", of making them matters of public discourse on collective ends (De Leonardis 1997, p. 174). The public character of goods may thus emancipate itself from features inhering to the good itself, rather emphasizing the process-like relationship the public establishes to it: in this sense, potentially everything may be assumed as a public good, and the concept may break the ethnocentric borders into which opposition to markets has confined it, unveiling its cultural-political dimension (Douglas 1992). A public sphere is thus given "in any social process in which goods and interests are developed, and acknowledged as such, to the extent they are shared, and in which action is constitutively interaction" (De Leonardis 1997, p. 169).

Consensus-Building, Collective Action and Institutionalization

Consensus-building, as long as such an understanding of the notion of consensus is shared, may be assumed as part of a strategy for addressing the possible conditions of emergence, evolution and institutionalization of collective action. In that, the relationships between the notions of consensus-building, collective action, and institutionalization and the possible actual coevolution of their processes become more clear.

'Collective action' is the emergent outcome of incremental processes in the building of commitments towards a practical definition of the 'public good'. Institutional settings in this perspective may be seen as *public goods of a second order* (Ostrom 1990; Donolo 1997), mediating the production of public goods: "institutions do 'matter' not only as far as [...] they predefine the framework for the actors' choices, but also – and, in the context of this reflection, even more – as they shape the intersubjective texture of interactions and communications on matters of collective relevance" (De Leonardis 1997, p. 185). As such, institutional settings act as 'formative contexts' (Lanzara 1993) for public choices while being themselves regenerated through their very enactment, in a constitutive relationship between regulatory patterns and social action (Giddens 1984) established through processes of communicative and strategic interaction.

The contingent and path-dependent outcomes of forms of collective action are thus defined by the position they assume along the continuum between the enabling and constraining dimensions of institutionalization, between the capability of reproduction and persistence and the capability of innovation and change. This duality of institutionalization processes constitutes the challenge for interactive planning approaches, inasmuch as planning – as a set of social practices – is both *embedded* in a pluralist field of practices, defined by different degrees of institutional properties and of institutionalization, and itself *constitutive* of institutional properties and of institutionalization processes.

An interactive perspective on the institutional dimension of planning should be framed with reference to the challenge represented by the hermeneutic involvement of planning practices in the definition of institutional properties and institutionalization processes. It should hence not so much address the formal consistency and legitimacy of institutional settings for interaction and public involvement, but rather explore the ability of the actual forms of interaction through which social practices unfold and develop to keep in tension between the aim of effectiveness and the integrative and innovative potential of institutional settings. It should not mainly focus on the formal features of institutional settings, placing the whole burden of innovation on inputs of institutional design, but rather on the coevolutive dimension of institutional settings, in terms of their practices of self-policing and regulation as well as of their practices of collective construction of meanings: in terms – to recall three influent conceptual references adopted in our discussion – of their ability to foster practices of collective *framing*, *probing*, and *sensemaking*.

As such, consensus-building as an interactive planning strategy may be conceived as a dual strategy.

On the one hand, as it has to address the 'pluralist challenge' represented by the development of social practices that are changing the policy-making and planning agenda of our territories, its institutional focus should be traced back to some fundamental questions. It should not be assumed as an end, and thus fall into the contradictions of an 'institutionalization of ends'; rather, it should be assumed as a way to explore the social mechanisms which may generate and sustain collective forms of action in face of increasing pluralism. It should contribute to explore the processes through which the mobilization and commitment of individuals, the contingent unity of meanings, and the constitution of collective forms of action become possible (Melucci 1987), as well as their dynamics of institutionalization. We may call this a focus on the dimension of *institution-building*.

On the other hand, such a strategy should address the conditions for collective action through experimental inquiries into their dimension of institutional embeddedness. It should address, in other terms, the *enabling* dimension of institutions, inquiring into "how [...] institutional arrangements affect the interpersonal, inter-temporal and substantive quality of policy choices", assuming institutions "as configurations of organizational capabilities (assemblies of personal, material and informational resources that can be used for collective action) and of sets of rules or normative

constraints structuring the interaction of participants in their deployment",
creating "the power to achieve purposes that would be unreachable in their
absence" (Scharpf 1989, p. 152). We may call this a focus on the dimension of *institutional design*.

The Virtuous Circle of Institutional Design and Institution-Building: Towards an Agenda for Action and Research

Beyond an obsession for a status – be it an ideal of 'consensus', or a formal
institutional setting or regime – interactive planning approaches need to be
grounded on a realist methodological attitude, on an experimental interplay
of *institutional design* and *institution-building*. Only by linking actual
forms of social interaction to experimental institutional arrangements
enabling their development and self-sustainment, a virtuous circle may be
established between the 'practical reasons' and the procedural rationales of
policies involved in the constitution of shared action-orientations.

But how can reference be made to tasks of institutional 'design', and
to their inherent normative perspective, without breaking this virtuous
circle? Let us make reference to three idealtypical levels for the 'design' of
settings for interactive processes as to 'nested' experimental opportunities
for an action-research perspective.

At a *microlevel* of interaction, in the 'everyday life of consensus-
building', planning processes face the interactional microfoundations of
institutionalization. Their design should be directed to the conduct of
experimental inquiries into the social mechanisms through which shared
commitments towards action are built and sustained and through which the
emergence of forms of collective action becomes possible.

Reference to 'consensus', in this perspective, becomes a tentative and
experimental task. It may be first defined, as such, in rather negative terms.
First, a critical and pragmatic commitment to interactive processes as
means for the effectiveness of planning and policy-making and for their
potential of innovation and discovery should not be intended as
'consensualism'. That is, it should neither conceive interaction as a means
to its own ends and 'consensus' as a condition for their achievement, nor
conceive 'consensus' as an end in itself: it should rather express a
commitment to an interactive definition of its ends, and consensus-building
as a means for pursuing this commitment. The 'design' of patterns of
interaction, accordingly, should not be conceived as discrete, and as
instrumental to distinct 'phases' of the policy process: it should rather be
intended as a means – in the words of two 'reflective practitioners' of

consensus-building – for "extending a broad interaction to all the phases of the decision making process, starting from the definition of the problem" (Balducci and Fareri 1996, p. 2).

Second, the frame of reference for such an assumption as well as for an evaluation of 'consensus' cannot be defined in static or one-dimensional terms: 'consensus' is not given, at a single point in time, in form of an 'exhaustion' of strategic-communicative practices and – possibly – of conflicts, but rather in the form of contingent and path-dependent agreements towards shared courses of action, defined through strategic-communicative practices and through conflict. Consensus-building approaches are ways of defining conditions for collective forms of action through the symbolic-cognitive and strategic-communicative means which frame everyday practices of social interaction.

New-institutionalist analysis points to the development of networks, to the definition of rules, and to the setting of social roles as some of the most crucial social mechanisms involved in interactive practices and in their processes of institutionalization. Networking, collective rule-sharing, and the mutual positioning of multiple social roles – as determinant dimensions of social practices, of the development of forms of collective action and of their processes of institutionalization – represent hence key references for a practice-oriented and interaction-based planning strategy.

The perspective involved in addressing such mechanisms as possible means for constituting collective forms of action is, as we have noted, necessarily path-dependent and iterative. It is thus not only or mainly directed towards the cooperative resolution of discrete problems, based on intensive and occasional forms of involvement, but towards the rooting of networks of action and communication enabling the extensive and diffuse cooperative treatment of problems over time.

The ways for addressing this *mesolevel* of interactive processes by means of institutional design are multiple, but all stress the paramount importance of settings for the development of negotiating abilities and procedures. Negotiation, in this sense, bears a different meaning from that of a technique for the discrete resolution of defined conflicts, as it is assumed in the theory and practice of alternative dispute resolution and in some of its applications to 'consensus-building' (e.g. Susskind *et al.* 1999). It should be rather developed in the direction of anticipatory and ongoing modes of public deliberation, amenable of rooting into embedded and stable forms of social reflexivity. Negotiating practices, in other terms, should evade a primary focus on defined ends and outcomes, and thus – as noted by Schön and Rein (1994) – the paradoxical consequence of an implicit assumption of the constancy of interests involved, which threatens to reduce the integrative redefinition of issues at stake to a mere strategic device of predefined actors; they should avoid constituting a 'pact of inclusion' among actors already endowed with decision-making power and resources, and rather become means for favoring the emergence of social preferences and their access to decision-making arenas.

The main features of negotiation practices thus redefined are their openness, their iterative dimension, and their flexibility to the contingency and dynamics of involvement. In the course of our discussion, several intrinsical contradictions of policy-making approaches based on extended negotiating procedures have been highlighted: the rise of complexity and of transaction costs involved, the 'paradoxes of participation', the 'fallacy of pure inclusiveness', and the marginal effectiveness of their results. However, adjustments to such basic contradictions are conceivable only and precisely within the 'extensive' sphere of relationships which defines negotiating practices: they may find a solution through the extension of negotiation as to constitute a set of practices embedded in multiple, interconnected action situations, as diffuse means for anticipating and framing problems and for enabling their cooperative solution.

The embeddedness of negotiating practices within consensus-building approaches from the very stage of problem formulation onwards is therefore a crucial design feature in defining trajectories of institutionalization. The dimension of the embeddedness of negotiations within a multidimensional field of social practices bears in fact a specific coevolutive potential: that of the constitution of mutual commitments, of practices of mutual monitoring, and – possibly – of self-policing and 'nested' forms of cooperation among the actors, whereby incentives and sanctions to cooperation may emerge in a non-hierarchical environment as low-cost by-products of ongoing patterns of interaction.

The constitution of incentives to cooperation – even if explicitly intended in a non-substantialist way – is nonetheless the dimension in which conditions of power more clearly emerge. Access to conditions of strategic-communicative symmetry among the actors involved in embedded negotiating settings is apparently to be achieved only in a 'conventional' way. For this very reason, a crucial factor is represented by the extent to which patterns of interaction allow for the shared definition of rules, i.e. for the sharing of opportunities for *(micro-)constitutional, collective,* and *operational choice* stratified in the configuration of rules which define action situations. Transparency on the contingent, conventional and configurational nature of rules is hence a condition for building mutual commitments and trust, for a perspective of iteration, and for constituting a 'critical tradition', a 'collective memory' of processes.

This latter aspect involves – to keep to the metaphor – a *macrolevel* of institutional design, responding to the territorial challenges of 'democratic governance', and setting conditions for the development of a policy-making environment defined in both *proactive* and *reflexive* terms.

Let us conclude on this by referring to this endeavor idealtypically as to the development of *active territorial policies*. *Active policies* may be conceived as 'policies of policies', as policies meant "to conceive their intervention on real processes not only in terms of handling them as objects, but also of as far as possible valorizing their actual dynamics and their latent resources" (Donolo 1997, pp. 97-98). Their action-orientation overarches a plurality of policy rationales as well as of policy areas: it combines regulatory settings, incentives, supply- and demand-orientations, intersecting the emergent aims of social, economic, and institutional policies and casting them into complex 'environments of scope', rectroacting on the constitution of preferences and valorizing unintended consequences as well as traded and untraded interdependencies. Their design rationality is thus at the service of effective outcomes as well as of an ongoing heuristic experience.

Territorial policies may become 'active' by acting on the growing interdependence between areas and actors of policy initiatives at the level of their place-boundedness, and by clustering their emergent configurations into self-policing local systems of action.

In a perspective of planning committed to interactive practices of institution-building, innovative inputs of institutional design and active territorial policies are set in a dialectic relationship. In the framework of active policies, processes of problem-setting and -solving may develop into arenas of microconstitutional, collective and operational choice, displaying reciprocal forms of commitment among the actors, and enhancing self-organizing and self-policing abilities. In the framework of innovative measures of institutional design, these arenas may be enabled to develop reflexive attitudes – to establish a 'triple-loop' of action-learning – linking everyday issues to a consciousness of their institutional determinations. But as the active dimension of public policy is where institutional design and social reflexivity may fruitfully hybridize, so their territorial dimension is where processes of institution-building – grounding the construction of consensus and legitimation on place-bound processes of identification and on concrete and situated balances among social claims – may recover their embeddedness in locales. And, as far as 'policies' define the nature of 'politics', 'active' territorial policies may become a primary domain in which a perspective of *generative politics* may develop.

Bibliography[*]

Primary Sources

Anglin, R. (1991), *The State Development and Redevelopment Plan: An Assessment*, Working Paper No. 22, Center for Urban Policy Research, Rutgers University, New Brunswick NJ

Beauregard, R.A. (1991), *Comments on "Economic Development" and "Urban Revitalization", "Communities of Place: The Interim State Development and Redevelopment Plan"*, unpublished paper

Bierbaum, M.A. and Nowicki, L.M. (1991), *The History of State Planning in New Jersey, 1935-1991*, unpublished paper presented at the Fourth National Conference on American Planning History

Buchsbaum, P.A. (1989a), *Summary of the Differences between League of Municipalities Draft of County Municipal Planning Partnership Bill*, unpublished paper

Buchsbaum, P.A. (1989b), *Summary Statement: New Jersey League of Municipalities' Draft and DOT's Draft of the County Municipal Planning Partnership Bill*, unpublished paper

Buchsbaum, P.A. (1993a), *Toward a Planned Center: A Procedural Guide to the Development of a Center*, Center for Analysis of Public Issues, Princeton NJ

Center for Urban Policy Research, Rutgers University (CUPR) (1992), *Impact Assessment of the New Jersey Interim Development and Redevelopment Plan*, 2 Vols., 28 February, New Brunswick NJ

Cidon, M. (1993), *Review of the Efficacy and the Efficiency of the Cross-Acceptance Process*, unpublished paper

DeGrove, J.M. (1991), *A Review of "Communities of Place: Interim State Development and Redevelopment Plan"*, unpublished paper

Liberty, R.L. (1991), *A Critique of "Communities of Place: The Interim State Development and Redevelopment Plan" for the State of New Jersey*, unpublished paper

Middlesex-Somerset-Mercer Regional Council (MSM) (1987), *Regional Forum: An Action Agenda for Managing Regional Growth*, MSM, Princeton NJ

Middlesex-Somerset-Mercer Regional Council (MSM) (1989), *The Growth Management Handbook: A Primer for Citizen and Government Planners*, MSM, Princeton NJ

Middlesex-Somerset-Mercer Regional Council (MSM) (n.d.), *Planning for Agriculture in New Jersey*, MSM, Princeton NJ

[*] Unless otherwise stated, translations of non-English citations are by Enrico Gualini.

New Jersey Department of Community Affairs, Division of State and Regional Planning (1977), *State Development Guide Plan*, Trenton NJ

New Jersey Future (1990a), *The State Plan: Realizing the Vision*, Highland Park NJ

New Jersey Future (1990b), *The Cities' Stake in a State Plan: An Urban Agenda for the 1990s*, Highland Park NJ

New Jersey Office of State Planning (OSP) (1987), *Growth Management Approaches*, Technical Reference Document prepared by: Freilich, Leitner, Carlisle & Shortlidge, Kansas City KA, February 25, Trenton NJ

New Jersey Office of State Planning (OSP) (1988a), *Cross-Acceptance Manual*, Trenton NJ

New Jersey Office of State Planning (OSP) (1988b), *Peer Review Technical Advisory Committee Report on the Draft Preliminary State Development and Redevelopment Plan Submitted to the Office of State Planning*, Trenton NJ

New Jersey Office of State Planning (OSP) (1988c), *Regions of Communities: A Regional Design System for Developing and Redeveloping Communities of Place*, Trenton NJ

New Jersey Office of State Planning (OSP) (1989), *Description of the Negotiation Phase of Cross-Acceptance*, Trenton NJ

New Jersey Office of State Planning (OSP) (1990), *Appendices to Cross-Acceptance Issues: Preliminary Staff Analysis*, Trenton NJ

New Jersey Office of State Planning (OSP) (1991), *Peer Review Planning Advisory Committee*, Trenton NJ

New Jersey Office of State Planning (OSP) (1992), *Summary of the Cross-Acceptance Debriefing Conducted on November 13, 1992*, Trenton NJ

New Jersey Office of State Planning (OSP) (1993a), *Guidelines for Establishing an Urban Complex, Technical Memorandum 2*, Trenton NJ

New Jersey Office of State Planning (OSP) (1993b), *Guidelines for Preparing a Strategic Revitalization Plan and Program, Technical Memorandum 1*, Trenton NJ

New Jersey Office of State Planning (OSP) (1993c), *The Centers Designation Process*, Trenton NJ

New Jersey Office of State Planning (OSP) (1994), *Cross-Acceptance II: Revising the Cross-Acceptance Process for Triennial Revision and Readoption of the State Development and Redevelopment Plan*, Trenton NJ

New Jersey State and Local Expenditure and Revenue Policy Commission (SLERP) (1988), *Final Report*, Trenton NJ

New Jersey State Legislature (1983), *Public Hearing before Senate Legislative Oversight Committee on State Response to the Mount Laurel II Supreme Court Decision*, Trenton NJ

New Jersey State Legislature (1984), *Public Hearing before State Senate Government, Federal and Interstate Relations and Veterans' Affairs Committee on Senate Bill 1464*, Trenton NJ

New Jersey State Planning Commission (SPC) (1987), *Trends and Hard Choices: Setting Objectives for New Jersey's Future*, Trenton NJ

New Jersey State Planning Commission (SPC) (1988a), *Communities of Place: The Preliminary State Development and Redevelopment Plan for the State of New Jersey, Vol. I: A Legacy for the Next Generation*, Trenton NJ

New Jersey State Planning Commission (SPC) (1988b), *Communities of Place: The Preliminary State Development and Redevelopment Plan for the State of New Jersey, Vol. II: Strategies and Policies*, Trenton NJ

New Jersey State Planning Commission (SPC) (1989), *Communities of Place: The Preliminary State Development and Redevelopment Plan for the State of New Jersey, Vol. III: Planning Standards and Guidelines*, Trenton NJ

New Jersey State Planning Commission (SPC) (1991), *Communities of Place: The Interim State Development and Redevelopment Plan for the State of New Jersey*, Trenton NJ

New Jersey State Planning Commission (SPC) (1992a), *Communities of Place: The New Jersey State Development and Redevelopment Plan*, Trenton NJ

New Jersey State Planning Commission (SPC) (1992b), *New Jersey State Planning Commission Report on Implementation Issues*, Trenton NJ

New Jersey State Planning Commission (SPC) (1992c), *Statement of Agreements and Disagreements of the State Development and Redevelopment Plan, prepared by: The New Jersey Office of State Planning*, Trenton NJ

Seskin, S.N. (1991), *The Interim State Plan from an Economic Perspective*, unpublished paper

State of New Jersey County and Municipal Government Study Commission (1985), *Functional Fragmentation and the Traditional Forms of Municipal Government in New Jersey*, Trenton NJ

State of New Jersey County and Municipal Government Study Commission (1986), *The Structure of County Government: Current Status and Needs*, Trenton NJ

State of New Jersey Executive Department (1994), *Executive Order No. 114*, Trenton NJ

Codes and Periodicals

MSM Review, quarterly, Middlesex-Somerset-Mercer Regional Council (MSM), Princeton NJ

New Jersey Administrative Codes (N.J.A.C.)

New Jersey Municipalities, monthly, New Jersey State League of Municipalities, Trenton NJ

New Jersey Register (N.J.R.)

New Jersey Reporter, monthly, Center for the Analysis of Public Issues, Princeton NJ

New Jersey Statutes Annotated (N.J.S.A.)

OSP Planning Memo, monthly, New Jersey Office of State Planning (OSP), Trenton NJ

State Planning Notes, biannual, New Jersey State Planning Commission (SPC) and New Jersey Office of State Planning (OSP), Trenton NJ

The New Jersey Planner, bimonthly, New Jersey Planning Officials, Watchung NJ

The State Planning Bullettin, monthly, New Jersey State Planning Commission (SPC) and New Jersey Office of State Planning (OSP), Trenton NJ

Secondary Sources

Alexander, E.R. (1982), 'Design in the Decision-Making Process', *Policy Sciences*, Vol. 14, No. 3, pp. 279-92

Alexander E.R. (1992), 'A Transaction Cost Theory of Planning', *Journal of the American Planning Association*, Vol. 58, No. 2, pp. 190-200

Alexander, E.R. (1994), 'To Plan or Not To Plan, That is the Question: Transaction Cost Theory and its Implications for Planning', *Environment and Design B: Planning and Design*, Vol. 21, No. 3, pp. 341-52

Alexander, E.R. (2000), 'Inter-organizational Coordination and Strategic Planning: the Architecture of Institutional Design', in Salet, W. and Faludi, A. (eds.) (2000)

Alexander, E.R. and Faludi, A. (1996) 'Planning Doctrine: its Uses and Implications', *Planning Theory*, 16, pp. 11-61

Altshuler, A. (1965a), *The City Planning Process*, Cornell University Press, Ithaca NY

Altshuler, A. (1965b), 'The Goals of Comprehensive Planning', *Journal of the American Institute of Planners*, Vol. 31, No. 3, pp. 186-97

Amin, A. and Thrift, N. (1995), 'Institutional Issues for European Regions: from Markets and Plans to Socioeconomics and Powers of Association', *Economy and Society*, Vol. 24, No. 1, pp. 41-66

Amin, A. (ed.) (1994), *Post-Fordism: A Reader*, Blackwell, Oxford

Amin, A. and Hausner, J. (1998), 'Interactive Governance and Social Complexity', in idem (eds.) (1998), *Beyond Market and Hierarchy: Interactive Governance and Social Complexity*, Edward Elgar, Aldershot

Amin, A. and Thrift, N. (1994), 'Living in the Global', in idem (eds.) (1994), *Globalization, Institutions, and Regional Development in Europe*, Oxford University Press, Oxford

Anderson, B. (1991), *Invented Communities: Reflections on the Origin and Spread of Nationalism*, Verso, London

Anglin, R. (1990), 'Diminishing Utility: The Effect on Citizen Preferences for Local Growth', *Urban Affairs Quarterly*, Vol. 25, No. 4, pp. 684-96

Anglin, R. (1994), 'Searching for Justice: Court-Inspired Housing Policy as a Mechanism for Social and Economic Mobility', *Urban Affairs Quarterly*, Vol. 29, No. 3, pp. 432-53

Argyris, C. (1993), *Knowledge for Action: A Guide to Overcoming Barriers to Organizational Change*, Jossey-Bass, San Francisco CA

Argyris C. and Schön, D.A. (1974), *Theory in Practice: Increasing Professional Effectiveness*, Jossey-Bass, San Francisco CA

Argyris, C. and Schön, D.A. (1978), *Organizational Learning: A Theory of Action Perspective*, Addison-Wesley, Reading MA

Arrow, K.J. (1963), *Social Choice and Individual Values*, Wiley & Sons, New York NY

Axelrod, R. (1984), *The Evolution of Cooperation*, Basic Books, New York NY

Axelrod, R. (1986), 'An Evolutionary Approach to Norms', *American Political Science Review*, Vol. 80, No. 4, pp. 1095-111

Babcock, R.F. (1966), *The Zoning Game: Municipal Practices and Policies*, University of Wisconsin Press, Madison WI

Babcock, R.F. and Siemon, C.L. (1985), *The Zoning Game Revisited*, Oelschlager Gunn & Hain, Boston MA

Bachrach, P. and Baratz, M.S. (1970), *Power and Poverty*, Oxford University Press, New York NY

Bachrach, P. and Baratz, M.S. (1962), 'Two Faces of Power', *American Political Science Review*, Vol. 56, pp. 947-52

Bagnasco, A. (1988), *La costruzione sociale del mercato. Studi sullo sviluppo della piccola impresa in Italia*, Il Mulino, Bologna

Bagnasco, A. (1994), *Fatti sociali formati nello spazio. Cinque lezioni di sociologia urbana e regionale*, Franco Angeli, Milano

Baldassare, M. (1986), *Trouble in Paradise: The Suburban Transformation in America*, Columbia University Press, New York NY

Balducci, A. and Fareri, P. (1996), 'Consensus Building as a Strategy to Cope with Planning Problems at Different Territorial Levels: Examples from the Italian Case', paper presented at the ACSP-AESOP Joint International Congress *Local Planning in a Global Environment*, Toronto, July 25-28

Bardach, E. (1977), *The Implementation Game: What Happens When a Bill Becomes a Law*, MIT Press, Cambridge MA

Barret, S. and Fudge, C. (eds.) (1981), *Policy and Action: Essays on the Implementation of Public Policy*, Methuen, London

Bateson, G. (1972), *Steps to an Ecology of Mind*, Ballantine, New York NY

Baum, H.S. (1980a), 'Analysts and Planners Must Think Organizationally', *Policy Analysis*, Vol. 6

Baum, H.S. (1980b), 'Sensitizing Planners to Organizations', in Clavel, P., Forester, J. and Goldsmith, W.W. (eds.) (1980) *Urban and Regional Planning in an Age of Austerity*, Pergamon Press, Oxford

Baum, H.S. (1983), *Planners and Public Expectations*, Schenkman, Cambridge MA

Baum, H.S. (1995), 'Book Review' of Schön, D.A., and Rein, M., "Frame Reflection", *Journal of the American Planning Association*, Vol. 61, No. 2, p. 277

Baum, J.A. and Oliver, C. (1991), 'Institutional Linkages and Organizational Mortality', *Administrative Science Quarterly*, Vol. 36, No. 2, pp. 187-218

Bebout, J. and Grele, R. (1964), *Where Cities Meet: The Urbanization of New Jersey*, Princeton University Press, Princeton NJ

Benhabib, S. (1990), 'Afterword: Communicative Ethics and Contemporary Controversies in Practical Philosophy', in Benhabib, S. and Dallmayr, F. (eds.) (1990), *The Communicative Ethics Controversy*, MIT Press, Cambridge MA

Benveniste, G. (1977), *The Politics of Expertise*, Boyd and Frazer, San Francisco CA

Berger, P.L., Berger, B. and Kellner, H. (1973), *The Homeless Mind: Modernization and Consciousness*, Random House, New York NY

Berger, P.L. and Luckmann, T. (1966), *The Social Construction of Reality: A Treatise in the Sociology of Knowledge*, Doubleday, Garden City NY

Blanco, H. (1994), *How to Think about Social Problems: American Pragmatism and the Idea of Planning*, Greenwood Press, Westport CN

Blau, J.R. (1991), 'When Weak Ties Are Structured', in Blau, J.R. and Goodman, N. (eds.) (1991)

Blau, J.R. and Goodman, N. (eds.) (1991), *Social Roles and Social Institutions: Essays in Honor of Rose Laub Coser*, Westview Press, Boulder CO

Blau, P.M. (1964), *Exchange and Power in Social Life*, John Wiley and Sons, New York NY

Bolan, R. (1991), 'Planning and Institutional Design', *Planning Theory*, Vol. 5-6, pp. 7-34

Bolan, R. (2000), 'Social Interaction and Institutional Design: The Case of Housing in the United States', in Salet, W. and Faludi, A. (eds.) (2000)

Bollens, S.A. (1992), 'State Growth Management: Intergovernmental Frameworks and Policy Objectives', *Journal of the American Planning Association*, Vol. 58, No.4, pp. 454-66

Börzel, T.A. (1997), *Policy Networks. A New Paradigm for European Governance*, EUI Working Papers No. 19, Robert Schuman Centre, European University Institute, San Domenico di Fiesole

Boudon R. (1977), *Effets pervers et ordre social*, Presses Universitaires de France, Paris

Boudon, R. (1984), *La place du désordre. Critique des théories du changement social*, Presses Universitaires de France, Paris

Boudon, R. (1987), 'Razionalità e teoria dell'azione', *Rassegna Italiana di Sociologia*, Vol. 28, No. 2, pp. 175-203

Bourdieu, P. (1994), *Raisons pratiques. Sur la théorie de l'action*, Editions de Seuil, Paris; engl. transl. (1998) *Practical Reason: On the Theory of Action*, Polity, Cambridge

Bromley, D.W. (1989), *Economic Interests and Institutions: The Conceptual Foundations of Public Policy*, Basil Blackwell, New York NY

Browne, A. and Wildavsky, A. (1983a), 'Implementation as Mutual Adaptation', in Pressmann, J.L. and Wildavsky, A. (1984)

Browne, A. and Wildavsky, A. (1983b), 'Implementation as Exploration', in Pressmann, J.L. and Wildavsky, A. (1984)

Bryson, J.M. and Crosby, B.C. (1992), *Leadership for the Common Good: Tackling Public Problems in a Shared-Power World*, Jossey-Bass, San Francisco CA

Bryson, J.M. and Einsweiler, R.C. (eds.) (1992), *Shared Power*, University Press of America, Canham MD

Buchsbaum, P.A. (1985), 'No Wrong Without Remedy: The New Jersey Supreme Court's Effort to Bar Exclusionary Zoning', *The Urban Lawyer*, Vol. 17, No. 1, Winter; now in Porter, D.R. (ed.) (1986)

Buchsbaum, P.A. and Smith, L.J. (eds.) (1993), *State and Regional Comprehensive Planning: Implementing New Methods for Growth Management*, American Bar Association, Chicago IL

Buchsbaum, P.A. (1977), 'The Irrelevance of the "Developing Municipalities" Concept', in Rose, J.G. and Rothman, R.E. (eds.) (1977)

Buchsbaum, P.A. (1991), 'The Courts', *New Jersey Reporter*, Vol. 21, No. 3, pp. 40-1

Buchsbaum, P.A. (1993b), 'Mount Laurel II: A Ten Year Perspective', *New Jersey Lawyer*, October, pp. 13-7

Buchsbaum, P.A. (1993c), 'The New Jersey Experience', in Buchsbaum, P.A. and Smith, L.J. (eds.) (1993)

Burchell, R.W. (1993), 'Issues, Actors, and Analysis in Statewide Comprehensive Planning', in Buchsbaum, P.A. and Smith, L.J. (eds.) (1993)

Burchell, R.W. (1995), *Mount Laurel and its Progeny after Twenty Years: Examples of Strategies for Affordable Housing Implementation*, unpublished paper prepared for "The 20th Anniversary of Mount Laurel", 1995 Annual Planning Conference, Toronto, Canada, April 8-12

Burchell, R.W. and Listokin, D. (1985), *The New Practitioner's Guide to Fiscal Impact Analysis*, Center for Urban Policy Research, Rutgers University, New Brunswick NJ

Burchell, R.W., Beaton, W.P., Listokin, D., Sternlieb, G., Lake, R.W. and Florida, R.L. (1983), *Mount Laurel II: Challenge and Delivery of Low-Cost Housing*, Center for Urban Policy Research, Rutgers University, New Brunswick NJ

Butzin, B. (1996), *Bedeutung kreativer Milieus für die Regional- und Landesentwicklung*, Lehrstuhl Wirtschaftsgeographie und Regionalplanung Universität Bayreuth, Bayreuth

Camagni, R. (ed.) (1991), *Innovative Networks: Spatial Perspectives*, Belhaven, London/New York

Chisholm, D. (1989), *Coordination Without Hierarchy*, University of California Press, Berkeley CA

Christensen, K.S. (1985), 'Coping with Uncertainty in Planning', *Journal of the American Planning Association*, Vol. 51, No.1, pp. 63-73

Cohen, M.D., March, J.G. and Olsen, J.P. (1972), 'A Garbage Can Model of Organizational Choice', *Administrative Science Quarterly*, Vol. 17, No. 1, pp. 1-25

Coleman, J. (1990), *Foundations of Social Theory*, Harvard University Press, Cambridge MA

Commons, J.R. (1957), *Legal Foundations of Capitalism*, University of Wisconsin Press, Madison WI

Connors, R.J. and Dunham, W.J. (1993), *The Government of New Jersey: An Introduction*, University Press of America, Lanham MD

Cook, K.S. and Levi, M. (eds.) (1990), *The Limits of Rationality*, University of Chicago Press, Chicago IL

Cooke, P. and Morgan, K. (1993), 'The Network Paradigm: New Departures in Corporate and Regional Development', *Environment and Planning D: Society and Space*, Vol. 11, pp. 543-64

Coser, L.A. (1991), 'Role-Set Theory and Individual Autonomy', in Blau, J.R. and Goodman, N. (eds.) (1991)

Coser, R.L. (1975), 'The Complexity of Roles as a Seedbed of Individual Autonomy', in Coser, L.A. (ed.) (1975), *The Idea of Social Structure: Papers in Honor of Robert K. Merton*, Harcourt Brace & Co., New York NY

Coser, R.L. (1991), *In Defense of Modernity: Role Complexity and Individual Autonomy*, Stanford University Press, Stanford CA

Council on New Jersey Affairs (1989a), *Land-Use Planning and Growth Management in New Jersey: A Policy Conversation*, Program for New Jersey Affairs, Working Paper No. 15, Woodrow Wilson School of Public and International Affairs, Princeton University, Princeton NJ

Council on New Jersey Affairs (1989b), *Prospects for New Jersey's Cities: A Conference Report*, Program for New Jersey Affairs, Working Paper No. 7, Woodrow Wilson School of Public and International Affairs, Princeton University, Princeton NJ

Council on New Jersey Affairs (1989c), *Prospects for New Jersey's Cities: A Policy Conversation*, Program for New Jersey Affairs, Working Paper No. 17, Woodrow Wilson School of Public and International Affairs, Princeton University, Princeton NJ

Cremaschi, M. (1994), *Esperienza comune e progetto urbano*, Franco Angeli, Milano

Crespi, F. (1982), *Mediazione simbolica e società*, Franco Angeli, Milano

Crespi, F. (1989), *Azione sociale e potere*, Il Mulino, Bologna; engl. transl. (1992) *Social Power and Action*, Blackwell, Cambridge

Crespi, F. (1986), 'Osservanza delle regole e rapporto con le regole', *Rassegna Italiana di Sociologia*, Vol. 27, No. 3, pp. 399-407

Crosta, P.L. (1990), *La politica del piano*, Franco Angeli, Milano

Crozier, M. and Friedberg, E. (1977), *L'acteur et le système*, Paris, Seuil; engl. transl. (1980) *Actors and Systems: The Politics of Collective Action*, University of Chicago Press, Chicago IL

Cullingworth, J. (1993), *The Political Culture of Planning: American Land-Use Planning in Comparative Perspective*, Routledge, London

Cyert, R.M. and March, J.G. (1963), *A Behavioral Theory of the Firm*, Prentice-Hall, Englewood Cliffs NJ

Dahl, R.A. (1961), *Who Governs? Democracy and Power in an American City*, Yale University Press, New Haven CT

Dahl, R.A. (1971), *Polyarchy: Participation and Opposition*, Yale University Press, New Haven CT

Danielson, M.N. (1976), *The Politics of Exclusion*, Columbia University Press, New York NY

Danielson, M.N. and Doig, J.W. (1982), *New York: The Politics of Urban Regional Development*, University of California Press, Berkeley CA

Davidoff, P. (1965), 'Advocacy and Pluralism in Planning', *Journal of the American Institute of Planners*, Vol. 31, No. 4, pp. 331-38

De Leonardis, O. (1997), 'Declino della sfera pubblica e privatismo', *Rassegna Italiana di Sociologia*, Vol. 38, N. 2, pp. 169-93

De Neufville, J.I. (1983), *Planning Theory and Practice: Bridging the Gap*, Working Paper No. 402, Institute of Urban and Regional Development, University of California, Berkeley CA

De Neufville, J.I. and Barton, S.E. (1987), 'Myths and the Definition of Policy Problems. An Exploration of Home Ownership and Public-Private Partnerships', *Policy Sciences*, Vol. 20, No. 3, pp. 181-206

De Rita, G. and Bonomi, A. (1998), *Manifesto per lo sviluppo locale*, Bollati Boringhieri, Torino

DeGrove, J.M. (1984), *Land, Growth, and Politics*, American Planning Association, Chicago IL

DeGrove, J.M. (1986), 'Creative Tensions in State/Local Relations', in Porter, D.R. (ed.) (1986)

DeGrove, J.M. (1993), 'The Emergence of State Planning and Growth Management Systems: An Overview', in Buchsbaum, P.A. and Smith, L.J. (eds.) (1993)

DeGrove, J.M. and Miness, D.A. (1992), *Planning and Growth Management in the States: The New Frontier for Land Policy*, Lincoln Institute of Land Policy, Cambridge MA

DeLeon, P. (1988), 'The Contextual Burdens of Policy Design', *Policy Studies Journal*, Vol. 17, No. 2, pp. 297-309

Dematteis, G. (1989), 'Regioni geografiche, articolazione territoriale degli interessi e regioni istituzionali', *Stato e mercato*, No. 27, pp. 115-37

Dematteis, G. (1995), *Progetto implicito. Il contributo della geografia umana alle scienze del territorio*, Franco Angeli, Milano

Dente, B. (1990), 'Introduzione', in idem (ed.) (1990), *Le politiche pubbliche in Italia*, Il Mulino, Bologna

Dente, B. (1998), 'Fabbricare istituzioni: alcune regole pratiche', in IRS - Istituto per la Ricerca Sociale, *Fare ricerca economica e sociale*, IRS, Milano

Dickens, P. (1990), *Urban Sociology: Society, Locality and Human Nature*, Harvester Wheatsheaf, New York NY

DiMaggio, P.J. (1988), 'Interest and Agency in Institutional Theory', in Zucker, L.G. (ed.) (1988)

DiMaggio, P.J. and Powell, W.W. (1991), 'Introduction', in Powell, W.W. and DiMaggio, P.J. (eds.) (1991)

DiMaggio, P.J. and Powell, W.W. (1983), 'The Iron Cage Revisited: Institutional Isomorphism and Collective Rationality in Organizational Fields', *American Sociological Review*, Vol. 48, No. 2, pp. 147-60; now in: Powell, W.W. and DiMaggio, P.J. (eds.) (1991)

Donolo, C. (1997), *L'intelligenza delle istituzioni*, Feltrinelli, Milano

Donolo, C. and Fichera, F. (eds.) (1988), *Le vie dell'innovazione*, Feltrinelli, Milano

Douglas, M. (1986), *How Institutions Think*, Syracuse University Press, Syracuse NY

Douglas, M. (1992), 'The Normative Debate and the Origins of Culture', in *Risk and Blame*, Routledge, London/New York NY

Douglas, M. and Wildavsky, A. (1982), *Risk and Culture: An Essay on the Selection of Technical and Environmental Dangers*, University of California Press, Berkeley CA

Dryzek, J.S. (1982), 'Policy Analysis as a Hermeneutic Activity', *Policy Sciences*, Vol. 14, No. 4, pp. 309-29

Dryzek, J.S. (1990), *Discursive Democracy: Politics, Policy and Political Science*, Cambridge University Press, Cambridge

Dugger, W.M. (ed.) (1989), *Radical Institutionalism*, Greenwood Press, New York NY

Dunn, W.N. (1993), 'Policy Reform as Arguments', in Fischer, F. and Forester, J. (eds.) (1993)

Elias, N. (1985), *Engagement und Distanzierung: Arbeiten zur Wissenssoziologie I*, Suhrkamp, Frankfurt a. M.

Elster, J. (1991) *Arguing and Bargaining in the Federal Convention and the Assemblée Constituante*, Working Paper, Center for the Study of Constitutionalism in Eastern Europe, School of Law, University of Chicago, Chicago IL

Emirbayer, M. and Goodwin, J. (1994), 'Network Analysis, Culture, and the Problem of Agency', *American Journal of Sociology*, Vol. 99, No. 6, pp. 1411-54

Epling, J.W. (1992), 'The New Jersey State Planning Process: An Experiment in Intergovernmental Negotiations', in Stein, J. (ed.) (1992)

Evans, P.B., Rueschemeyer, D. and Skocpol, T. (eds.) (1985), *Bringing the State Back In*, Cambridge University Press, New York NY

Faludi, A. (1996), 'Framing with Images', *Environment and Planning C*, 23, 1, pp. 93-108

Faludi, A., (ed.) (1973), *A Reader in Planning Theory*, Pergamon Press, Oxford

Faludi, A. and van der Walk, A. (1994), *Rule and Order: Dutch Planning Doctrine in the Twentieth Century*, Kluwer, Dordrecht

Feldman, M.S. and March, J.G. (1981,) 'Information in Organizations as Signal and Symbol', *Administrative Science Quarterly*, Vol. 26, No.1, pp. 171-86

Ferrari, G. (1986), 'Problemi di epistemologia delle regole', *Rassegna Italiana di Sociologia*, Vol. 27, No. 3, pp. 369-97

Fischer, F. and Forester, J. (eds.) (1993), *The Argumentative Turn in Policy Analysis and Planning*, Duke University Press, Durham NC

Fischer, F. and Forester, J. (1993), 'Editors' Introduction', in Fischer, F. and Forester, J. (eds.) (1993)

Fischer, R. and Ury, W.L. (1981), *Getting to Yes: Negotiating Agreement without Giving In*, Penguin, New York NY

Fishman, R. (1987), *Bourgeois Utopias: The Rise and Fall of Suburbia*, Basic Books, New York NY

Florida, R. (1995), 'Towards the Learning Region', *Futures*, Vol. 27, No. 5, pp. 527-36

Forester, J. (1980), 'Critical Theory and Planning Practice', in Clavel, P., Forester, J. and Goldsmith, W.W. (eds.) (1980)

Forester, J. (1982), 'Planning in the Face of Power', *Journal of the American Planning Association*, Vol. 48, No. 1, pp. 67-80

Forester, J. (1989), *Planning in the Face of Power*, University of California Press, Berkeley CA

Forester, J. (1993a), 'Critical Etnography: On Fieldwork in an Habermasian Way', in Alvesson, M. and Malmott, H. (eds.) (1992), *Critical Management Studies*, Sage, Newbury Park CA

Forester, J. (1993b), *Critical Theory, Public Policy and Planning Practice: Toward a Critical Pragmatism*, State University of New York Press, New York NY

Forester, J. (1994), 'Bridging Interests and Community: Advocacy Planning and the Challenges of Deliberative Democracy', *Journal of the American Planning Association*, Vol. 60, No. 2, pp. 153-58

Forester, J. and Krumholz, N. (1988), 'L'urbanistica tra la pressione del potere e l'urgenza del bisogno', in Triennale di Milano, *Le città del mondo e il futuro delle metropoli. Partecipazioni internazionali*, vol. II, ed. by L. Mazza, Electa, Milano

Forester, J. and Krumholz, N. (1990), *Making Equity Planning Work: Leadership in the Public Sector*, Temple University Press, Philadelphia PA

Forester, J. (1999), *The Deliberative Practitioner: Encouraging Participatory Planning Processes*, MIT Press, Cambridge MA

Friedland, R. and Alford, R. (1991), 'Bringing Society Back In: Symbols, Practices, and Institutional Contradictions', in Powell, W.W. and DiMaggio, P.J. (eds.) (1991)

Friedmann, J. (1973), *Retracking America: A Theory of Transactive Planning*, Doubleday and Anchor Books, New York NY

Friedmann, J. (1987), *Planning in the Public Domain: From Knowledge to Action*, Princeton University Press, Princeton NJ

Friedmann, J. (1990), 'Human Territoriality and the Struggle for Place', *Giornale del Dottorato in Pianificazione Territoriale*, No. 1, p. 10

Friedmann, J. (1992), *Empowerment: The Politics of Alternative Development*, Blackwell, Oxford

Friedmann, J. and Abonyi, G. (1976), 'Social Learning: A Model for Policy Research', *Environment and Planning A*, Vol. 8, No. 8, pp. 927-40

Friedmann, J. and Weaver, C. (1979), *Territory and Function: The Evolution of Regional Planning*, Arnold, London

Friend, J.K. and Hunter, J.M.H. (1970), 'Multi-Organizational Decision Processes in the Planned Expansion of Towns', *Environment and Planning*, Vol. 2, No. 1, pp. 23-54

Gale, D.E. (1992), 'Eight State Sponsored Growth Management Programs: A Comparative Analysis', *Journal of the American Planning Association*, Vol. 58, No. 4, pp. 425-39

Garfinkel, H. (1967), *Studies in Ethnomethodology*, Prentice Hall, Englewood Cliffs NJ

Garreau, J. (1991), *Edge City: Life on the New Frontier*, Doubleday, New York NY

Geertz, C. (1973), *The Interpretation of Cultures*, Basic Books, New York NY

Geertz, C. (1980), *Negara: The Theatre State in Nineteenth-Century Bali*, Princeton University Press, Princeton NJ

Geertz, C. (1983), *Local Knowledge: Further Essays in Interpretive Anthropology*, Basic Books, New York NY

Giddens, A. (1984), *The Constitution of Society*, Polity Press, Cambridge

Giddens, A. (1990), *The Consequences of Modernity*, Stanford University Press, Stanford CA

Giddens, A. (1991), *Modernity and Self-Identity: Self and Society in the Late Modern Age*, Polity, Cambridge

Giglioli, P.P. (1989), 'Teorie dell'azione', in Panebianco, A. (ed.) (1989), *L'analisi della politica*, Il Mulino, Bologna

Goertz, M.E. (1988), *Excerpts from the Initial Decision of the Office of Administrative Law in the Case Abbott v. Burke*, Council on New Jersey Affairs, Program for New Jersey Affairs, Working Paper No. 16, Woodrow Wilson School of Public and International Affairs, Princeton University, Princeton NJ

Gottlieb, H. (1988), 'For State Planners, It's Policy (and Politics)', *New Jersey Law Journal*, Vol. 122, No. 10, September

Gouldner, A.W. (1984), *Against Fragmentation: The Origins of Marxism and the Sociology of the Intellectuals*, Oxford University Press, Oxford

Grabher, G. (ed.) (1993), *The Embedded Firm: On the Socioeconomics of Industrial Networks*, Routledge, London/New York

Granovetter, M.S. (1973), 'The Strength of Weak Ties', *American Journal of Sociology*, Vol. 78, No. 6, pp. 1360-80

Granovetter, M.S. (1985), 'Economic Action and Social Structure: The Problem of Embeddedness', *American Journal of Sociology*, Vol. 91, No. 3, pp. 481-510

Granovetter, M.S. and Swedberg, R. (eds.) (1992), *The Sociology of Economic Life*, Westview Press, Boulder CO

Gray, B. and Wood, D.J. (1991a), 'Collaborative Alliances: Moving From Practice to Theory', *Journal of Applied Behavioral Science*, Vol. 27, No. 1, March, pp. 3-22

Gray, B. and Wood, D.J. (1991b), 'Toward a Comprehensive Theory of Collaboration', *Journal of Applied Behavioral Science*, Vol. 27, No. 2, June, pp. 139-62

Gruber, J. (1993), *Coordinating Growth Management through Consensus Building: Incentives and the Generation of Social, Intellectual and Political Capital*, IURD Working Paper 617, University of California, Berkeley CA

Gualini, E. (forthc.) 'Institutional Capacity Building as an Issue of Collective Action and Institutionalisation: Some Theoretical Remarks', in Cars, G., Healey, P., Madanipour, A., and de Magalhaes, C. (eds.) (forthc.), *Institutional Capacity and Urban Governance*

Haas, P. (1990), *Saving the Mediterranean*, Columbia University Press, New York NY

Habermas, J. (1981), *Theorie des kommunikativen Handelns*, Suhrkamp, Frankfurt a.M.; engl. transl. (1984) *The Theory of Communicative Action*, Polity, Cambridge

Habermas, J. (1983), *Moralbewusstsein und kommunikatives Handeln*, Suhrkamp, Frankfurt a.M.; engl. transl. (1990) *Moral Consciousness and Communicative Action*, Polity, Cambridge

Hajer, M.A. (1993), 'Discourse Coalitions and the Institutionalization of Practice: The Case of Acid Rain in Britain', in Fischer, F. and Forester, J. (eds.) (1993)

Hajer, M.A. (1995), *The Politics of Environmental Discourse: Ecological Modernisation and the Policy Process*, Oxford University Press, Oxford

Hajer, M.A. (2000), 'Transnational Networks as Transnational Policy Discourse: Some Observations on the Politics of Spatial Development in Europe', in Salet, W. and Faludi, A. (eds.) (2000)

Hall, P.A. (1986), *Governing the Economy: The Politics of State Intervention in Britain and France*, Polity Press, Cambridge

Hall, P.A. (1993), 'Policy Paradigms, Social Learning, and the State: The Case of Economic Policy Making in Britain', *Comparative Politics*, Vol. 25, No. 3, pp. 275-96

Hall, F.W. (1977), 'An Orientation to Mount Laurel', in Rose, J.G. and Rothman, R.E. (eds.) (1977)

Hall, P.A. and Taylor, R.C.R. (1996), 'Political Science and the Three New Institutionalisms', *Political Studies*, Vol. 44, pp. 936-57

Hardin, G. (1968), 'The Tragedy of the Commons', *Science*, Vol. 162, pp. 1243-48

Hardin, R. (1982), *Collective Action*, Johns Hopkins University Press, Baltimore MD

Harding, A. (1994), 'Urban Regimes and Growth Machines: Towards a Cross-National Research Agenda', *Urban Affairs Quarterly*, Vol. 29, No. 3, pp. 356-82

Harding, A. (1997), 'Urban Regimes in a Europe of the Cities?', *European Urban and Regional Studies*, Vol. 4, No. 4, pp. 291-314

Harding, A. and Garside, P. (1993), *Globalization, Urban Political Economy, and Community Power*, European Institute for Urban Affairs, John Moores University, Liverpool

Harvey, J.M. and Katovich, M.A. (1992), 'Symbolic Interactionism and Institutionalism: Common Roots', *Journal of Economic Issues*, Vol. 26, No. 3, pp. 791-812

Hassink, R. (1997), 'Die Bedeutung der Lernenden Region für die regionale Innovationsförderung', *Geographische Zeitschrift*, Vol. 85, No. 2-3, pp. 159-73.

Häussermann, H. und Siebel, W. (1994), 'Wie organisiert man Innovation in nichtinnovativen Milieus?', in Kreibich, R., Schmid, A.S., Siebel, W., Sieverts, T. and Zlocnicky, P (ed.) (1994), *Bauplatz Zukunft. Dispute über die Entwicklung von Industrieregionen*, Klartext Verlag, Essen

Healey, P. (1992), 'A Planner's Day: Knowledge and Action in Communicative Practice', *Journal of the American Planning Association*, Vol. 58, No. 1, pp. 9-20

Healey, P. (1993a), 'The Communicative Turn in Planning Theory', in Fischer, F. and Forester, J. (eds.) (1993)

Healey, P. (1993b), 'The Communicative Work of Development Plans', *Environment and Planning B: Planning and Design*, Vol. 20, No. 1, pp. 83-104

Healey, P. (1996), *Collaborative Planning: Shaping Places in Fragmented Societies*, Macmillan, Basingstoke and London

Healey, P., Khakee, A., Motte, A., and Needham, B. (eds.) (1997), *Making Strategic Spatial Plans*, University College of London Press, London

Hechter, M. (1987), *Principles of Group Solidarity*, University of California Press, Berkeley CA

Heclo, H. (1974), *Modern Social Policy in Britain and Sweden*, Yale University Press, New Haven CT

Heclo, H. and Wildavsky, A. (1974), *The Private Government of Public Money*, Macmillan, London

Heinelt, H. and Wollmann, H. (eds.) (1991), *Brennpunkt Stadt. Stadtpolitik und lokale Politikforschung in den 80er und 90er Jahren*, Birkhäuser, Basel/Berlin/Boston

Heinelt, H. and Smith, R. (eds.) (1996), *Policy Networks and European Structural Funds*, Avebury, Aldershot

Heiner, R.A. (1983), 'The Origin of Predictable Behavior', *American Economic Review*, Vol. 73, No. 4, pp. 560-95

Heiner, R.A. (1985), 'Rational Expectations when Agents Imperfectly Use Information', *Journal of Post-Keynesian Economics*, Vol. 8, No. 2, pp. 201-7

Heinze, R.G. and Schmid, J. (1997), 'Industrial Change and Meso-Corporatism: A Comparative View on Three German States', *European Planning Studies*, Vol. 5, No. 5, pp. 597-617

Heinze, R.G. and Voelzkow, H. (1991), 'Kommunalpolitik und Verbände. Inszenierter Korporativismus auf lokaler und regionaler Ebene?', in Heinelt, H. and Wollman, H. (eds.) (1991)

Hirschman, A.O. (1994) Social Conflicts as Pillars of Democratic Market Society, *Political Theory*, Vol. 22, pp. 203-18

Hirst, P. (1994), *Associative Democracy: New Forms of Economic and Social Governance*, Polity, Cambridge

Hodgson, G.M. (1967), *The Consequences of Utilitarism*, Clarendon Press, Oxford

Hodgson G.M. (1988), *Economics and Institutions: A Manifesto for a Modern Institutional Economics*, Polity, Cambridge MA

Hodgson, G.M., (ed.) (1993), *The Economics of Institutions*, Edward Elgar, Aldershot

Hooghe, L. (ed.) (1996), *Cohesion Policy and European Integration: Building Multi-Level Governance*, Clarendon Press, Oxford

Hugues, M.A. and Vandoren, P.M. (1990), 'Social Policy through Land Reform: New Jersey's Mount Laurel Controversy', *Political Science Quarterly*, Vol. 105, No. 1, pp. 97-111

Hunter, F. (1953), *Community Power Structure*, University of Carolina Press, Chapel Hill NC

Innes, J.E. (1990), *Knowledge and Public Policy: The Search for Meaningful Indicators*, Transaction Books, New Brunswick NJ

Innes, J.E. (1991), *Implementing State Growth Management in the U.S.: Strategies for Coordination*, IURD Working Paper 542, University of California, Berkeley CA

Innes, J.E. (1992a), 'Group Processes and the Social Construction of Growth Management: Florida, Vermont and New Jersey', *Journal of the American Planning Association*, Vol. 58, No. 4, pp. 440-53

Innes, J.E. (1992b), 'Implementing State Growth Management Programs', in: Stein, J. (ed.) (1992)

Innes, J.E. (1995a), 'Planning is Institutional Design', *Journal of Planning Education and Research*, Vol. 14, No. 2, pp. 140-3

Innes, J.E. (1995b), 'Planning Theory's Emerging Paradigm: Communicative Action and Interactive Practice', *Journal of Planning Education and Research*, Vol. 14, No. 3, pp. 183-9

Innes, J.E. (1996), 'Planning through Consensus Building: a New View of the Comprehensive Planning Ideal', *Journal of the American Planning Association*, Vol. 62, No. 4, pp. 460-72

Innes, J.E. and Booher, D.E. (1999), 'Consensus-building as Role Playing and Bricolage: Toward a Theory of Collaborative Planning', *Journal of the American Planning Association*, Vol. 63, No. 3, pp. 9-26

Innes, J.E. and Booher, D.E. (2000), 'Planning Institutions in the Network Society: Theory for Collaborative Planning', in Salet, W. and Faludi, A. (eds.) (2000)

Innes, J.E., Gruber, J., Neuman, M. and Thompson, R. (1994), *Coordinating Growth and Environmental Management through Consensus Building*, CPS-California Policy Seminar, University of California, Berkeley CA

Jackson, K.T. (1985), *Crabgrass Frontiers: The Suburbanization of the United States*, Oxford University Press, New York NY

Jepperson, R.L. (1991), 'Institutions, Institutional Effects, and Institutionalism', in Powell, W.W. and DiMaggio, P.J. (eds.) (1991)

Jessop, B. (1990), 'Regulation Theories in Retrospect and Prospect', *Economy and Society*, Vol. 19, No. 2, pp. 153-216

Jessop, B. (1995a), 'The Regulation Approach, Governance and Post-Fordism: Alternative Perspectives on Economic and Political Change?', *Economy and Society*, Vol. 24, No. 3, pp. 307-33

Jessop, B., (1995b), 'Towards a Schumpeterian Workfare Regime in Britain? Reflections on Regulation, Governance, and Welfare State', *Environment and Planning* A, Vol. 27, No. 10, pp. 1613-26

Jordan, G. (1990), 'Policy Community Realism versus "New Institutionalist" Ambiguity', *Political Studies*, Vol. 38, No. 3, pp. 470-84

Judge, D., Stoker, G. and Woolman, H. (eds.) (1995), *Theories of Urban Politics*, Sage, London

Kanige, J. (1988), 'The Best Houses Money Can Buy: Builders Dig Deep for Campaign Cash', *New Jersey Reporter*, Vol. 17, No. 10, May

Kanige, J. (1987), 'Off the Drawing Board, Into the Fire: The Best Laid Plans...', *New Jersey Reporter*, Vol. 17, No. 3, pp. 8-14

Kean, T.H. (1988), *The Politics of Inclusion*, Collier Macmillan, London

Keating, M. (1993), 'The Politics of Economic Development: Political Changes and Local Development in the United States, Britain, and France', *Urban Affairs Quarterly*, Vol. 28, No. 3, pp. 373-96

Keating, M. (1997), 'The Invention of Regions: Political Restructuring and Territorial Government in Western Europe', *Environment and Planning C*, Vol. 15, No. 4, pp. 383-98

Keating, M. and Hooghe, L. (1996), 'By-Passing the Nation State? Regions and the EU Policy Process', in Richardson J. (ed.) (1996), *European Union: Power and Policy-making*, Routledge, London

Kelman, S., Clavel, P., Forester, J. and Goldsmith, W.W. (1980), 'New Opportunities for Planners', in Clavel, P., Forester, J. and Goldsmith, W.W. (eds.) (1980), *Urban and Regional Planning in an Age of Austerity*, Pergamon Press, Oxford

Keohane, R.O. and Ostrom, E. (1994), 'Introduction. Local Commons and Global Interdependence: Heterogeneity and Cooperation in Two Domains', *Journal of Theoretical Politics*, Vol. 6, No. 4, pp. 403-28

Kingdon, J.W. (1984), *Agendas, Alternatives and Public Choices*, Little Brown, Boston MA

Kiser, L.L. and Ostrom, E. (1982), 'The Three Worlds of Action: A Metatheoretical Synthesis of Institutional Approaches', in Ostrom, E. (ed.), *Strategies of Political Inquiry*, Sage, Beverly Hills CA

Kolesar, J. (1992), 'Battle for the Boondbooks: Farmland Preservation Was Once an Option. Now, It's a Must', *New Jersey Reporter*, Vol. 22, No. 1, May-June

Koelble, T.A. (1995), 'The New Institutionalism in Political Science and Sociology', *Comparative Politics*, Vol. 27, No. 2, pp. 231-43

Kooiman, J. (1993a), 'Governance and Governability: Using Complexity, Dynamics and Diversity', in idem (ed.) (1993c)

Kooiman, J. (1993b), 'Findings, Speculations, and Recommendations', in idem (ed.) (1993c)

Kooiman, J. (ed.) (1993c), *Modern Governance: New Government-Society Interactions*, Sage, London

Krumholz, N. and Clavel, P. (1994), *Reinventing Cities: Equity Planners Tell Their Stories*, Temple University Press, Philadelphia PA

Lake, R.W. (1981), *The New Suburbanities: Race and Housing in the Suburbs*, Center for Urban Policy Research, Rutgers University, New Brunswick NJ

Lanzalaco, L. (1995), *Istituzioni organizzazioni potere. Introduzione all'analisi istituzionale della politica*, La Nuova Italia Scientifica, Roma

Lanzara, G.F. (1993), *Capacità negativa. Competenza progettuale e modelli di intervento nelle organizzazioni*, Il Mulino, Bologna

Le Galès, P. (1998), 'Regulation and Governance in European Cities', *International Journal of Urban and Regional Research*, Vol. 22, No. 3, pp. 482-506

Le Galès, P. and Thatcher, M. (eds.) (1995), *Les reseaux de politique publique*, L'Harmattan, Paris

Lederman, S.S. (ed.) (1989), *The SLERP Reforms and Their Impact on New Jersey Fiscal Policy*, Program for New Jersey Affairs, Woodrow Wilson School of Public and International Affairs, Princeton University, Princeton NJ

Lenz, S. (1985), *Keeping the Garden State Green*, Council on New Jersey Affairs, Program for New Jersey Affairs, Working Paper No. 8, Woodrow Wilson School of Public and International Affairs, Princeton University, Princeton NJ

Lewin, K. (1948), *Resolving Social Conflicts: Selected Papers on Group Dynamics*, Harper & Bros., New York NY

Lewin, K. (1951), *Field Theory in Social Science: Selected Theoretical Papers*, Harper & Bros., New York NY

Lewis, D.K. (1968), *Convention: A Philosophical Study*, Harvard University Press, Cambridge MA

Libecap, G.D. (1994), 'The Conditions for Successful Collective Action', *Journal of Theoretical Politics*, Vol. 6, No. 4, pp. 563-624

Liepitz, A. (1993), 'The Local and the Global: Regional Individuality or Interregionalism', *Transactions of the Institute of British Geographers: New Series*, Vol. 18

Lindblom, C.E. (1965), *The Intelligence of Democracy*, The Free Press, New York NY

Lindblom, C.E. (1990), *Inquiry and Change: The Troubled Attempt to Understand and Shape Society,* Yale University Press/Russell Sage Foundation, New Haven and London

Lindblom, C.E. and Cohen, D.K. (1979), *Usable Knowledge: Social Science and Social Problem Solving*, Yale University Press, New Haven CN

Listokin, D. (1976), *Fair Share Housing Allocation*, Center for Urban Policy Research, Rutgers University, New Brunswick NJ

Logan, J.R. and Molotch, H.L. (1987), *Urban Fortunes: The Political Economy of Place*, University of California Press, Berkeley CA

Low, N. (1997), 'What Made it Happen? Mapping the Terrain of Power in Urban Development', *Planning Theory*, Vol. 17, pp. 88-112

Lowi, T.A. (1972), 'Four Systems of Policy, Politics, and Choice', *Public Administration Review*, Vol. 32, No. 4

Luberoff, D. (1993), *State Planning in New Jersey*, A. Alfred Taubman Center for State and Local Government, John F. Kennedy School of Government, Harvard University, Cambridge MA

Luberoff, D. and Altshuler, A. (1998), *State Planning in New Jersey*, A. Alfred Taubman Center for State and Local Government, John F. Kennedy School of Government, Harvard University, Cambridge MA

Lukes, S. (1974), *Power: A Radical View*, Macmillan, London

Majone, G. (1989), *Evidence, Argument, and Persuasion in the Policy Process*, Yale University Press, New Haven and London

Majone, G. and Wildavsky, A. (1979), 'Implementation as Evolution', in Pressman, J.L. and Wildavsky, A. (eds.) (1984)

Malkin, J. and Wildavsky, A. (1991), 'Why the Traditional Distinction Between Public and Private Goods Should Be Abandoned', *Journal of Theoretical Politics*, Vol. 3, No. 4, pp. 355-78

Mandelbaum, S.J. (1985), 'The Institutional Focus of Planning Theory', *Journal of Planning Education and Research*, Vol. 5, No. 1, pp. 3-9

Mandelbaum, S.J. (1990), 'Reading Plans', *Journal of the American Planning Association*, Vol. 56, No. 3, pp. 350-6

Mandelbaum S.J. (1991), 'Telling Stories', *Journal of Planning Education and Research*, Vol. 10, No.3, pp. 209-14

Mandelbaum, S.J. (2000), *Open Moral Communities*, MIT Press, Cambridge MA

March, J.G. (1991), 'Exploration and Exploitation in Organizational Learning', *Organization Science*, Vol. 2, No. 1, pp. 71-87

March, J. G. and Olsen, J.P. (1976), *Ambiguity and Choice in Organizations*, Universitetsforlaget, Bergen

March, J. G. and Olsen, J.P. (1983), 'Organizing Political Life: What Administrative Reorganization Tells Us About Government', *American Political Science Review*, Vol. 77, No. 2, pp. 281-96

March, J. G. and Olsen, J.P. (1984), 'The New Institutionalism: Organizational Factors in Political Life', *American Political Science Review*, Vol. 78, No. 3, pp. 734-49

March, J. G. and Olsen, J.P. (1989), *Rediscovering Institutions: The Organizational Basis of Politics*, The Free Press, New York NY

March, J. G. and Olsen, J.P. (1995), *Democratic Governance*, The Free Press, New York NY

March, J.G. and Simon, H.A. (1958), *Organizations*, John Wiley and Sons, New York NY

Marin, B. and Mayntz, R. (eds.) (1991), *Policy Networks: Empirical Evidence and Theoretical Considerations*, Campus, Frankfurt a.M. and New York

Marks, G. (1992), 'Structural Policy in the European Community', in Sbragia, A. (ed.) (1992), *Europolitics: Institutions and Policymaking in the New European Community*, The Brookings Institution, Washington DC

Markusen, A.R. (1978), 'Class and Urban Social Expenditure: A Marxist Theory of Metropolitan Government', in Tabb, W.K. and Sawers, L. (eds.) (1978), *Marxism and the Metropolis: New Perspectives in Urban Political Economy*, Oxford University Press, New York NY

Marshall, J. and Peters, M. (1985), 'Evaluation and Education: The Ideal Learning Community', *Policy Sciences*, Vol. 18, No. 2, pp. 263-88

Martha Lamar Ass. (1988), *Affordable Housing in New Jersey: The Results o Mount Laurel II and the Fair Housing Act*, unpublished paper for The Fund for New Jersey and The Alliance for Affordable Housing Education Fund

Martin, L.L. (1994), 'Heterogeneity, Linkage and Commons Problems', *Journal of Theoretical Politics*, Vol. 6, No. 4, pp. 473-94

Massey, D. (1991), 'A Global Sense of Space', in Open University (1991), *The Making of the Regions*, Open University Press, Milton Keynes; now in Massey, D. (1994)

Massey, D. (1993), 'Power Geometry and a Progressive Sense of Space', in Bird, J., Curtis, B., Putnam, T., Robertson, G. and Tickner, L., (eds) (1993), *Mapping the Futures: Local Cultures, Global Change*. London: Routledge; now in Massey, D. (1994)

Massey, D. (1994), *Space, Place and Gender*, Polity, Cambridge

Matthiesen, U. (ed.) (1998), *Die Räume der Milieus. Neue Tendenzen in der sozial- und raumwissenschaftlichen Milieuforschung, in der Stadt- und Raumplanung*, Sigma, Berlin

Mayer, M. (1991), '"Postfordismus" und "lokaler Staat"', in Heinelt, H. and Wollman, H. (eds.) (1991)

Mayer, M. (1996), 'Postfordistische Stadtpolitik. Neue Regulationsformen in der lokalen Politik und Planung', *Zeitschrift für Wirtschaftsgeographie*, Vol. 40, No. 1-2, pp. 20-7

Mayntz, R. (1977), 'Die Implementation politischer Programme: Theoretische Überlegungen zu einem neuen Forschungsgebiet', *Die Verwaltung*, Vol. 10, No. 1, pp. 51-66

Mayntz, R. (1993), 'Modernisation and the Logic of Interorganisational Networks', in Child, J. *et al.* (eds.) (1993), *Societal Change between Market and Orgasnisation*, Avebury, Aldershot

Mayntz, R. and Scharpf, F.W. (1975), *Policy-making in the German Federal Bureaucracy*, Elsevier, Amsterdam

Mayntz, R. and Scharpf, F.W. (eds.) (1995), *Gesellschaftliche Selbstregelung und politische Steuerung*, Campus, Frankfurt a.M.

McCay, B.J. and Acheson, J.M. (1987), *The Question of the Commons: The Culture and Ecology of Communal Resources*, University of Arizona Press, Tucson AZ

McDougall, H.A. (1987), 'From Litigation to Legislation in Exclusionary Zoning Law', *Harvard Civil Rights-Civil Liberties Law Review*, Vol. 22, pp. 623-663

Melucci, A. (1987), 'Sul coinvolgimento individuale nell'azione collettiva', *Rassegna Italiana di Sociologia*, Vol. 28, No. 1, pp. 29-53; engl. transl. (1988), 'Getting Involved: Identity and Mobilization in Social Movements', in: Klandermans, B., Kriesi, H., and Tarrow, S. (eds.) (1988), *From Structure to Action: Comparing Social Movement Research across Cultures*, JAI Press, Greenwich CN

Melucci, A. (1991), *Il gioco dell'io. Il cambiamento di sé in una società globale*, Feltrinelli, Milano; engl. transl. (1996) *The Playing Self*, Cambridge University Press, Cambridge

Merton, R.K. (1957), *Social Theory and Social Structure*, Free Press, Glencoe IL

Meyer, J.W. and Rowan, B. (1977), 'Institutionalized Organizations: Formal Structure as Myth and Ceremony', *American Journal of Sociology*, Vol. 83, No. 2, pp. 340-63; now in Powell, W.W. and DiMaggio, P.J. (eds.) (1991)

Meyer, J.W. and Scott, R.W. (eds.) (1983), *Organizational Environments: Ritual and Rationality*, Sage, Beverly Hills CA

Meyer, M.W. and Zucker, L. (1989), *Permanently Failing Organizations*, Sage, Newbury Park CA

Michael, D.N. (1973), *On Learning to Plan and Planning to Learn: The Social Psychology of Changing Toward Future-Responsive Societal Learning*, Jossey Bass, San Francisco CA

Moe, T.M. (1984), 'The New Economics of Organization', *American Journal of Political Science*, Vol. 28, No. 4, pp. 739-77

Moore, T.D. (1986), 'Saving the Pinelands', in Porter, D.R. (ed.) (1986)

Morgan, K. (1997), 'The Learning Region: Institutions, Innovation and Regional Renewal', *Regional Studies*, Vol. 31, No. 5, pp. 491-503

Muller, P. and Surel, Y. (1998), *L'analyse des politiques publiques*, Montchrestien, Paris

Mumford, L. (1938), *The Culture of Cities*, Harcourt Brace & Co., New York NY

Mutti, A. (1986), 'Le regole della cooperazione strategica tra Stato e gruppi di interesse', *Rassegna Italiana di Sociologia*, Vol. 27, No. 3, pp. 223-47

Nathan M.L., Mitroff I.I. (1991), 'The Use of Negotiated Order Theory as a Tool for the Analysis and Development of an Interorganizational Field', *Journal of Applied Behavioral Science*, Vol. 27, No. 2, June, pp. 163-80

Neuman, M.C. (1992), *A New Planning Style: Governing Growth in New Jersey*, unpubl. paper, University of California, Berkeley CA

Neuman, M.C. (1996), *The Imaginative Institution: Planning and Institutions in Madrid*, unpubl. doctoral dissertation, Department of City and Regional Planning, University of California, Berkeley CA

Nielsen, R.P. (1993), 'Woolman's "I Am We" Triple-Loop Action Learning: Origin and Application in Organizational Ethics', *Journal of Applied Behavioral Science*, Vol. 29, No. 1, pp. 117-38

North, D.C. (1981), *Structure and Change in Economic History*, Norton, New York NY

North, D.C. (1990a), *Institutions, Institutional Change and Economic Performance*, Cambridge University Press, New York NY

North, D.C. (1990b), 'A Transaction Cost Theory of Politics', *Journal of Theoretical Politics*, Vol. 2, No. 4, pp. 355-67

Olson, M. (1965), *The Logic of Collective Action: Public Goods and the Theory of Groups*, Harvard University Press, Cambridge MA

Ostrom, E. (1986), 'An Agenda for the Study of Institutions', *Public Choice*, Vol. 48, No. 1, pp. 3-25

Ostrom, E. (1990), *Governing the Commons: The Evolution of Institutions for Collective Action*, Cambridge University Press, Cambridge

Ostrom, E. (1991), 'Rational Choice Theory and Institutional Analysis: Toward Complementarity', *American Political Science Review*, Vol. 85, No. 1, pp. 237-43

Ostrom, E. (1992), 'Community and the Endogenous Solution of Commons Problems', *Journal of Theoretical Politics*, Vol. 4, No. 3, pp. 343-51

Ostrom, E. (1994a), *Neither Market nor State: Governance of Common Pool Resources in the Twenty-First Century*, International Food Policy Research Institute, Washington DC

Ostrom, E. (1994b), 'Constituting Social Capital and Collective Action', *Journal of Theoretical Politics*, Vol. 6, No. 4, pp. 527-62

Ostrom, E. (1995), 'New Horizons in Institutional Analysis', *American Political Science Review*, Vol. 85, No. 1, pp. 174-78

Ostrom, V., Feeny, D. and Picht, H. (1988), *Rethinking Institutional Analysis and Development: Issues, Alternatives, and Choices*, Institute for Contemporary Studies Press, San Francisco CA

Painter, J. and Goodwin, M. (1995), 'Local Governance and Concrete Research: Investigating the Uneven Development of Regulation', *Economy and Society*, Vol. 24, No. 3, pp. 334-56

Parsons, T. (1990), 'Prolegomena to a Theory of Social Institutions' (orig. 1934), *American Sociological Review*, Vol. 55, No. 3, pp. 319-33

Parsons, T. (1951), *The Social System*, The Free Press, Glencoe IL

Payne, J.M. (1983), 'Starting Over: Mount Laurel II', *Real Estate Journal*, Vol. 12, No. 1, pp. 85-96.

Pedersen, O.E. (1991), 'Nine Questions to a Neo-Institutional Theory in Political Science', *Scandinavian Political Studies*, Vol. 14, pp. 125-48

Pittenger, J.C. (1986), 'The Courts', in Pomper, G. (ed.) (1986)

Pizzorno, A. (1986), 'Sul confronto intertemporale delle utilità', *Stato e mercato*, No. 16

Pizzorno, A. (1989), 'Spiegazione come reidentificazione', *Rassegna Italiana di Sociologia*, Vol. 30, No. 2, pp. 161-84

Plotkin, S. (1987), *Keep Out: The Struggle for Land-Use Control*, University of California Press, Berkeley CA

Polanyi, K. (1944), *The Great Transformation*, Beacon Press, Boston MA

Polanyi, K., (ed.) (1957), *Trade and Markets in the Early Empires*, Free Press, New York NY

Polsby, N. (1980), *Community Power and Democratic Theory*, 2nd ed., Yale University Press, New Haven CT

Pomper, G. (ed.) (1986), *The Political State of New Jersey*, Rutgers University Press, New Brunswick NJ

Porter, D.R. (1992), *About Growth Management: Defining the Issues, Assessing the Techniques*, Association for Commercial Real Estate, Arlington VA

Porter, D.R. (ed.) (1986), *Growth Management: Keeping on Target?*, The Urban Land Institute, Washington DC

Powell, W.W. (1990), 'Neither Markets nor Hierarchy: Network Forms of Organisation', *Research in Organizational Behaviour*, Vol. 12, pp. 295-336; now in Thompson, G., Frances, J., Levancic, R. and Mitchell, J. (eds.) (1991), *Markets, Hierarchies and Networks: The Co-ordination of Social Life*, Sage, London

Powell, W. W. and DiMaggio, P.J. (eds.) (1991), *The New Institutionalism in Organizational Analysis*, University of Chicago Press, Chicago IL

Pressmann, J.L. and Wildavsky, A. (1984), *Implementation*, exp. ed., University of California Press, Berkeley CA

Putnam, R.D. (1993), *Making Democracy Work: Civic Traditions in Modern Italy*, Princeton University Press, Princeton NJ

Rabinovitz, F.F. (1989), 'The Role of Negotiation in Planning, Management, and Policy Analysis', *Journal of Planning Education and Research*, Vol. 8, No. 2, pp. 87-95

Rahenkamp, J. and Rahenkamp, C. (1986), 'Fair Share Housing in New Jersey', in Porter, D.R. (ed.) (1986)

Raiffa, H. (1982), *The Art and Science of Negotiation*, Harvard University Press, Cambridge MA

Regonini, G. (1993), 'Il principe e il povero. Politiche istituzionali ed economiche negli anni 80', *Stato e mercato*, No. 39, pp. 361-403

Rein, M. and Laws, D. (2000), 'Controversy, Reframing and Reflection', in Salet, W. and Faludi, A. (eds.)(2000)

Rein, M. and Schön, D.A. (1993), 'Reframing Policy Discourse', in Fischer F., Forester J. (eds.) (1993)

Rhodes, R.A.W. (1996), 'The New Governance: Governing without Government', *Political Studies*, Vol. 64, pp. 652-67

Richardson, J.J. (ed.) (1982), *Policy Styles in Western Europe*, Allen & Unwin, London

Rittel, H.W.J. and Webber, M.M. (1973), 'Dilemmas in a General Theory of Planning', *Policy Sciences*, Vol. 4, No. 2, pp. 155-69

Robichaud, B. and Russel, E.W.B. (1988), *Protecting the New Jersey Pinelands: A New Direction in Land-Use Management*, Rutgers University Press, New Brunswick NJ

Roper, R.W., Lago, J.R., Beer, N.G. and Bierbaum M.A. (1986), *Federal Aid in New Jersey*, Council on New Jersey Affairs, Program for New Jersey Affairs, Working Paper No. 9, Woodrow Wilson School of Public and International Affairs, Princeton University, Princeton NJ

Rose, J.G. (1977), 'The Courts and the Balanced Community', in Rose, J.G. and Rothman, R.E. (eds.) (1977)

Rose, J.G. and Rothman, R.E. (eds.) (1977), *After Mount Laurel: The New Suburban Zoning*, Center for Urban Policy Research, Rutgers University, New Brunswick NJ

Rosenau, J.N. and Czempiel, E.-O. (eds.) (1992), *Governance without Government: Order and Change in World Politics*, Cambridge University Press, Cambridge

Rositi, F. (1986), 'Tipi e dimensioni dei sistemi normativi', *Rassegna Italiana di Sociologia*, Vol. 27, No. 3, pp. 347-67

Rowe, N. (1989), *Rules and Institutions*, Philip Allan, New York NY

Rubin, J.I., Seneca, J.J. and Stotsky, J.G. (1990), 'Affordable Housing and Municipal Choice', *Land Economics*, Vol. 66, No. 3, pp. 325-40

Rusconi, G.E. (1992), 'Agire strategico e agire comunicativo. Un contrasto apparente nella spiegazione della politica', in Cecchini, A. and Indovina, F. (eds.) (1992), *Strategie per un futuro possibile*, Franco Angeli, Milano

Sabatier, P.A. and Jenkins Smith, H. (eds.) (1993), *Policy Change and Learning: An Advocacy Coalition Approach*, Westview Press, Boulder CO

Sabel, C.F. (1989), 'Flexible Specialization and the Reemergence of Regional Economies', in Hirst, P. and Zeitlin, J. (eds.) (1989), *Reversing Industrial Declines*, St. Martin's Press, New York NY

Sabel, C.F. (1992), 'Studied Trust: Building New Forms of Cooperation in a Volatile Economy', in Romo, F. and Swedberg, R. (eds.) (1992), *Readings in Economic Sociology*, Russel Sage, New York NY

Salet, W. (2000), 'The Institutional Approach to Strategic Planning', in Salet, W. and Faludi, A. (eds.) (2000)

Salet, W. and Faludi, A. (2000), 'Three Approaches to Strategic Planning', in Salet, W. and Faludi, A. (eds.) (2000)

Salet, W. and Faludi, A. (eds.) (2000), *The Revival of Strategic Spatial Planning*, Royal Netherlands Academy of Arts and Sciences, Amsterdam

Salmore, B. and Salmore, S.A. (1993), *New Jersey Politics and Government: Suburban Politics Comes of Age*, University of Nebraska Press, Lincoln NE

Savitch, H. (1997), 'Meeting the Global Challenge through Institutional Capacity and Social Capital', paper presented at the International Seminar *Governing Cities: International Perspectives*, Bruxellels, 18-19 September 1997

Scharpf, F.W. (1973), *Planung als politischer Prozess. Aufsätze zur Theorie der planenden Demokratie*, Suhrkamp, Frankfurt a.M.

Scharpf, F.W. (1978), 'Interorganizational Policy Studies: Issues, Concepts, and Perspectives', in Hanf, K. and Scharpf, F.W. (eds.), *Interorganizational Policy Making: Limits to Coordination and Central Control*, Sage, London

Scharpf, F.W. (1989), 'Decision Rules, Decision Styles and Policy Choices', *Journal of Theoretical Politics*, Vol. 1, No. 2, pp. 149-176

Scharpf, F.W. (1994a), 'Community and Autonomy: Multi-Level Policy-Making in the European Union', *Journal of European Public Policy*, Vol. 1, No. 2, pp. 219-242

Scharpf, F.W. (1994b), 'Games Real Actors Could Play: Positive and Negative Coordination in Embedded Negotiations', *Journal of Theoretical Politics*, Vol. 6, No. 1, pp. 27-53

Scharpf, F. W. (1997), *Games Real Actors Play: Actor-Centered Institutionalism in Policy Research*, Westview Press, Boulder CO

Schattschneider, E.E. (1960), *The Semi-Sovereign People: A Realist's View on Democracy in America*, Holt Rinehart and Winston, New York NY

Schelling, T.C. (1960), *The Strategy of Conflict*, Oxford University Press, Oxford

Schelling, T.C. (1978), *Micromotives and Macrobehavior*, Norton, New York NY

Schmitter, P.C. and Grote, J.R. (1997), *The Corporatist Sisyphus: Past, Present & Future*, EUI Working Papers SPS 97/4, European University Institute, Florence

Schön, D.A. (1983), *The Reflective Practitioner: How Professionals Think in Action*, Basic Books, New York NY

Schön, D.A. (1987), *Educating the Reflective Practitioner*, Jossey-Bass, San Francisco CA

Schön, D.A. and Rein, M. (1994), *Frame Reflection: Toward the Resolution of Intractable Policy Controversies*, Basic Books, New York NY

Schotter, A. (1981), *The Economic Theory of Social Institutions*, Cambridge University Press, Cambridge

Schütz, A. (1932), *Der sinnhafte Aufbau der sozialen Welt*, Springer, Wien; engl. transl. *The Phenomenology of the Social World*, Northwestern University Press, Evanston IL 1967

Scott, A.J. (1996), 'Regional Motors of the Global Economy', *Futures*, Vol. 28, pp. 391-411

Scott, R.W. (1987), 'The Adolescence of Institutional Theory', *Administrative Science Quarterly*, Vol. 32, No. 4, pp. 493-511

Scott, R.W. and Meyer, J.W. (1983), 'The Organization of Societal Sectors', in Meyer, J.W. and Scott, R.W. (eds.) (1983); now in Powell, W.W. and DiMaggio, P.J. (eds.) (1991)

Scott, R.W. and Meyer, J.W. (eds.) (1994), *Institutional Environments and Organizations: Structural Complexity and Individualism*, Sage, London

Selig, J.M. (1988) *Implementing Mount Laurel: An Assessment of Regional Contribution Agreements*, Council on New Jersey Affairs, Program for New Jersey Affairs, Working Paper No. 14, Woodrow Wilson School of Public and International Affairs, Princeton University, Princeton NJ

Selznick, P. (1949), *TVA and the Grass Roots: A Study of Politics and Organization*, University of California Press, Berkeley CA

Selznick, P. (1957), *Leadership in Administration*, Harper & Row, New York NY

Sen, A.K. (1977), 'Rational Fools: A Critique of the Behavioral Foundations of Economic Theory', *Philosophy and Public Affairs*, Vol. 6, No. 4, pp. 317-44

Seneca, J.J. and Rubin, J.I. (1988), *Economic Issues in State Planning*, Economic Policy Council and Office of Economic Policy, unpubl. paper, Trenton NJ

Shepsle, K.A. (1986), 'Institutional Equilibria and Equilibrium Institutions', in Weisberg, H.F. (ed.) (1986), *Political Science: The Science of Politics*, Agathon Press, New York NY

Shepsle, K.A. (1989), 'Studying Institutions: Some Lessons from the Rational Choice Approach', *Journal of Theoretical Politics*, Vol. 1, No. 2, pp. 131-47

Shepsle, K.A. and Weingast, B.R. (1987), 'The Institutional Foundations of Committee Power', *American Political Science Review*, Vol. 81, pp. 85-104

Simmons, P. (1977), 'Introduction', in Rose, J.G. and Rothman, R.E. (eds.) (1977)

Simon, H.A. (1947), *Administrative Behavior*, Macmillan, New York NY

Simon, H.A. (1972), *Human Problem Solving*, Prentice-Hall, Englewood Cliffs NJ

Simon, H.A. (1973), 'Organizational Man: Rational or Self-Actualizing?', *Public Administration Review*, Vol. 33, No. 4, pp. 346-53

Sjoblom, G. (1993), 'Some Critical Remarks on March and Olsen's "Rediscovering Institutions"', *Journal of Theoretical Politics*, Vol. 5, No. 3, pp. 397-407

Sjöstrand, S.E. (1992), 'On the Rationale Behind "Irrational" Institutions', *Journal of Economic Issues*, Vol. 26, No. 4, pp. 1007-40

Snidal, D. (1994), 'The Politics of Scope: Endogenous Actors, Heterogenity and Institutions', *Journal of Theoretical Politics*, Vol. 6, No. 4, pp. 449-72

Snow, D.A., Burke Rochford Jr., E., Worden, S.K. and Benford, R.D. (1986), 'Frame Alignement Processes, Micromobilization, and Movement Participation', *American Sociological Review*, Vol. 51, No. 3, pp. 464-81

Snow, D.A., Zurcher, L.A. and Ekland-Olson, S. (1980), 'Social Networks and Social Movements: A Microstructural Approach to Differential Recruitment', *American Sociological Review*, Vol. 45, No. 4, pp. 787-801

Stein, J., (ed.) (1992), *Growth Management: The Planning Challenge of the Nineties*, Sage, Newbury Park CA

Steinberg, M.K. (1989), *Adaptations to an Activist Court Ruling: Aftermath of the Mount Laurel II Decision for Lower-Income Housing*, Lincoln Institute of Land Policy, Cambridge MA

Steinmo, S., Thelen, K. and Longstrethm F. (eds.) (1992), *Structuring Politics: Historical Institutionalism in Comparative Analysis*, Cambridge University Press, Cambridge

Sternlieb, G. and Schwartz, A. (1986), *New Jersey Growth Corridors*, Center for Urban Policy Research, Rutgers University, New Brunswick NJ

Stoker, G. (1990), 'Regulation Theory, Local Government, and the Transition from Fordism', in King, D. and Pierre, J. (eds.) (1990), *Challenges to Local Government*, Sage, London

Stoker, G. and Mossberger, K. (1994), 'Urban Regime Theory in Comparative Perspective', *Environment and Planning C: Government and Policy*, Vol. 12, pp. 195-212

Stone, C.N. (1989), *Regime Politics: Governing Atlanta, 1946-1988*, University of Kansas Press, Kansas City KA

Storper, M. (1995), 'The Resurgence of Regional Economies, Ten Years Later: The Region as a Nexus of Untraded Interdependencies', *European Urban and Regional Studies*, Vol. 2, pp. 191-221

Streeck, W. and Schmitter, P.C. (1985), 'Community, Market, State - and Associations? The Prospective Contribution of Interest Governance to Social Order', *European Sociological Review*, No. 1, pp. 119-38

Sullivan, T. (1984), *Resolving Development Disputes through Negotiation*, Plenum Press, New York NY

Sureman, S.R. (1986), *Mount Laurel II and the Fair Housing Act*, New Jersey Institute for Continuing Legal Education, Newark NJ

Susskind, L. and Cruickshank, J. (1987), *Breaking the Impasse: Consensual Approaches to Resolving Public Disputes*, Basic Books, New York NY

Susskind, L., McKearnan, S. and Thomas-Larmer, J. (eds.) (1999), *The Consensus-Building Handbook: A Comprehensive Guide to Reaching Agreement*, Sage, Los Angeles CA

Susskind, L. and Ozawa, C. (1984), 'Mediated Negotiation in the Public Sector: The Planner as Mediator', *Journal of Planning Education and Research*, Vol. 4, No. 1, pp. 5-15

Swidler, A. (1986), 'Culture in Action: Symbols and Strategies', *American Sociological Review*, Vol. 51, No. 2, pp. 272-86

Tarrow, S. (1998) *Power in Movement: Social Movements and Contentious Politics*, Cambridge University Press, Cambridge

Taylor, M. (1976), *Anarchy and Cooperation*, Wiley & Sons, London

Taylor, M. (1982), *Community, Anarchy and Liberty*, Cambridge University Press, New York NY

Taylor, M. (1987), *The Possibility of Cooperation*, Cambridge University Press, Cambridge

Thelen, K. and Steinmo, S. (1992), 'Historical Institutionalism in Comparative Politics', in Steinmo, S., Thelen, K. and Longstreth, F. (eds.) (1992)

Thompson, J.D. (1967), *Organizations in Action: Social Science Basis of Administrative Theory*, McGraw Hill, New York NY

Thrift, N. (1983), 'On the Determination of Social Action in Space and Time', *Environment and Planning D: Society and Space*, Vol. 1, No. 1, pp. 23-57

Throgmorton, J.A. (1993), 'Survey Research as Thetorical Trope: Electric Power Planning Arguments in Chicago', in Fischer, F. and Forester, J. (eds.) (1993)

Webber, M.M. (1969), 'Planning in an Environment of Change, Part II: Permissive Planning', *Town Planning Review*, Vol. 39, No. 4, pp. 277-95

Webber, M.M. (1978), 'A Difference Paradigm for Planning', in Burchell, R.W. and Sternlieb, G. (eds.) (1978), *Planning Theory in the 1980's: A Search for Future Directions*, Center of Urban Policies Research, Rutgers University Press, New Brunswick NJ

Webber, M.M. (1983), 'The Myth of Rationality: Development Planning Reconsidered', *Environment and Planning B: Planning and Design*, Vol. 10, No. 1. pp. 89-99

Weick, K.E. (1979), *The Social Psychology of Organizing*, 2nd ed., Addison-Wesley, New York NY

Weick, K.E. (1976), 'Educational Organizations as Loosely Coupled Systems', *Administrative Science Quarterly*, Vol. 21, No. 1, pp. 1-19

Weick, K.E. (1995), *Sensemaking in Organizations*, Sage, Thousand Oaks CA

Weiss, S. (1989), *Decision of the Commissioner of Education in the Case of Abbott v. Burke: A Summary*, New Jersey State Department of Education, Trenton NJ

White, M.J. (1978), 'Self-Interest in the Suburbs: The Trend toward No-Growth Zoning', *Policy Analysis*, Vol. 4, No. 2, pp. 185-203

Whyte, W.F. (1991), *Social Theory for Action: How Individuals and Organizations Learn to Change*, Sage, Newbury Park CA

Wildavsky, A. (1979), *The Art and Craft of Policy Analysis*, Macmillan, London and Basingstoke

Wildavsky, A. (1987), 'Choosing Preferences by Constructing Institutions: A Cultural Theory of Preference Formation', *American Political Science Review*, Vol. 81, No. 1, pp. 3-22

Wildavsky, A. (1994), 'Why Self-Interest Means Less Outside of a Social Context: Cultural Contributions to a Theory of Rational Choices', *Journal of Theoretical Politics*, Vol. 6, No. 2, pp. 131-59

Williams, N. Jr. (1984), 'The Background and Significance of Mount Laurel II', *Journal of Urban and Contemporary Law*, Vol. 26, No. 3, pp. 3-23

Williamson, O.E. (1985), *The Economic Institutions of Capitalism: Firms Markets, Relational Contracting*, Free Press, New York NY

Wisman, J.D. and Rozanski, J. (1991), 'The Methodology of Institutionalism Revisited', *Journal of Economic Issues*, Vol. 25, No. 3, pp. 709-37

Wood, R.C. (1961), *1400 Governments: The Political Economy of the New York Metropolitan Region*, Harvard University Press, Cambridge MA

Woolman, H. and Goldsmith, M. (eds.) (1993), *Urban Politics and Policy: A Comparative Approach*, Blackwell, Oxford.

Wright, E.O. (ed.) (1995), *Associations and Democracy*, Verso, London

Young O.R. (1994), *International Governance*, Cornell University Press, Ithaca NY

Znaniecka Lopata, H. (1991), 'Role Theory', in Blau, J.R. and Goodman, N. (eds.) (1991)

Zonnefeld, W. (2000) 'Discoursive Aspects of Strategic Planning: a Deconstruction of the "Balanced Competitiveness" Concept in European Spatial Planning', in Šalet, W. and Faludi, A. (eds.) (2000)

Zucker, L.G. (1977), 'The Role of Institutionalization in Cultural Persistence', *American Sociological Review*, Vol. 42, No. 5, pp. 726-43; now in Powell, W.W. and DiMaggio, P.J. (eds.) (1991)

Zucker, L.G. (1987), 'Institutional Theories of Organizations', *Annual Review of Sociology*, Vol. 13, pp. 443-64

Zucker, L.G. (ed.) (1988), *Institutional Patterns and Organizations: Culture and Environment*, Ballinger, Cambridge MA

Zukin, C. (1986), 'Political Culture and Public Opinion', in Pomper, G. (ed.) (1986)

Index

Printed and bound by CPI Group (UK) Ltd, Croydon, CR0 4YY

22/10/2024

01777627-0008